Neurobiological Evolution of Humans

Our perception of the brain structure and function as an organ full of secrets and mysteries must change, and it is necessary to consider it as a part of the body that is constantly evolving and developing to maintain homeostasis for the entire human organism. New fossils, imaging methods, molecular genetics, comparative anatomy, evolutionary psychology, cognitive concepts in neuroscience, and neurology have made it possible to address questions of the neurological evolution of humans. This book integrates disparate findings from those disciplines in a succinct history by explaining how evolutionary process occurred, reshaped our brain, reconstructed our mind, and finally changed our "selves."

Key Features

- Explains complex scientific concepts in neurobiology using lucid prose and maintaining a clear narrative
- Provides well-documented arguments supported by hundreds of references and citations
- Documents a complex history which unfolded over millions of years
- Offers new approaches intended to transform ideas, opinions, and intellectual foundations of the brain

Neurobiological Evolution of Humans

From Neurons to Narratives

Ali Moghtaderi

CRC Press
Taylor & Francis Group
Boca Raton London New York

CRC Press is an imprint of the
Taylor & Francis Group, an **informa** business

Designed cover image: www.sciencephoto.com

First edition published 2025
by CRC Press
2385 NW Executive Center Drive, Suite 320, Boca Raton FL 33431

and by CRC Press
4 Park Square, Milton Park, Abingdon, Oxon, OX14 4RN

CRC Press is an imprint of Taylor & Francis Group, LLC

ISBN: 9781032885537 (hbk)
ISBN: 9781032885544 (pbk)
ISBN: 9781003538400 (ebk)

DOI: 10.1201/9781003538400

Typeset in Times
by KnowledgeWorks Global Ltd.

To my family,

Roya

Nazanin

Navid

Those who were with me all the way

Contents

About the Author

Dr. Ali Moghtaderi is a professor of clinical neurology at a local university (ZAUMS) in Iran. He had been trained as a neurologist at Tehran University (TUMS) and had also worked as a fellow for more than four years in different disciplines in neurology; vascular intervention fellowship for one year in 2001 in Freiburg Albert Ludwig University, Germany, under the supervision of Martin Schumacher; and in 2007, he spent a year working with Louis Cuevas at the University of Liverpool (LSTM), UK, on tuberculous meningitis. In 2014–2015, Dr. Moghtaderi studied Neuroimmunology and Multiple Sclerosis with Luanne Metz at the University of Calgary, Canada. Although the fellowship courses expanded his scientific knowledge in neuroscience, he never found answers to the basic questions of the mind about how the mechanism of functional processes of the brain, such as consciousness, thinking, and memory. During his career, Dr. Moghtaderi has trained neurology to many students, interns, residents, and physicians; in addition doing academic research, writing articles and continue to do so. Recently, he had an opportunity to reconsider his old passion for writing a book about the neurobiological evolution of the humankind. Therefore, in the past four years, Dr. Moghtaderi has dedicated more time to it, resulting in the present book with more than 800 references. He hopes this is a path to the dream of truth.

Preface

For years, the process of the evolution of the nervous system in *Homo sapiens* has occupied my mind. However, the development of such a concept required extensive study and experience. While piecing together the puzzle over the past years, I encountered a new problem. Some fundamental concepts in neuroscience, such as consciousness, memory, thinking, emotions, and morality, seem simple in appearance, but when we attempt to define them, we face difficulties. We know that the evolutionary process mostly occurs in organs that are intricate and complex like skull and brain. In this way, at the same time as the ability and function of the relevant organ rises, energy consumption drops down. The most important and also intricate system in the evolution of humans is the nervous system, especially the brain, firstly to maintain our homeostasis. In fact, the story of "how we became us" pertains to the evolutionary scenario of the brain in which consciousness, mind, memory, intelligence, thinking, etc., being installed on it over the years. Despite this, we not only cannot define them correctly, but we also do not know what they are made of or what their nature is.

Although this problem is not new, for many years, scientists in the experimental sciences have had to reach out to philosophers, who have tried to find suitable answers to our questions. Of course, they couldn't. Generally, philosophers work on questions that have no answers or for which an answer can never be found, and are likely to be debated and scrutinized forever. Perhaps the thought process or problem-solving method to reach answers to those questions is very ambiguous or completely wrong. It seems that concepts such as consciousness, memory, and self, all fall within the realm of experimental sciences and thus should be addressed using scientific methods. Even if we do not find answers, we still will not reach the correct answer using the wrong method. Although the content presented in this book has a strong scientific foundation and utilizes numerous academic sources, some questions still remain in ambiguity. Unlike many existing books, this book strives to be written in a relatively simple and understandable language so that it can be comprehensible to those without extensive experience in life sciences. I hope that reading each page will accompany the reader on a winding path toward the dream of truth.

Ali Moghtaderi
Calgary, August 2024

Introduction

The story of *Homo narrans*, the "*Speaking Man*," is the story of "*How We Became Us*." It is the story of the life of ancient humans and their transformation into intelligent beings. It is the story of a human species that has outlasted and eradicated other species to establish itself as the dominant reign in the present world. It is also the story of the rise of neuroscience, the gradual retreat of philosophy, and other metaphysical sciences that have yet to provide satisfactory answers to the fundamental questions of humanity. Although mixing and comparing these two disciplines is not a logical endeavor, philosophers seek answers to questions such as the nature of the human soul, ethics, reason, mind, love, beauty, the existence or nonexistence of God, and immortality. However, many of these concepts are now being analyzed clearly and prominently with the help of neuroscience.

Unveiling the secrets locked within ancient fossils, new methods of imaging the nervous system, molecular genetics, comparative anatomy, evolutionary psychology, cognitive concepts in neuroscience, and neurology have greatly increased our ability to unravel the mysteries of these questions. This perspective provides us with a fresh opportunity to redefine many diseases that were previously labeled as functional, mental, or psychological, based on the findings of neurobiology. It can be thought of as the story of our improved and enhanced understanding of things that philosophy had been unable to answer for years, and many considered them to be an elusive, invisible, and uncertain concept, that is, the human psyche. In fact, our mystique perception of the brain and the central nervous system as an organ full of secrets and mysteries must change, and it is necessary to consider it as a part of the body that is constantly changing, evolving, and developing to maintain crucially homeostasis for the entire human organism.

Today, we are almost certain that the closest creatures to the primates, including tree mice (squirrels) and ring-tail lemurs, are our ancestors. About 6–8 million years ago, our common history with great apes existed in Africa, and a branch of them, called *Homo sapiens*, demonstrated their dominance over the Earth around 300,000 years ago in that region. Basically, evolution based on genetic diversity and adaptation to the environment ensures the survival of a species of living organisms, which has been able to sustainably maintain its reproductive capacity. The evolutionary process of this creature's mental structure, including toolmaking and various methods of communication, especially speech, self-awareness, awareness of the others and the surrounding environment, memory formation, and the development of culture, religion, and civilization, significantly improved.

The mechanisms and methods governing this evolutionary structure have long been debated and discussed by experts. This book primarily aims to explain how these evolutionary modifications occurred in the neurobiological structure of humans. Fundamental and foundational changes in the human brain have allowed for remarkable abilities. A species that can walk on two legs, make effective use of its hands, speak different languages, and ultimately consider itself the noblest of creatures.

However, before discussing the changes in the nervous system, I found it necessary to introduce the basic anthropological concepts to provide the reader with a better understanding of the evolutionary process of what we call the nervous system. In some discussions, reference has also been made to anatomical concepts and their changes over time. Therefore, this book is divided into two main sections. The first part, consisting of five chapters, mainly deals with the anthropological history of man and focuses on the two topics of Darwinian evolution and the ability of man to move on two legs (bipedalism). The second part, which includes the remaining eleven chapters, emphasizes the significance of brain size and function, intelligence, mind, emotions, consciousness, and ultimately, the ability to speak. The ability to speak and communicate is the most prominent and distinguishing human characteristic through which human consciousness finds meaning, thinks, reflects on its thoughts, and expresses them to others. Therefore, this topic forms the largest part of the book. However, three fundamental concepts, namely, mind, self, and consciousness, form the basis of language and speech, and understanding them is of utmost importance.

It seems that the ability to walk upright on two legs has been the most significant and major event in human evolution according to Darwinian evolution. To undergo such an event, humans had to modify their skeletal structure. Simultaneously with the freedom of hands, the ability to create tools also evolved in *Homo erectus*. Toolmaking aided their cognitive self-awareness, forming a mental image of the body, constructing consciousness, developing the concepts of "self" and "the other," and ultimately gaining awareness of "self." Historically, the mind and the brain have been considered as two separate entities. Many neuroscientists disagree with the Cartesian dualistic view that the mind is a non-material product of the physical brain. This dualistic Cartesian perspective is now evident in the duality of the brain substance (res extensa) and spiritual or mental substance (res cogitans). While it is true that our brain plays a significant role in controlling our body, our body also influences our sensory perceptions, thoughts, and emotions, and both play essential roles in constructing the concepts of "self" and "the other." It appears that we do not perceive the reality of the world accurately; rather, our brain tries to infer the best guess about the external reality. This best guess is the result of the interaction between the sensory system of the nervous system and the limited information it receives from the surrounding world. The same process occurs in the brain's perception of the body. Through our sensory system, we construct models of "self" and the surrounding world in our minds. The brain, as an interface, uses the body as a tool or object to receive and/or intervene in our sensations from the external world. However, our perception of ourselves is not merely an objective phenomenon but rather an active and subjective selection. The result of this sensory perception gives rise to the concept of "me." In fact, our body, with the help of our brain, plays an active role in how we perceive the external world and represents it in the form of thoughts, ideas, insights, and emotions, ultimately contributing to the creation of self-consciousness, and awareness of ourselves and the surrounding world. Body awareness is the foundation of constructing consciousness. This phenomenon may be the greatest and most significant unknown of human beings.

Primitive humans attempted to teach their fellow beings different ways of survival through body language, marking the gradual beginning of cultural evolution. With the discovery and control of fire, as well as increased consumption of protein (meat), their brain started to grow and gradually became larger. However, the optimal control and consumption of energy resources led to a halt in this growth after a while. Then, they tried to adopt a specific structure and cellular communication in the brain. Neurons gathered in different areas of the brain, carrying out their tasks more specialized and establishing cellular intermediary centers to communicate with each other when necessary. This process laid the foundation for the asymmetry and lateralization of the brain. Now, the intelligent human was born in an asymmetric way, a *lopsided ape*.*

Alongside changes in skeletal structure and the nervous system, the human lifespan also underwent transformation. Their embryonic period in the mother's womb became shorter, and they had to spend more time outside the mother's body to complete their growth, becoming more dependent on their mother. This continuous attachment changed the framework of their emotions and feelings. Their nervous system underwent fundamental changes to perceive the concepts of external and internal emotions. Continuous reception of sensory information in the framework of the central nervous system, like a collection of frames in a movie, brought forth the concept of the human mind. Consciousness, cognition, and distinguishing "self" from "the others" were established based on the constructed human mind.

During childhood and adolescence, humans invest in brain growth and then, with considerable effort, begin reproducing after sexual maturity. During this period, especially with the cessation of sexual capability, they utilize their cognitive capacity to transmit information to their

* I borrowed this term from Michael Corballis: The lopsided ape (1993, Oxford University Press).

offspring. This required communication to be established. Humans gradually expanded body language and learned language by transforming gestures and manual–facial expressions into sound symbols around 70,000 years ago. Language played a crucial role in the coevolution of the brain and increased lifespan, making 60 years, the time for the onset of cognitive decline. Learning language alone was not sufficient for them, so they used their freed hands to record sound symbols. Five thousand years ago, in ancient Mesopotamia (a region between to rivers: Tigris and Euphrates), humans successfully invented writing. The ability to walk upright, toolmaking, language invention, and the recording of human speech through symbols and inscription allowed humans to transcend boundaries of power and create their own culture based on aesthetics, such as writing, painting, music, art, and more. The intelligent human had now become modern.

Overall, an effort has been made to maintain a scientific structure in the book, so I have tried to utilize credible sources in writing it. But in no way have I tried to teach the anatomy and physiology of the nervous system; however, many medical concepts are simply expressed along with evolutionary alterations. Unfortunately, it is difficult for people who are unfamiliar with medical concepts to understand most of the sources given at the end of the book. At the same time, I have aimed to make the text understandable and comprehensible for individuals who have not studied in medical or related fields. For this reason, I have redefined certain concepts in different chapters of the book in order to enhance understanding. However, this may make some topics appear slightly more complex. In many cases, to avoid repeating the word *H. sapiens*, I have used wise, intelligent, or modern man. The content throughout the book is relatively interconnected, and it is advisable not only for people familiar with biological sciences but also for readers unfamiliar with such concepts to read the book in order, although it is not mandatory.

Part I

Human History

1 Genesis

CREATION

Approximately 500 million years ago, a remarkable event, the emergence of the first vertebrates unfolded, marking a significant milestone roughly four billion years after the Earth's formation. They set foot on the stage of existence. Due to the fact that ultraviolet rays destroy the structure of DNA, with the explosion of the Precambrian (age of early life) period, 500 million years ago, a large number of light-loving single cells appeared. The movement of the Earth's position and the circadian cycle gradually increased to twenty hours, and after that, living beings gradually got used to the 24-hour cycle. Mammals lived under the dominance of dinosaurs due to their small size, starting from around 185 million years ago. The largest among them was the size of a cat. They were mostly warm-blooded and had to constantly eat to regulate their body temperature and meet other metabolic needs. However, the process of discovering and consuming what they needed only occurred at night. They gradually covered themselves with a layer of fur and a layer of fat to control their body temperature. Dental modifications, the presence of sweat glands, and ultimately the existence of mammary glands all contributed effectively to the struggle for survival. Initially, monotremes (with their origins in the Mesozoic era) like platypus emerged, which were the only extant egg-laying mammals capable of breastfeeding their young[1]. In the next stage, pouch-bearing animals like kangaroos and opossums, which kept their babies in pouches, entered the realm of life. The final stage belonged to placental mammals, which, due to having a placenta, could nurture their embryos in the womb until they could live outside the womb with the help of the mother.

Perhaps about 65 million years ago, the impact of a large asteroid with a diameter of over 10 km managed to wipe out the reign of dinosaurs, their airborne counterparts (pterosaurs), and their marine counterparts (ichthyosaurs, plesiosaurs). Smaller creatures like bees, which were responsible for pollinating plants, also did not escape this catastrophe. This period marked the largest mass extinction event in the history of the Earth, known as the K/T extinction, as it occurred during the Cretaceous-Tertiary (Paleogene) geological era. This description was the result of joint research by a physicist father, named Luis Alvarez and a geologist son, named Walter. In the 1970s, the son was working on a layer of the Earth that dated back to 66 million years ago. They discovered that a layer of iridium was scattered on the Earth's surface as well as deep into the Earth[2,3]. Furthermore, it was determined that the amount of iridium was more than ninety times the expected amount in that region. Based on subsequent studies, it was revealed that this heavy metal had arrived on Earth from space. It is estimated that the sudden influx of half a million tons of iridium from space was required for such a distribution of this substance. The most probable scenario was the impact of a 10–15-km asteroid at a speed exceeding 100,000 km/h on the Yucatan Peninsula[4]. Initially, the impact wave, earthquake, and tsunami destroyed many dinosaurs. Afterward, clouds of ash and dust in the sky and widespread fires on the Earth's surface eliminated many living organisms. Gradually, over the course of many years, the temperature decreased, and the surviving organisms began to reproduce and expand[5].

The era of freedom after the extinction of dinosaurs began. The early and primitive primates (a biological order that comprises monkeys, apes, humans, …) such as prosimians, emerged with fox-like eyes, forward-facing gaze, binocular vision, and fingers capable of grasping and hanging[6]. About 23–33 million years ago, monkeys capable of harming Prosimians entered the realm of existence. Except in Madagascar, where there are no monkeys, various Prosimians like lemurs are abundant; otherwise, their numbers decreased in other parts of the world. Gradually, some of the monkeys underwent transformation, lost their tails, became more intelligent, and with the

DOI: 10.1201/9781003538400-2

ability to freely rotate their shoulder joints, they could easily hang from tree branches. They were called apes (tailless monkeys). This evolutionary process was completed approximately 23–25 million years ago. Apes continued their existence in two Asian groups, including gibbons and orangutans, and an African group consisting of gorillas and chimpanzees. Gibbons have the least resemblance to current humans. They are also skilled acrobats in trees. Orangutans mostly live in Eastern Asia among trees. African gorillas prefer living on the ground and feeding on plants. The largest body size among primates belongs to gorillas, and like orangutans, they have a harem of females (one alpha male and several females) and have the closest genetic relationship to humans after chimpanzees.

Chimpanzees have longer hands compared to their legs. The span of their open hands is one and a half times their height, and they generally live in social groups. Their inclination for social life is so intense that they cannot survive alone and die. An alpha male dominates others and is constantly engaged in fighting and conflict to maintain his position. They have the ability to use simple tools and can use a piece of wood to extract termites from their nests and eat them. Large testicles and aggressive behavior to protect their territory and life area differentiate chimpanzees from gorillas. Alpha males and females have free sexual relationships, but they prevent lower-ranking individuals from engaging in sexual relations to prevent the development of their sperm. They are mostly herbivorous, occasionally consuming meat, and live in trees and, if necessary, on the ground.

Bonobos have many similarities to chimpanzees, but they are slightly smaller, more delicate, and hang from branches with their long arms. They live socially, but the female gender dominates the group. Instead of using violence to solve problems, they employ a strategy of providing sexual services to other males. Additionally, they sexually stimulate other females through masturbation. People living around the Congo River in Africa have witnessed the stark difference in behavior between aggressive alpha male chimpanzees in the north of the river and the calmer and female-dominated behavior of bonobos in the south. Perhaps easier food availability in the southern regions is one of the potential explanations for this type of behavior[7].

THE LINK

Since the early Permian period, which takes its name from the *Perm* region located in the Eastern Europe of Russia, roughly 300 million years ago, vertebrates have held sway and served as the dominant predators on Earth. Large amphibians which live both in land and water, had to reproduce and raise their offspring in water, while a new group emerged that deposited their fertilized eggs in moist soil instead of water. A protective membrane called the amnion surrounds the fertilized eggs for oxygen and nutrient exchange. Two well-known groups of animals, mammals and birds, evolved from these organisms. It is not known precisely when mammals acquired characteristics such as lactation, parental care, hair covering, constant body temperature, and neocortex (new brain cortex) during their evolutionary journey[8,9]. Some of them had not only a stirrup bone in the ear but also two additional bones in the middle ear, the hammer and anvil, which improved the efficiency of their auditory system. The evolutionary journey of these mammals continued for approximately 250–65 million years under the shadow of dinosaurs.

The most significant feature of these animal species was the presence of the neocortex. Cortical layers in the brain of reptiles formed a three-layered cortex, but mammals continued their existence with a six-layered neocortex, which is considered a form of "tinkering" in their evolutionary path. The neocortex not only brought about a versatile and computationally powerful structure but also easily expanded[10]. The neocortex ultimately created sensory maps for understanding visual, auditory, and sensory-motor inputs accurately. Simultaneously, the motor cortex and its association with the sensory system emerged; although the mechanisms are not precisely known, it ultimately led to "cognitive innovation." Although no neural tissue was added to it, its function and size underwent changes. These changes involved not only size variations but also different ratios of the neocortex to other brain regions and different ratios of the sensory-motor cortex to the associated cortex[11].

It appears that this structure was preserved as an expandable foundational structure, supporting the development of a complete language through basic neural tissue evolution[12].

The increased number of mammals was not the result of any internal or inherent superiority but simply a result of a chance event: the extinction of dinosaurs. This event led to an explosion in mammalian diversity, especially primates, and an increase in their numbers. As mentioned earlier, this event occurred approximately 65 million years ago when the Cretaceous geological period was giving way to the ancient Paleogene (Tertiary) period. After a prolonged era of being overshadowed, mammals suddenly took over the rule of dinosaurs. An extraterrestrial event, which likely resulted in the dispersion of iridium, was the cause of this, and with the beginning of the Eocene geological period around 56 million years ago, many mammalian lineages, whose fossils exist today, spread across the Earth's surface. In fact, the impact of the K/T boundary changes signifies a conventional external chance event that has had an irreversible and fundamental impact on other major evolutionary patterns.

Primates gradually transitioned from insect-eating (insectivore) to plant-eating (herbivore) animals, indicating cognitive changes in living organisms following ecological and environmental changes known as the cognitive niche[13]. In other words, these creatures had to know which area of the forest had ripe fruits and where they were better and then move in that direction. It seems that this analog cognitive force drove them to be a plant-eater. The dependency on fruits also led to an improvement in their color vision. Colorful flowers are a means by which plants utilize for better pollination and reproduction. Colorful fruits assist in the dispersal of plant seeds by animals. Plants, in their evolutionary route, have used red color for flowers and fruits and green color for tree leaves. Many fruit-eater (frugivorous) birds can detect red color quite well. Unlike primates; birds, fish, and even some reptiles have better color vision and sometimes, contrary to humans, they have four sensory receptors instead of three for color perception. Since primates initially had a nocturnal lifestyle, they had only two color receptors, and gradually, in their evolutionary path, they incorporated three receptors into their genetic structure. The extinction of dinosaurs resulted in primates becoming mammals that continued their lives by consuming large, colorful fruits while possessing color vision and depth perception.

With the cooling of the air over two million years, the surface temperature of the seas and oceans also decreased from 23°C to 17°C, and the tropical zone shifted toward the equator. Primates began to evolve from the early stages of the Miocene (means less recent) geological period and gave rise to the last common ancestor. Gradually, their hands, elbows, fingers, and teeth became similar to those of early humans. They mostly had small bodies, with an average weight ranging from 10 to 15 kg. From the middle of the Miocene period, species with larger bodies and larger brains emerged. Some of them had larger skeletal structures and weighed up to 70 kg. The presence of thick enamel teeth indicates a common characteristic between humans and other primates. The structure of their teeth suggests that their main diet consisted of fruits and plant stems, and they were likely arboreal (living among trees) creatures. They experienced slower growth, longer lifespans, and larger brains accompanied by larger bodies. Color vision indicated a higher hunting ability, and their communal living demonstrated social intelligence. Climate changes over the past five million years led to increased reproduction and breeding among monkeys, while apes were forced to inhabit more limited forest areas. The predominance of the ecological niche in monkeys was due to their ability to survive in unstable environments, have higher reproductive rates, and not rely heavily on intelligence. In summary, the last common ancestor of modern humans is an exceptional mammal with a large body, long lifespan, and large brain. Furthermore, the genetic pattern of their nervous system is borrowed from a bilaterian (symmetrically two-sided) vertebrate and insect ancestral structure, loading it into a fish model. Finally, about 65 million years ago, it displayed its genetic codes in a new mammal. Afterward, the most significant modifications have been in form and size, rather than genetic structure[14].

About seven million years ago, what we call early humans diverged from chimpanzees. Thus, our last common ancestor, the last ancestor of intelligent humans, was created, although a good fossil of it is not yet available. There are several shared characteristics between modern humans, the last common ancestor, and chimpanzees. African monkeys are explicitly social and live in relatively close-knit groups or have local social groups. Contact between social groups is prohibited, and offenders are disciplined with aggressive methods. Apes usually create a nest for sleeping and mating. Perhaps our last common ancestor also had similar behavior. Another shared characteristic is the use of simple tools, usually to extract termites and ants from their nests. Another common feature is their use of plants for personal healing. Self-medication is observed in both humans and monkeys. However, the methods of treatment vary depending on the type of soil and plants in the region. Violence is the fourth shared behavior, which in chimpanzees can lead not only to the death of adult animals but also to infanticide. In most cases, this behavior is used to demonstrate the dominance of male monkeys and increase the probability of fatherhood for their offspring[15].

Jane Goodall, a British primatologist in 1986, witnessed the first battle between two groups of chimpanzees, where the defeated males were all killed, and the females came under the control of the victorious group. Many of them act as guards for their territories and attack any lone individual they see near their boundaries. Typically, the victim is held by several monkeys while others strike or bite their bodies or reproductive organs and ultimately leave them half-dead and bloody[16]. Many of them assess their capabilities seriously before engaging in an attack. Bonobos, on the other hand, have less aggressive behavior, and although severe clashes are observed among them, their predominant behavior is peaceful intervention. Unlike chimpanzees, they have lower levels of testosterone and higher levels of serotonin, which enables them to demonstrate higher social tolerance in resolving conflicts. They often attempt to resolve their issues with rival monkeys through contact and sexual gratification. What we can call humans attempted to separate themselves from their ape-like ancestors, come down from the trees, walk upright on two legs, and make optimal use of their hands about 4–6 million years ago[17].

In 1925, an Australian anatomist and anthropologist named Raymond Dart from the Witwatersrand School of Medicine in Johannesburg, South Africa, reported the discovery of a fossil, which he later called the missing link in the evolution between humans and primates. A year before that, he managed to discover the skull of a child in the Taung region, 320 km southwest of Johannesburg. The special feature of the skull was its vertical position, small canine teeth, and the volume inside the skull was relatively small, and it was clearly placed between humans and apes. He named it *Australopithecus africanus*, meaning South African monkey, which was a combination of two Latin and Greek words. It is not clear to me whether he adapted the word *Australis* from its Latin root meaning south, or whether it was related to the word Australia, the country of Dart's birth, by chance. Anyway, from now on in this book, we call it *Austra*. Although his theory was not approved, later it became clear that his theory has created a fundamental revolution in the science of anthropology[18,19].

Mary Leakey, a British paleontologist in 1976, found the footprints of two adult *Austras* and one juvenile in a thirty-meter span in the Laetolil (now Laetoli) beds of northern Tanzania[20]. They were walking side by side in an ash-rich volcanic area. The footprints quickly hardened due to the crystallization of dissolved salts in the ash, and later, another layer of ash rapidly covered them. Mary Leakey discovered those footprints 3.5 million years later. It appears that the two adult individuals were thirty meters ahead and were leaning forward. The presence of smaller footprints around them indicates the presence of smaller animals. Their curved hands indicate their arboreal nature, and they were probably able to sleep more soundly by constructing shelters on top of trees at night.

Austras, with their long arms, would engage in fruit gathering from trees, while using their shorter legs, which allowed for greater speed and less energy consumption during walking, they roamed the continent of Africa. The slightly bent knee and hip joints gave it a somewhat stooped posture. The angle between the upper surface of the coccygeal bone and the horizon line, known as the promontory, which has an acute angle in modern humans, still resembled that of a chimpanzee or an ape and was close to being upright in *Austras*. This overall pattern allowed *Austras* to have the ability to both climb the trees and move on two feet, without the need to knuckle-walk like a chimpanzee. It seems that standing upright and the ability to move forward had greater importance for *Austras* compared to life in trees. Although it reduced energy consumption through this method, *Austras* still primarily relied on fruits, plant roots, small animals, and carrion for its diet. Therefore, its teeth had shorter canines compared to chimpanzees, and gradually, the enamel thickness of its teeth increased. The thickening of the enamel prevented tooth damage caused by consuming nutrient-rich roots buried in the soil, grains, and edible seeds of plants[21,22].

As a rule, the upsurge in *Austras*' capabilities were accompanied by an increase in brain volume and improved brain function. The average brain volume of *Austras* was about 460 cc, which is approximately 30–35% larger than that of chimpanzees with a cranial capacity of 350 cc. Although meat consumption played a fundamental role in brain development *Austras*' weight remained between 35 and 42 kg, and its height did not exceed 4–5 feet, equivalent to 120–130 cm. Additionally, the upright stance resulted in a narrower pelvic outlet, leading to more difficult childbirth. Consequently, the duration of pregnancy was shorter, and the time period of childhood and dependence on the mother escalated. The freedom of both hands in *Austras*, while walking on two legs, allowed the mother to carry her child with both hands. Gradually, due to the pressure exerted on the lower limbs, the angle of the femur bone (bicondylar) increased by ten degrees, resulting in a reduction in energy consumption during walking and hopping from one foot to the other[23]. Contrary to chimpanzees, whose leg (tibia) and ankle (talus) bones deviate outward, the angle of the shin and ankle did not change, and the big toe in chimpanzees is in a perpendicular position to other toes, in *Austras* were parallel to them apparently. Moreover, the foot arch had a noticeable curve to better withstand the pressure on the sole. Collectively, these changes improved the speed and precision of forward movement for bipedal humans[24].

To maintain balance in the spinal column, the large opening of the skull (foramen magnum) that connects the brain to the spinal cord shifted slightly to the center of the skull base. Unlike animals that walk on all fours, where the foramen magnum remains posteriorly (toward the back), in *Austras*, it moved toward the anterior (front). Like chimpanzees, *Austras* were bisexual and the male's body size was twice that of the female. This difference in body size is reaching up to 15% maximally for modern humans, which is very similar to bonobos. *Austras* had various extinct species. Lucy, unlike Taung, has a longer history and dates back to 3.9 million years ago. These events may be the result of genetic changes. Studies have revealed that two out of three SRGAP2 genes on chromosome one have undergone duplication. These genes are likely associated with neuron migration in the brain.

2 Darwin's View

THE JOURNEY

Following his completion of a bachelor's degree at the University of Cambridge, Charles Darwin received an invitation from Captain Robert Fitzroy to join him on his voyage as a biologist. Although his father initially opposed this journey, his maternal grandfather Josiah Wedgwood changed his father's mind. Initially scheduled for November 4th, the trip was delayed until December 27, 1831. Finally, the HMS Beagle set sail from the shores of England toward South America. Darwin was supposed to embark on a four-year sea voyage around South America with Captain Fitzroy. When their ship reached the southern regions of Chile, Darwin's perspective on the wildlife of that area, particularly the native Fuegians of Tierra del Fuego, who were willing to kill someone else to preserve food for their dogs during times of famine, underwent a change. He pondered on the fundamental differences in the evolutionary process between a civilized English citizen and, in his view, the savages of that region. They circumnavigated South America and eventually arrived at the Galapagos Islands near Ecuador. This region, located approximately six miles off the western coast of South America, consists of around one hundred islands. It was in this area, practically on the Earth's surface, that Darwin's theory of evolution began to take shape in his mind[25]. With a slight delay, his rival and codiscoverer of the theory of natural selection, Alfred Russel Wallace, who had traveled to the islands of the Malay Archipelago in the vicinity of Indonesia, reached similar conclusions[26,27].

Darwin had a very simple concept of natural selection, which he called the "struggle for survival." However, this behavior of selection is not necessarily achieved rapidly and through aggressive behavior, but rather through small changes accompanied by collective cooperation. Darwin considered natural selection as the transfer of better traits for adaptation to the environment. He believed that these better traits are transmitted to future generations through mating and reproduction. Natural selection affects the gene frequency and, of course, the phenotypic characteristics of organisms. In other words, the genomes that create phenotypic traits that have better compatibility with the environment leave more offspring and persist for a longer time. Natural selection or different reproduction ensures that an organism's structures and functions are in accordance with environmental conditions. In other words, Darwin practically did not talk about natural selection, but rather focused on artificial selection and sexual selection. The former refers to the performance of human individuals in breeding better specimens, similar to what animal breeders do, while the latter refers to mate choice. In the latter case, males usually engage in conflict with each other to mate with a female. They often establish territories or, in some cases, male birds choose their mate or vice versa. In essence, males fight over a mate who has the power and the right to choose. In his book "*The descent of man*," published in 1871, Darwin does not fully explain the reason for this aesthetic preference for finding a mate. He believed that this esthetic preference could be for reasons other than the mate's visible characteristics. It seems that the fundamental force in the evolutionary process depends on sexual selection rather than natural selection. Overall, the theory of natural selection is not the cause and driving force for biological adaptation, but rather its effect and result on populations, resulting from various factors, for survival control. However, Wallace had a different view than Darwin about the evolution of the brain and mind[26]. He had no justification for how the indigenous people of Australia have brains of the same size as civilized English urbanites but lack mathematical knowledge and abstract thinking, and all of this has supposedly arisen through natural selection.

The theory of evolution added two factors to natural selection to solve many of its problems. The first was sexual selection, meaning that females are generally looking for better males, and

DOI: 10.1201/9781003538400-3

the second was kin selection and altruism, which reflects a friendly reaction, meaning an organism can even sacrifice its life for the survival of its own kind[28]. Initially, the effort is to preserve and raise their offspring, but they are not reluctant to sacrifice their lives to preserve the lives of their children. Perhaps this second phenomenon has also been influential in the evolutionary path of language. These two concepts, however, were accepted with great delay. Acceptance of sexual selection was very difficult in a Victorian society dominated by males. Ultimately, another concept of the two aforementioned selections is proposed, and that is group selection, which is interpreted as intragroup competition of species for progress and survival.

Charles Darwin returned to England, after nearly five years, in October 1936. He was influenced by prominent figures such as Jean Lamarck (1744–1829), who first introduced the concept of evolution, and individuals like Comte de Buffon and Charles Lyall. Prior to Darwin, the concept of creation had been expressed in "*Genesis, Old Testament*" which was probably the first religious book. Before that, there were many stories about the creation of humans in Persian and/or Greek myths. God created everything in six days, and if we consider Saturday, which is a day of rest, it adds up to seven days (the seventh day being the Sabbath, a day of rest for Jews). However, other theories gradually gained ground in scientific circles. George Combe (1788–1858), a Scottish lawyer, introduced the theory of phrenology in the early 19th century. He believed that the shape of the skull could reveal the function of the human brain and interpret it accordingly. He saw the brain as a factory, with each part engaged in a specific task. He published his theories in the book "*The constitution of man*"[29] in 1828, and by the beginning of the 20th century, around 350,000 copies were purchased, while Darwin's book "*On the origin of species*," first published in 1859, had sold about 50,000 copies by that time. The influence of Combe's theory can still be seen in the thoughts and writings of many laypeople. Prior to him, Franz Joseph Gall (1758–1828), a German anatomist, in 1796 enumerated 27 organs that comprised the human brain[30]. Darwin also lived in a period where there were no satisfactory answers to many complex and significant scientific questions, and only religious books provided simple and relatively childish answers to those issues. These answers were usually presented in a way that individuals had to accept them. Darwin, however, probably did not properly acknowledge the theories of his paternal grandfather, Erasmus, who had written the book "*Zoonomia*"[31]. Erasmus expressed the connection between ecosystems and evolution. It seems that he somewhat overlooked the contributions of his father, Robert, who had helped him gain entry into the Royal Society of London. Perhaps this phenomenon, using political power to reach higher positions, also helped him understand the behavior of chimpanzees to become an alpha male. He believed that natural selection never occurred through significant leaps but always progressed with small steps forward. He believed that natural selection was not associated with violence and that many living beings tried to find a way to survive with each other's help. He introduced the concept of "recycling," which refers to the transformation and modification of an organ's structure in a living organism to achieve a new functional endpoint. The use of the terms "evolution" or "transformation" at that time seemed radical, therefore, he cleverly avoided stating the evolutionary process for intelligence, consciousness, and language, a problem that later his opponents, such as Friedrich Max Muller, professor of linguistics at Oxford University, insisted on[32]. Darwin tried to discuss this problem in his next book, "*The descent of man*" (1971), more clearly. This theory was contrary to the theory of species fixity, which was accepted by renowned biologists of that era, such as Sir Richard Owen and Baron Georges Cuvier, and their main foundation could be found in the Bible. Additionally, it was considered politically dangerous, as it could disrupt traditional rights in the judicial and social systems of that time. Ultimately, this new intellectual invention made it impossible to draw a clear line between different species, and it diminished the position of humans as the noblest creatures. The main significance of his theory was not only demonstrating the unity of organisms from a common ancestry but also justifying their diversity due to living in different environments.

Approximately two years after his return, he primarily worked on his theory, but practically did not publish it until 20 years later, in 1859. He only expressed his ideas among friendly scientific circles. The main reason for his silence was the pressure from the Church regarding the correctness of theories and ideas presented in religious books, and the fear of religious punishment kept him cautious. Wallace's letter to him in early 1858 indicated that similar results had been obtained in the Ternate Islands[33,34]. However, upon learning that his rival had achieved similar results, the scientific pressure outweighed the religious pressure, and he published his book titled "*On the origin of species.*" But it took another seven years for Gregor Mendel (1822–1884), Austrian priest and biologist, to present his own theories on the principles of genetics and the transmission of traits from parents to offspring. The issue of trait inheritance was one of Darwin's fundamental and unresolved problems. Prior to that, his grandfather Erasmus had been disregarded due to his research on the theory of human embryology, which contradicted the Orthodox Church's belief that humans and other living beings were created in the Garden of Eden about 4000 years ago and were expelled after eating the forbidden fruit. Erasmus believed that generally, the strongest and most active animal could reproduce and increase its species to ensure its own survival. He referred to this as the "law of nature," which was based on divine will. Furthermore, Jean Lamarck expressed the same idea as Erasmus in his book "*Philosophie zoologique.*" The elongation of giraffes' necks to reach higher branches and leaves and its survival in the next generation support this theory, which had been presented by Erasmus Darwin and Jean Lamarck. As I mentioned before, although Darwin drew from his famous grandfather's book "*Zoonomia*" and had intellectual influence from it, he never made direct references to it in his writings. In his letters to his children, he claimed to have read the book, but it did not have a significant impact on him.

With the disappearance of dinosaurs, the environment underwent fundamental changes. Small mammals started reproducing and expanding in various regions. Claws turned into feet. Eyes that were located on the sides of the head transformed into forward-facing eyes. Binocular and three-dimensional vision emerged. The brains of these creatures increased their ability to perceive a range of colors from black and white to color. Many nocturnal animals transitioned to being diurnal. The most distinguishing feature that set primates apart from other mammals was the presence of opposable thumbs, positioned opposite the other fingers, which played a crucial role in grasping objects and eventually in toolmaking and manual dexterity[35]. A smaller number of mammals that used all four limbs for locomotion gradually began to stand on their hind legs. Bipedalism became a prominent characteristic that distinguished *hominids* from other primates[36]. Perhaps three major events during the Permian geological period, approximately 40 million years ago, caused a fundamental transformation in mammals that lived on the tree of life. First, the ability to manipulate objects with hands, mainly due to the opposable thumb. Second, the emergence of three-dimensional vision, and third, the acquisition of color vision as a result of changes in the neuroanatomy of light-receiving nerve cells (neurons). Prosimians were small mammals, often insectivorous, that lived in trees. Some of them gradually developed the ability to walk on two feet and occasionally use their hands, while their teeth became stronger with an enamel layer for consuming fruit. They were now called monkeys and had tails. A portion of them, mainly in South America with flat noses, were known as New World monkeys, while a larger portion in Africa with downward-curving eagle-like noses were referred to as Old World monkeys, including macaques and baboons, continuing their own existence. This process took about two million years. About 28 million years ago, another branch of monkeys called *hominids* emerged. After that, gibbons, orangutans, gorillas, and chimpanzees separated from their common ancestry approximately 16, 13, 8, and 6 million years ago, respectively. For a long time, it was unclear what had happened between the Miocene (means less recent) period around eight million years ago and the Pliocene (means continuation of recent) period about four million years ago. The main reason was the lack of fossils that could establish a connection with *hominids* (human ancestors between great apes and *sapiens*). The discovery of *Ardipithecus* fossils, dating back to around 4.5 million years ago, which walked on two legs with a height of about one meter and weighed approximately 50 kg, partially resolved this issue. Sarich and his colleague[37,38] in 1967 demonstrated that based on the analysis of human albumin, the genetic difference between humans and chimpanzees is only about 1%. Previous studies showed a 1.6% difference and confirmed that chimpanzees are more similar to humans than gorillas. Recent studies[39] revealed 1.23% alteration in specific human nucleotides. Eighteen out of the twenty-three current human chromosomes are nearly identical to those of orangutans, gorillas, and chimpanzees. This indicates that the last common ancestor of humans and chimpanzees existed approximately 4.5–5 million years ago. However, this time is just a blink of an eye in geological and evolutionary scales.

EXAPTATION

In any case, these changes have occurred very slowly from a human perspective and over multiple generations. Genetic mutations (variations in genes) and the production of subsequent generations based on these genetic changes are significant. Isolation and the separation of a species in a particular environment can lead to the growth and development of a species. For example, Darwin used the modification in beak shapes of cormorants on the Galapagos Islands, and Wallace used the physiological changes in iguanas to support the theory of natural selection in a limited area[40]. Additionally, female bonobos in the Congo are likely dominant over males due to the abundance of plant resources, and they often resolve intra-group conflicts through sexual services. However, in the north of the river where food resources are scarce, male chimpanzees possess power and consider the use of violence as a normal means of problem-solving. Hunger is also an influential factor in the emergence of new species. Warm-blooded animals must constantly eat to maintain their metabolism and body heat, whereas cold-blooded animals can remain hungry for up to ten times longer than warm-blooded animals. Penguins are birds but have learned to swim, catch fish, and continue their own species. However, bats are mammals and unlike other birds that hunt during the day, they find their prey at night using their ultrasonic system. Severe stress can weaken and ultimately lead to the extinction of a species. Vital obstacles, artificial changes, and severe climate changes such as the Ice Age have caused changes in food and the status of animal species. Both the hunter and the prey had to improve themselves every day and adapt to the environment, otherwise survival would be impossible for them. When either the hunter or the prey becomes stronger, smarter, or more cunning, the other is doomed to extinction if they cannot improve their abilities. This phenomenon is known as the "Red Queen effect." Breeding and sexual reproduction are very helpful in this regard. In 1959, Russian researcher Dmitry Belyayev[41] attempted to breed tame foxes by continuous reproduction and raising them in a calm environment in Novosibirsk. After about 50 years of ongoing research, now conducted by Lyudmila Trut[42], they were able to breed foxes that were tamer and less aggressive toward humans, with a greater inclination to play. This collection of changes, including reduced pigmentation of the skin, smaller teeth, smaller snouts and ears, longer juvenile period, and nonseasonal increase in the sexual cycle, is referred to as the "domestication syndrome." However, it appears that these foxes in the wild are returning to some extent to their initial characteristics[43]. Perhaps initial epigenetic environmental changes are transformed into genetic hereditary changes. Commensal behaviors and mutual performance can facilitate this process in intelligent humans[44]. These changes have been manifested in the form of skull shape changes, depressed eyebrows, smaller teeth, and jaw size.

The question that arises here is whether evolution can also go backward. It seems that this phenomenon is possible but so rare that it can be said not to exist. The irreversibility of the evolutionary process is called "Dollo's Law"[45,46] after Belgian paleontologist Louis Antoine Dollo (1857–1931) first described it. In general, a good theory of evolution should be able to explain the course of changes well and potentially succeed in delivering a good end product. The classic example is the transformation of gills into air-filled sacs (swim bladder) for buoyancy in fish and then its transformation into lungs in other vertebrates. Another example is the existence of a small upper limb that has been used for temperature control after the addition of feathers to it, or the jaw bones have turned into two of the three middle ear bones and one hyoid bone in the larynx. This change of function was initially called pre-adaptation. However, Stephen J. Gould and Elisabeth Vrba revived and reconstructed it in 1982 and chose the term "exaptation" instead of the concept of adaptation[47]. This issue later received more attention in Gould's cognitive evolution. The study of animal behavior has been of interest since the time of Aristotle, but knowledge about their evolution began in detail with Darwin's publication of "*The descent of man*" in 1871. This start was accompanied by several fundamental problems. First, there were no fossils to study behavior. Second, genetic alterations have secondary effects on behavior. For example, genetic changes affect the brain, and the brain affects behavior, but the fundamental and ultimate problem was how the brain, being a highly complex organ, is capable of adapting to rapid environmental changes.

Behavior can have a specific impact on the evolutionary trajectory. First, it can guide the course of evolution. For example, many aquatic mammals that fed in water and had no specialized adaptations for aquatic life gradually influenced their evolutionary process through their behavior. The "Baldwin effect" refers to a type of evolutionary mechanism that results from the incorporation of a newly acquired or learned behavior from the environment into a living organism[48] and can later be transmitted through genetic or epigenetic means, resulting in homogenization which was first proposed in 1896 by American child psychologist called James Mark Baldwin. He was of the opinion that when a new feature created in a living being makes it flexible and improves its abilities[49,50], and it will eventually be integrated into the evolutionary process. A classic example of this is the learning of new behaviors by prey in response to predators, which gradually gets transmitted across generations. Another example is the learning of various social skills initially performed by a few individuals in a society, such as web surfing, which gradually leads to a large number of individuals in the community acquiring it, strengthening it from generation to generation. Calligraphy is another similar behavior where only a few individuals possess the skill initially, but gradually, with training, more individuals learn it. However, behavior can also have reverse effects. For example, early humans, after migrating from Africa to cold regions, utilized various coverings. This gradually led to a reduction in their body hair growth. This phenomenon is evident in Inuit Eskimos or certain nocturnal animals that have adapted to using the sun's light, resulting in the loss or reduced activity of many enzymes, particularly in their skin, which functions in cold weather.

One of the main criticisms that Wallace, a contemporary rival of Darwin, raised against the theory of natural selection was that abstract intelligence and thinking, including sciences, philosophy, mathematics, and law, all had little value and utility for early humans. Darwin believed that the phenomenon of intelligence can be explained by natural selection, but according to Wallace, abstract intelligence had no value for early humans. In his opinion, only intelligent design could justify such a paradox. However, intelligence in animals may be an adaptation to cognitive capacities[51] resulting from environmental changes. This makes human behavior, unlike other animals, not based on fixed defensive methods, and due to the existence of reasoning. He was able to design trap, make weapons, or detoxify edible plants which have toxic compounds. The term "cognitive niche" was first introduced by John Tooby and Irven DeVore[52]. Essentially, one should consider that animals evolve in response to each other. Any non-plant food source essentially comes from another organism's body. As a result, animals learn new behaviors for self-defense; they learn to hide, develop new defensive strategies, and learn how to defend themselves. However, plants cannot learn new behaviors, so they undergo chemical evolution, and produce chemical compounds in their leaves and stems that are toxic, repellent, bitter, or distasteful to deter herbivores from consuming them. This prompts animals or predators to learn that in an ongoing arms race, they must present new offensive and defensive methods. Humans also have their own cognitive niche. In biology, a "niche" refers to the role an organism plays in its environment (ecosystem). Humans use their cognitive abilities to adapt to their environment. This phenomenon[53] allows humans to invent tools, traps, and weapons. This evolutionary behavior arises through what is called mental design, and with repeated execution, it becomes more precise and sophisticated from generation to generation. Thus, humans exhibit their aggressive behavior in the present, while the defensive reactions of other animals evolve over time. Therefore, humans are highly advanced and dangerously dominant in the arms race.

Unusual characteristics of intelligent humans, unlike other animals, include the ability to create tools, engage in social cooperation among individuals (beyond close relatives), and possess the ability to speak. It can be inferred that the cognitive niche is a result of two-way natural selection pressures, and intelligence resulting from this phenomenon[54] is not a cheap or free commodity. Early humans required larger brains with greater capabilities to obtain it, which also resulted in higher energy and metabolic needs compared to other animals. The large head also brought fundamental problems such as infection and death during childbirth for mothers. In other words, early humans had various traits that they considered as a set for their evolutionary progress. For example, having hands suitable for hanging from trees (prehensile hands) rather than walking on two feet, which provided better mobility. The discovery and cooking of scarce foods such as meat, which is a nutritious and concentrated protein source for the brain, have also contributed to the improvement of alertness and cognitive abilities in early humans. Another characteristic is group living, which leads to the sharing of social intelligence. Overall, it can be concluded that intelligence has evolved with the help of a larger brain, meat-eating, group living, and a longer period of childhood. In the midst of all this, the cognitive niche has been able to play a completely effective role in the formation of the mind's structure through the development of metaphorical and abstract thinking, and it has had an impact on the evolution of intelligence. Abstract thinking makes us combine different knowledge that was originally only for solving common physical problems. This ability causes abstract cognition to emerge without the presence of an external evolutionary force[53].

3 Bipedalism

HANDYMAN

In the summer of 1959, Louis Leakey made a significant discovery in Olduvai, Tanzania; the fossilized remains of a woman. This ancient individual stood approximately one meter tall and weighed between 25 and 35 kg, and had been deceased for about 2.3 million years ago[55–57]. However, the official report of this discovery was presented in April 1964, approximately 30 years after Leakey began his research in that area. Fifteen years later, his son, Richard, found a skull in the same region with a greater age and a brain volume of about 750 cc, nearly twice that of *Australopithecus* (we call them *Austras* in this book). They named this species *Homo habilis*[58,59], meaning "skilled human" or "handyman."

He consumed more meat compared to *Austras*. He would create simple but heavy tools with his own hands or use animal bone fragments as tools for cutting meat and butchery. Gradually, his hands became longer, which aided in climbing trees, cutting objects, and swimming. He no longer lived exclusively in trees but was not a proficient hunter when he came to the ground. The strength of his shorter lower limbs increased his ability to run, walk, and climb trees. His facial bones became more delicate and smaller, although the prominent and protruding cheeks were still evident in his face. Like *Austras*, he obtained a significant portion of his energy from chewing plants and fruits, so his lower jaw was also strong and large compared to modern humans to exert more force while chewing. After butchering and skinning animals, he learned to use stone tools for cutting wood and plant stems. He now used one hand to grasp objects[60]. Furthermore, he had learned to use his thumb and forefinger to precisely manipulate objects. In other words, he had transformed into an asymmetrical lopsided ape. It is not accurately known whether they were a refined version of *Austras* or had displayed a sudden genetic change, but various species of them with larger body sizes existed under the name *Homo rudolfensis*. The average height and weight did not show significant differences compared to *Austras*, but there was evident sexual dimorphism with larger body size in male individuals. He had now learned to live in groups of seventy to eighty individuals. The presence of multiple males was necessary for the protection of the group in the plains against predatory animals and wild cats. However, they had not yet learned social behavior. Males had to compete and fight to acquire females, similar to their predecessors, and usually each male attempted to possess multiple females for himself and produce offspring for the next generation[61].

The discovery of the *H. habilis* was an influential factor in shifting research focus from Asia toward Africa in search of early human. His three main characteristics were an upright stature, walking on two feet, and the predominant use of hands for tool manipulation[62]. The dominance of one side of the body in performing tasks (lateralization), the preference for the right hand (dexterity), and the increased precision in grasping objects with fingers were all not possible without simultaneous brain growth. The average volume inside the skull of this man was approximately 650–750 cc. The relatively square-shaped skull, along with the posterior recession of the occipital bone and parietal ridges, all indicated an increase in brain volume in the parietal regions. This increase in brain volume had an impact on learning, improved memory, and the ability to initiate speech, although he did not yet know how to speak. His long and thick hair indicated that he gradually migrated to colder regions and likely adopted a less mobile lifestyle through the use of tools. He crafted simple stone tools with similar shapes and structures. Due to his higher intelligence, he achieved better advancements in cultural evolution, and his way of life was clearly more complex than that of older relatives[63].

DOI: 10.1201/9781003538400-4

ON THE FEET

Becoming bipedal was actually the first revolution in the evolutionary path of humans. The initiation of human movement on two feet led to numerous, profound, and varied impacts on human functionality. Walking upright not only caused anatomical changes in the living organism compared to quadrupeds but also resulted in functional changes in the brain. The hands became more specialized in their functions, and their somatotopic representation of those body parts in the brain became more pronounced. Actually, the body surface is mapped onto the cerebral cortex very carefully in an ordered fashion. In many stages of walking on two feet, only one foot is in contact with the ground[64]. They not only walk but also run, jump, and dance, so improving the balance system in the cerebellum, basal ganglia, inner ear, depth perception, and binocular vision, all seem to be essential simultaneously. The evolution of the function of the semicircular canals in the ears not only adjusted the various and multiple movements of the enlarged skull on the body but also proved effective in stabilizing the image on the retina during motion. These canals have undergone modifications and alterations from their ape-like form in ancestors over millions of years to their current form in upright humans.

Those changes, which began around 3–4 million years ago, accompanied the enlargement of brain volume, reduction in canine teeth size, elongation of childhood time interval, and expansion of the human lifespan. The liberation of the hands, along with the increased ability to create and use tools, established non-verbal communication, and ultimately expanded cultural capacity, were all associated with this process. Previously, the early humans were knuckle walkers, meaning they walked on two feet in an unusual way, placing the back of one of their hands, usually the right one, on the ground instead of their palms while walking, similar to the posture of an ape[65]. They often used three limbs instead of four. Knuckle walking in the jungle was somewhat more efficient and practical due to the presence of trees. Other animals such as birds and dinosaurs also walked on two feet, so their respiratory system was completely separated from their movement process, such as running. However, in animals that moved on all fours, the speed of respiration was closely linked to the speed of running. The process of walking on two feet was considered an initial adaptation (pre-adaptation) for long-distance running on extended paths[66].

In the path of natural selection, taking a quiet step has not been the main focus of how humans stroll, but the main attention has been paid to walking in difficult conditions such as running, walking on rocks, when escaping from predators and in other difficult situations. In fact, our normal walking is the result of strengthening movement skills that were previously practiced in difficult conditions. Therefore, the examination of human walking ability should not be considered from the traditional clinical point of view, in which the standard type of strolling is the walking of healthy people, but it should be evaluated from the adaptive evolutionary and osteological point of view, which pays more attention to the morphology of a person vertically. In other words, until we engage with our locomotor apparatus in a fully active manner like athletes, we will never fully understand its functional importance[67].

To transition from quadrupedalism (walking on four limbs) to bipedalism (two limbs), anatomical changes were necessary in the human body. The torso and chest transitioned from a conical shape to a barrel shape, and the pelvis became bowl-shaped. The intestines, which had become shorter, took their place in this reshaped cavity. The increased length of the small intestine indicates a change in dietary habits, which began around half a million years ago with the use of fire for cooking, enabling better utilization of food. The abdominal muscles also strengthened and transitioned from a protruding shape to a more direct and upright form[68,69]. The iliac bones in the pelvis aligned opposite each other, and the curvature of the lumbar spine increased in the pelvic region. The angle of the femur and tibia bones changed. The femur became stronger, and ultimately, the pelvic outlet became smaller[70]. The presence of abductor muscles prevented excessive rotation of the body on the head of the femur, thus preventing over-twisting during mid-stride. The big toe, which in monkeys and apes, was almost perpendicular to the other toes, gradually became parallel to them (adducted hallux). The position of the foot-arch changed as well. The foot-arch reduced (plantigrade) as the sole of the foot made more contact with the ground and increased friction, enhancing mobility[71]. The toes of the foot became shorter, with the length of the third toe accounting for about 18% of the total foot length in humans, compared to approximately 35% in chimpanzees. Therefore, the greatest force exerted in humans falls on the medial part of the foot, that is big toe (the first metatarsal bone) and the ball of the foot (the second metatarsal bone)[72]. In non-human primates, such as monkeys and apes, the force is distributed across four external plantar bones, which results in more force (torque) exerted on the NHPs' ankle. Thus, if they were to walk like humans, it would put more pressure on their tendons and muscles[72].

At the same time, the lumbar spine became longer and more mobile compared to other great apes. This type of modification has caused the weight of the head to be placed along the spine in different positions of the body. In addition, the lumbar spine, having more curvature (lordosis), has been able to stabilize the head, arms, and trunk above the pelvis and hip joint so that they are completely straight and extended during movement. The most energy consumption when taking a step is related to the ability to keep the head, arms, and trunk upright. In general, the amount of energy consumption decreases when walking on two legs, and of course, the lengthening of the two lower limbs alone does not reduce energy consumption. Walking on two legs is actually a dangerous form of walking and intensifies the possibility of falling and physical injuries. This change of shape was probably the first step in the adaptation of early humans. At the same time, with the sacrum bone (a triangular bone below the lumbar spine) becoming wider and the iliac bone shortening in the pelvic structure, childbirth has become slightly easier for bipeds. A phenomenon that has been a constant problem for humans since the Pleistocene period which is the most recent episode of Earth cooling (Ice Age)[73].

The fundamental significance of walking on two feet was the liberation of the hands. One of the first benefits was the ability to gather small fruits from short trees and carry food, including hunted meat, to higher tree branches. Additionally, in times of food scarcity, they could relocate their food to another place and have a relative sense of privacy. Chimpanzees learn to wash food roots, a behavior that still exists in Southeast Asian macaques. Having free hands gave humans the ability to create and use tools while simultaneously caring for infants.

An upright posture not only reduced the amount of sunlight radiation on the body's surface and enhanced temperature control but also reduced side effects such as heat stress. Furthermore, it resulted in reduced energy consumption, especially at lower speeds, increased field of vision for traveling from one place to another, particularly in open plains, and of course, a reduction in the risk of encountering predatory animals, which was essential for the survival of early humans. While it helped improve the ability to walk and run in pursuit of hunting consuming less energy to some extent. Standing also made humans appear more formidable against other larger predators and demonstrated a better ability to throw stones for hunting or self-defense.

Two types of chimpanzees, of the *Pan* species, namely common chimpanzees and bonobos, were separated from each other for at least one million years due to the presence of the Congo River. Humans gradually showed more pronounced differences compared to their *Pan* ancestors. The most significant difference between *Austras* and chimpanzees was walking on two feet and an increase in cranial (brain) volume, while tool use by *Austras* was similar to modern chimpanzees. Unlike the *Pan* species, which have noticeable body hair, their bodies became hairless. *Austras* still had long arms and short legs, which allowed them to move faster but with higher energy consumption. However, their long arms indicated that they still relied on tree fruits for nutrition. Longer legs, which have been observed in paleontological studies since around 2.5 million years ago, indicate that early humans became more dependent on two-legged locomotion.

Gradually, their legs became longer than their arms, allowing them to walk upright more effectively. A larger skull and relatively flat face were among the distinguishing features of early humans compared to *Pan*. A more evolved group of *Austras* called *Paranthropus* had stronger chewing muscles, smaller incisors and canines, larger molars, and a stronger lower jaw for chewing[74]. It was during this period that the ability to create simple tools known as the Oldowan industry emerged. The first findings of this industry were discovered in the Olduvai Gorge in northern Tanzania and southern Kenya. In this industry, they mostly used stone tools and generally settled by the rivers. Using these simple tools to skin animals resulted in early humans obtaining a larger amount of meat and bone marrow, which were essential for body and brain growth. Additionally, it provided nourishment for a larger number of individuals in their small communities.

Climate change has caused increased humidity to be trapped in icy lakes, polar ice, and glaciers, resulting in forests transforming into vast plains. Walking upright in the plains helped to see farther distances. The toes became shorter, losing the ability to hang and grasp tree branches and leaves, and instead, the ability to move forward was enhanced. The large hole (foramen magnum) in the lower part of the skull shifted slightly forward to align the head along the correct axis of the spinal column[75]. Walking and leaping in the plains led early humans to live further away from forested areas, transitioning from a tree-dwelling life to a free-roaming life on the ground. Eventually, primates on trees evolved into two-legged humans on the ground. As the weather grew colder, the number of trees decreased while the number of bipeds on the ground increased.

Gibbons were more capable than other non-human primates in walking on two legs. Chimpanzees, on the other hand, walk on two legs for about 10–15% of their ground movement. *Austras*, who lived in this manner for three million years, divided into two branches: robust and gracile. The first group had large teeth with muscular jaws and focused on a diet of fibrous fruits and nutritious roots, while the second group showed a greater preference for small fruits and edible seeds. Eventually, with worsening weather conditions, they turned to consuming carrion[76]. Gradually, during the *H. habilis* period, they began using simple stone tools to hunt and consume meat. The ability to engage in this behavior improved in *H. erectus*, and ultimately, the process of making fire, maintaining it, and utilizing cooked meat led to a three-fold increase in their body size and, of course, their brain size.

The human vertebral column has an S-shaped curvature, providing greater flexibility. In contrast, chimpanzees have a rigid and curved vertebral column. Moving on two legs clearly increases the torque on the knees and ankles. This situation puts a lot of pressure on the lower limbs. General deformation of the lumbar spine and pelvis has been associated with a rise in diseases such as discopathy (disc herniation), low back pain, arthritis, joint injuries, and varicose veins of the lower limbs. Although chimpanzees have better walking abilities than humans; humans have superior running capabilities compared to primates.

The growth of the head, through the consumption of meat, and the narrowing of the pelvic canal due to walking on two legs, resulted in a shorter gestation period and earlier birth of human offspring[68]. Although it was difficult for the fetus's head to pass through the birth canal, the infant became heavily dependent on the parents due to incomplete growth of the head and inability to walk immediately after birth. The brain volume at birth was around 230–340 cc, approximately 510 cc at three months, and around 720 cc at first year. In other words, at three months, it reached approximately 55% of adult size, and at the end of first year, it reached around 70–85% of adult size. In comparison, the brain volume at birth for chimpanzees is about 40% of adult human volume for *H. erectus* and 35% for *sapiens*.

The ability to walk in infants begins after 12 months, indicating a shortened gestational period in humans and the need for premature delivery due to the small diameter of the birth canal. Improvement in walking skills and the use of limbs depends on the growth and maturation of the nervous system. Central nervous system damage that leads to local tension and contracture affects motor development, standing capability, and walking. At the same time, the evolution of the vestibular (balancing) system in the inner ear and improved vision contribute to more delicate and precise movements[77]. However, the completion and improvement of this situation take until the age of twelve. The development of the proprioceptive system for the perception of body position also plays an effective role in balance control and movement improvement. In fact, there is a clear and significant relationship between posture and motion. Understanding the movement of objects in the surrounding visual field helps adjust and effectively maintain balance, ultimately leading to independent movement in children. Environmental and seasonal factors are also likely to be necessary and influential in this matter. The weight of the lower limbs in infants is also essential for initiating effective movement. At birth, reflexive stepping movements can be observed in infants. As the weight of the lower limbs increases due to the augmentation of muscles and the strengthening of them, this reflex disappears. Studies by Thelen and colleagues[78] have shown that by placing the feet of infants in water, possibly due to the sensation of reduced limb weight, this reflex can be reactivated.

Approximately 30,000–40,000 years ago, significant and prominent evolutionary changes ceased, and small evolutionary changes at the cellular and molecular levels continued. The size of the human body became adapted to the environment[79], where individuals living closer to the equator became leaner, while those living farther away tended to be relatively more robust and stocky. In other words, Neanderthals, like intelligent humans, walked upright, but due to their shorter and more robust stature, they consumed more energy. The width of the pelvis in men can vary between about 20 and 30 cm. Leaner individuals have a greater capacity for heat exchange, whereas this capacity is lower in obese individuals. In essence, skeletal structure represents a balance between bipedal locomotion capabilities and the limitations imposed by environmental, climatic, and childbirth factors. Key changes in the modern skeletal structure include the reduced height of the iliac bone wings in the pelvis, spinal column curvature, foot arch, and the positioning of the big toe parallel to the other toes.

The learning process of walking usually takes several months in children. Although mastering this skill requires years of practice, overall it leads to improvement in motor skills. However, it should be noted that the skill of using both feet precedes the ability to use a dominant hand, and the use of oppositional thumb is before the ability to use one hand dominantly. In order for language to develop and for us to be able to communicate, bipedalism still requires more time. The use of a dominant hand and the ability to speak are almost exceptional skills of humans[80]. In other words, bipedalism is the root of what we call human. This evolutionary process is accompanied by significant changes in behavior. For example, visual attention to the surrounding environment is necessary for monitoring the environment and preventing collisions with other beings. Additionally, visual ability allows us to visualize the whereabouts of any being, perceive its presence or absence, and remember where it is hiding. We keep pathways in mind and learn not to get lost and find doorways, and ultimately, we can make decisions independently.

Modern humans shift their weight forward during movement, which allows them to move at an average speed of about 0.7 m/s. The movement on two feet has evolved from using the toes to using the heel and sole of the foot, although modern humans still maintain the ability to move on their toes to reduce noise while walking[81]. The most important needs of intelligent beings for bipedal movement are balance, forward propulsion, gravity support, and taking proper steps. At the same time, motor coordination, vision, and proper functioning of the vestibular system in the inner ear are crucial. The hip–knee flexion strategy is highly important for anticipating obstacles and timely adaptive reactions during fast movements. This set of abilities demonstrates that early humans had acquired the ability to receive and transform sensory information from vision, vestibular (balance in the inner ear) system, depth perception, and other internal and external senses into motor output signals.

UPRIGHT POSTURE

The moniker "*Java man*" is attributed to one of the earliest discovered fossils of an upright human, linked to the Kedung Brubus region in Java, Indonesia. This fossil was first discovered in 1890 by Eugene Dubois[82], a Dutch anthropologist, but the most well-preserved fossil, dubbed *Turkana boy*, belonged to a 12-year-old boy standing at 1.5 meters tall, who lived approximately 1.65 million years ago in what area is now called Kenya[83]. He has been given different names in different regions. In Asia and East Asia, he is known as *Peking man*, in Southeast Asia as *Java man*, in Africa as *Homo ergaster*, in Europe as *Tautavel man*, and in the Middle East and parts of Europe as *Homo erectus*.

With a short thoracic vertebral column, he reached a height of 1.8 meters and had a brain volume of up to 900 cc. His short stature shows that his level of breathing control was not complete and probably could not talk like a modern human being. Larger and bigger vertebral bones in *sapiens* is a marker of thicker spinal cord which was an adaptation for better breathing and ability to speak. He had learned to walk easily on two feet and even swim better than modern humans. His longer legs and shorter hands compared to *H. habilis* made it easier for him to do so. In addition, the structure of his chest changed from conical to barrel-shaped to improve respiratory capacity and volume.

In warmer regions close to the equator, long legs, tall stature, and narrow hips (width of 23–30 cm in men) helped them better regulate heat, but in *H. erectus*, in northern regions, they still maintained their robust characteristics. He was a successful species that continued to live in limited geographical areas of the earth until 50,000 years ago. Body size was directly related to geographic latitude. At the same time, the reduction of surface skin hair and improved sweating ability helped control body temperature. Additionally, melanocyte skin receptors increased. Darker skin, with better heat absorption from sunlight and increased vitamin D production, contributed to improved bone formation and walking on two feet. He was also a toolmaker, and in addition to using stone layers (lithic flakes), he used tools whose two-sided surfaces were more effective[84] for hitting and cutting. These tools, such as pickaxes, knives, cleavers, and hatchets, were usually larger, heavier, and multi-functional. Actually, he entered the Acheulean industry, named after Saint-Acheul in northern France. In fact, *H. erectus* lived in this period and in terms of structure, he had covered up to 90% of the way to become *sapiens*. These tools, which had two faces (bifacial), were often durable, versatile, and specialized for thousands of years. Making these tools require long-term training and learning. They were used for striking, cutting, and skinning, and are still used in some primitive tribes[85].

He knew fire from 1.5 million years ago in Kenya, but maintaining fire required time and the need for individuals who had knowledge and expertise in combustible materials. This process took more than a million years. Around 700,000 years ago, the use of stone tools and fire-making techniques to ignite fire became known, and around 300,000 years ago, the use of materials that burned slowly, such as tinder, was learned. Using various tools and utilization of fire improved the consumption of meat obtained from hunting large animals such as buffalo, boar, elephant, rhino and/or hippo. Increased consumption of meat, especially cooked meat, and decreased consumption of plants, root vegetables, and fruits led to a smaller lower jaw and a temporary reduction in tooth enamel thickness. This change in consumption patterns was accompanied by a reduction in the length of the intestine. Cooking food, especially meat, significantly increases the absorption of nutrients. The most important invention of *H. erectus*, the use of fire for cooking meat and food, led to an increase in human socialization. Additionally, improved nutrition resulted in a decrease in sexual dimorphism (the differences in size or appearance between males and females). These changes were observed not only in body size but also in the size of the larynx and a change in vocal frequency, ultimately leading to the development of two-way communication.

The height increased up to 180 cm, and weight increased up to 70 kg. Although males were still larger and stronger than females, the difference was less compared to their ancestors. Polygamy (having multiple spouses) and polygyny (having multiple wives) transformed into monogamy. Physical structure underwent changes due to environmental variations and genetic mutations. The skull became larger. The average volume inside the skull increased to 950 cc and sometimes reached 1200 cc with prominent eyebrows and for the first time, a protruding nose appeared on the face. Improving the structure of the nose not only helped to warm and humidify the air for the barrel-shaped chest but also made it learn the ability to navigate and eventually migrate long distances with the spatial perception of different smells. With increased socialization, division of labor and sharing of food and its storage became part of social behavior. Women engaged in household chores and farming, while men were occupied with hunting, warfare with other groups, and defending their territory and land. The cave-dwelling human transformed into a hunter–gatherer. The use of fire not only aided in cooking food and warming living spaces, as well as keeping away wild animals but also by creating light at night, it brought group members together for up to 16 hours a day. Although they still did not possess the ability to speak and used a form of proto-language, this ultimately led to the emergence of the most prominent human characteristic, language, and the ability to communicate. Furthermore, fire helped in keeping themselves warm and surviving during migration from tropical regions to cold and frigid regions. It has also enabled them to adapt to terrestrial and sparsely vegetated areas. Gradually, these individuals learned how to cooperate with each other to protect themselves against environmental hazards and provide care and support for the sick and injured.

At the same time as sea travel, painting, and art forms, the use of color, especially red color for wall paintings of animals, other people, and everyday events, became evident in social behavior. Sea voyages sometimes indicated the ability to map and plan an action before carrying it out. At the same time, the ability to create artistic objects and various paintings showcased their symbolic thinking[86]. This thinking was accompanied by and harmonized with modern behavior and understanding of various phenomena. The aesthetic-cognitive sense of humans likely aided in their survival. The coloring of the human body reflected a simpler understanding of self and others. However, they still lacked a well-developed vocal structure for speech and relied on what we call the primary or ancient language. Changes in consumption patterns, the use of different coverings for warmth, walking, and long-distance movement on two feet exposed them to new diseases such as tooth decay, oral diseases, skin parasites, back pain, scoliosis, and misalignment of the spine and lumbar discs.

It should be noted that most of the anthropologists in the last century were strongly influenced by Darwin's theories. He founded a hypothesis that with the release of the hands and the ability to walk on two legs to move, the dominance of one hand in doing daily tasks, toolmaking, the reduction of fangs, and making weapons for hunting and self-defense was justified. *Austra* transformed from an arboreal, omnivorous, and bipedal creature into a toolmaking human, generally right-handed, with the ability to grasp objects in the fist (power grip) or between the fingers (precision grip) with a small cheek (*H. habilis*) and turning it into a hunter–gatherer, with a bigger brain, a more complex technology and globally spread (*H. erectus*). Of course, we still need newer findings to prove it definitively, and there is still a sea of difference between ape-man and true-man.

ANCIENT HUMAN

The emergence of *archaic* (ancient) humans stemmed from the gradual and slow evolution of *H. erectus* which represented the early forms of both intelligent humans (*H. sapiens*) as well as *Homo neanderthalensis*. The most famous fossil of this species[87] was discovered in Swanscombe, near London, in 1935. Other findings gradually emerged in the following 20 years. Like their predecessors, they had a prominent brow ridge and a slightly sloping forehead. Approximately 200,000 years ago, they split into three main branches. The first branch was in East Asia, where they were able to produce stone tools. The second group formed in Europe, the Mediterranean region, and Central Asia, representing the Neanderthal humans, and eventually, modern humans emerged in the southern region of the African desert.

Neanderthals existed on Earth until 35,000 years ago, but the oldest fossil attributed to them dates back to 100,000 years ago. However, they commonly lived around 70,000 years ago. Nowadays, the term "Neanderthal" in the lexicon is used figuratively to mean someone brutish, uncivilized, and somewhat foolish. In 1856, their fossils were first discovered in the Neander valley, after the name of German theologian Joachim Neander, next to the Dussel River, 20 km far from Dusseldorf (in fact Neandertal written with a "T" in German but often spelled with "Th" in English texts). As John Pfeiffer[88] put it, the discovery of Neanderthals during the Victorian era was like encountering wild and naked savages thrown into the midst of the classy ladies' circle of that period[89]. Since they lived during the Ice Age in Europe, they were cold-resistant, characterized by larger skulls, thicker bones, and a muscular and robust physique. Their receding chin and relatively large nose served as a radiator to warm the air. Their muscular body and strong legs suited them for long walks. Meanwhile, they used their hard and durable teeth to crush and cut food, mainly consisting of hunted meat. The primary hunting method involved ambushing and launching collective attacks on large animals. However, their fundamental challenge was the lack of advanced intelligence to create weapons, hunting in small groups not large ones like modern human[90], and the inability to establish verbal communication with their fellow beings. It is likely that this factor prevented them from being able to confront modern humans and led to their extinction at the hands of their fellow intelligent beings. With the death of the last Neanderthal about 28,000 years ago[91], space was opened for the undisputed rule of wise man.

The migration of *sapiens* about 120,000 years ago from Africa to the Middle East, Far East, and Europe forced them to adapt to new biospheric conditions[92] including colder weather, higher altitude, and exposure to new microbes. Contact with microbes and susceptibility to diseases, especially after the transition of humans from hunter–gatherers to pastoralists in the last 3000–6000 years[93,94] has shown a clear increase. So, this period, known as the Bronze Age, can be considered a period of co-evolution between humans and microbes. Perhaps one of the most prominent characteristics of archaic humans was their understanding of the concept of death. Burial of the dead[95] was among their customs. Placing flowers beside the lifeless bodies of their fellow beings, as recently discovered in a site in Iraq dating back 60,000 years ago, and conducting specific burial rituals, indicate a symbolic manifestation of their understanding of the concept of death. These traditions sometimes went to the extent of consuming the brains of their deceased after crushing their skulls. They adapted to cold climates, and their primitive form was distinguished by dark hair and dark skin, but the Cro-Magnon branch had lighter hair and skin. They also had control over fire and were skilled hunters, often hunting collectively for medium-sized animal species such as deer or larger ones like mammoths. DNA analysis has shown a similarity of up to 99.7% between them and modern humans. Although they were not highly intelligent, their average brain volume was larger than that of modern humans, reaching up to 1600 cc. Some groups among them also had the ability to use needles and had a basic understanding of what we now call sewing.

Whether modern humans evolved directly from gradual modifications in Neanderthals or if they were a separate branch, we do not know for certain. Secondly, the details of how this group of humans dispersed are still not well established. It is possible that these human specimens of *H. erectus* originated in various parts of the Earth and then spread. However, in other studies, it has been proposed that they initially emerged in one region and gradually dispersed and expanded. Clark Howell[96,97] from Chicago University believes in an intermediate model. Ultimately, due to their better abilities in planning, employing improved hunting methods, training dogs-wolves for hunting, and having simple verbal communication skills through vocalization, modern humans (*sapiens*) caused the demise of Neanderthals[98]. In this way modern humans settled approximately 90,000 years ago in the Middle East, 65,000 years ago in Australia, 30,000 years ago in New Guinea, and 12,000 years ago in the New World[99].

4 Aquaman

SKELETAL STRUCTURE

In order to successfully take a single step, not it only is imperative that our nervous system undergoes meticulous formation and evolution, but we also need a well-developed, finely tuned, and proportionate skeletal-muscular structure. In bipedal (two limbs) organisms, the bones are located beneath the skull, unlike quadrupedal (four limbs) organisms. In humans, out of the 206 bones in the skeletal structure, 177 (86%) are located below the skull. Out of this number, 126 are related to the upper and lower limbs, which are almost equally divided (31% in upper limbs and 30% in lower limbs). The cross-sectional area on which quadruped animals stand is much wider than that of bipedal animals, indicating that the evolution of a balanced system for bipeds has been crucial in controlling their posture[100]. The center of gravity in humans is located in front of the sacrum bone in the pelvis, and the contraction of muscles and ligaments within the pelvis (iliopsoas and hip flexors), prevents the body from falling forward or backward. Increased oxygen consumption indicates that standing posture is associated with a 7% increase in energy expenditure compared to lying down, but the energy expenditure in humans for walking is lower compared to chimpanzees.

The lower limbs in humans make up about 30% of body weight, slightly more than in gorillas (28%) and orangutans (24%). In contrast, the weight of the upper limbs in modern humans is approximately 8% of the total body weight, less than in gorillas (12%) and orangutans (16%). The reason for the increase in weight and the stronger skeletal structure is due to the weight-bearing capacity of the trunk, pelvis, head, and upper limbs, which directly exert pressure on a small cross-sectional area[101]. However, it should be noted that different climatic conditions and types of manual labor can alter the shape of the upper limbs, especially the hands, as seen in Neanderthals due to the colder climate of their living environment. The combination of these factors results in the maximum walking speed of humans being 4.5–5 km/h and their running speed being up to 9 km/h. It should be noted that the energy consumption in the human model is lower compared to chimpanzees. The main reason for this is that the hip and knee joints of chimpanzees never fully extend, unlike in humans. This phenomenon is accompanied by increased energy consumption during walking. Non-human primates relatively conserve a lot of energy, and the most efficient and similar model of movement or walking in nonhuman primates is related to the gait of a spider monkey.

The foramen magnum is a large hole in the occipital bone that connects the brain and spinal cord to each other. In humans, this opening has moved slightly forward, causing the head to be positioned directly above the cervical spine. In other words, the anterior (front) edge of the foramen magnum is located right in the middle of the base of the skull. The placement of the foramen magnum depends on the positioning of the head, torso, and the locomotion method[102] of the living organism. Quadrupeds and ungulates, which move on two legs, have undergone evolutionary modifications, including elongation of the legs, and generally utilize a galloping gait, so their foramen magnum is slightly forward of the midline. It is possible that in humans, as the brain size expanded and the frontal lobes protruded, the head gradually rotated downward, causing the foramen magnum to move forward. Eventually, the tail completely disappeared, although the presence of a tail in primates effectively aids in maintaining their balance, it has lost its significance in the evolutionary process of humans[103].

The skeletal structure in animal models clearly shows that the head and torso are bent forward, and the tail helps with balance in the rear. As we approach the human model in the evolutionary process, the tail shrinks, and the head moves backward until ultimately humans stand upright on

DOI: 10.1201/9781003538400-5

two legs. The evolution of cellular structure in the cerebellum (literally means little brain) has also led to the coordination of fine and delicate hand movements with the eyes in humans, allowing for walking on two legs and efficient use of the hands, especially the fingers well.

LIFE AND ENERGY

Approximately two million years ago, the African tectonic plate encountered the Eurasian plate from the northeast; initiating a monumental collision. This geological event gave rise to towering mountain ranges, with some peaks transforming into active volcanoes. These tall mountains caused a significant reduction in rainfall in the region behind them, transforming forested areas into fertile plains and open lands. Some areas gradually became semi-arid and semi-desert. These mountain ranges have created the Great Rift Valley, which starts from the Beqaa Valley in Lebanon and extends to Mozambique[104] in southern Africa. In the path of this region, the Omo River in Ethiopia carries rich deposits of soft soil, containing fossils of early humans dating back several million years. The geographical changes on Earth clearly enhanced the visual ability of both predators and prey which was the major role of early humans at different times. The transition from a quadrupedal "arbori-life" form (living on trees) to a bipedal form was a result of natural selection pressure, compared to a four-legged creature that moves on the ground. Most likely, the fundamental significance of this natural selection was the liberation of hands. When hands were freed from the task of weight-bearing, humans could use their hands for grasping, climbing, pointing, and carrying objects. Pointing at objects played a crucial role in shaping verbal language. The use of language and memory also played an essential role in the reconstruction of the thinking process. In other words, we think with our speech ability, the language.

Bipedal humans had to reduce energy consumption for survival. This natural selection, originating from the function of the nervous system, had to coordinate the skeletal-muscular structure accordingly. There are 26 bones in the human foot. Unlike apes, the thumb is positioned parallel to the other fingers, which has changed since the existence of skillful humans (*H. habilis*). Additionally, human feet are shorter and lack the ability to hang from tree branches compared to apes. The ratio of the length of the third toe to the entire foot is 18% in humans, 33% in gorillas, and 35% in chimpanzees[101]. At the same time, the size of the human foot is longer than the human hand. The ratio of the upper limbs to the lower limbs is 0.7 in humans, while in the nearest non-human primates, the orangutans are 4.40, pygmy chimpanzees are 4.01, and in gorillas are 1.20. Scientific studies have shown that humans exert more pressure on the wrist and sole bones compared to non-human primates, where the highest pressure is applied to the outer fingers. This issue is caused if primates want to walk on two legs like humans, they put about twice the pressure on the leg muscles and ligaments, and of course, they get tired much faster. Given that the highest pressure is exerted inward on the foot when humans take a step, the change in the shape of the foot from a perpendicular position to the other fingers to a parallel position has been a very accurate and important process, which has made the inner bones of the foot stronger and more robust than the outer bones of that area. Ultimately, the arch of the foot has reduced the transfer of pressure to the inner part of the foot when taking a step.

The usual walking speed is about 0.7 m/s. Stride consists of a stance phase, which accounts for about 60% of the time, and the swinging phase, which makes up the remaining 40% of the time. This type of walking is commonly referred to as "plantar" walking, meaning that the heel and sole of the foot make contact with the ground. The characteristic of this walking style is the impact on the heel when the foot strikes the ground and the toe-off when the foot lifts off the ground. The necessary factors for this type of gait include initially maintaining body alignment against gravity, followed by stepping and ultimately maintaining balance[105,106]. When this system is formed, a forward propulsive force is generated by shifting the weight forward of the center of gravity. Typically, it creates a relatively constant speed of 0.7–1.25 m/s. This forward motion requires adaptation to the external environment. Reactive and predictive control mechanisms located in the upper and lower parts of the nervous system are involved in this process. Therefore, having a sense of depth, vision, and a well-functioning balance system will greatly impact walking. Additionally, the coordination of the visual-motor system is a key factor in successful navigation without assistance from other objects. In this situation, when the foot encounters an obstacle during movement, the predictive mechanisms initiate balance adjustments and gait regulation through reactive bending of the knee and hip.

In the evolutionary course of early humans, they were forced to migrate or engage in constant battles to preserve their living territory. Division of labor typically involved women carrying and caring for infants, while men searched for food and fought other wild animals. Generally, the *Homo* species utilized most of their acquired energy for brain growth until reaching adolescence. After that, with increased hunting abilities, the energy production rate increased significantly in the pursuit of discovering food sources or hunting other animals. Although their capabilities begin to decline around the age of 45, they still provide more than they consume to their group or tribe until the age of sixty. Hunting contributed significantly to their socialization, cognitive development, and ultimately the size of the human brain. The process of white matter degeneration and brain shrinking starts around the age of sixty, indicating a reduced need for innovative brain function. It seems that if innovation continues, the quantity of white matter does not change. This process can be considered as a kind of brain tissue adaptation, which of course is associated with high energy consumption. Free hands were very effective for both women and men in this regard. This greatly assisted in the survival of future generations and proved valuable. Furthermore, the hunter–gatherer humans had to start constructing tools to trap and kill prey and for solving other problems. Toolmaking, which began with skilled humans, required free hands. Gradually, stone tools were used for grinding plants and cutting meat from hunted animals. This made food consumption easier for them. Previously, early humans had to use their teeth, like carnivorous animals, to tear the skin of their prey or wait for scavengers to tear the skin before using the decaying meat. Stone tools were very helpful in this regard. Women invented agricultural tools at home to be able to share the increased food production with their children. This behavior could somewhat feminize male behavior and improve social relationships between women and men[107]. Many non-human primates could cover a relatively short distance of about 30 meters while carrying food. They were probably engaged in this activity for 2–3 hours[108] during the night. The best current nonhuman primates to perform this task are gibbons, although others are also capable. Chimpanzees perform about 10–15% of their ground movements in this manner. During periods of food scarcity, non-human primates prefer to transfer their food to a place that is out of reach in order to preserve their privacy. Additionally, food has also had an influential role in culture, to the extent that many societies are identified based on the combination and diversity of their diets. The agricultural revolution has caused the *Homo* branch to separate significantly from other primates, to the point where certain groups are named based on their dietary preferences and interests, such as calling themselves vegetarians.

In this regard, management of energy consumption becomes important. If quadrupeds were to stand on two legs, their energy consumption would increase[109]. The reason for this is the bending of the legs and the constant contraction of the abdominal, gluteal, and thigh muscles to maintain balance. In chimpanzees, this posture of standing on two legs is associated with high energy consumption. For a chimpanzee to be able to easily stand on two legs, the angle between the upper part of the pelvis (ilium) and the lower part of the pelvis (ischium), known as the fulcrum angle, must transition from an acute position to a divergent position to bear the weight. At the same time, the lumbar curve increases. This shows itself well with the increase in back pain and the exacerbation of different types of disc diseases in today's people. Ultimately, a reduction in pelvic height can be helpful. This alteration in the shape of the pelvis is also accompanied by a reduction in the diameter of the birth canal. Finally, modern humans (*sapiens*) were forced to reduce the duration of the life period of an unborn baby in the womb.

AQUATIC PHASE

The conversion from quadrupedalism to bipedalism remains a puzzle for and continues to baffle many researchers. Evidence supporting this transition has proven elusive, relying primarily on a range of theories. In the 1950s, two separate teams were working on the discovery of the structure of DNA. The first team, led by Linus Pauling in the United States, had two Nobel Prizes in his pocket. The second team, led by Francis Crick and James Watson[110,111], was based in England. In fact, they were interested in the origin of life and the puzzle of consciousness. Another crucial and important figure who assisted them in capturing images through X-ray diffraction was Rosalind Franklin[112], although her role in the discovery of DNA structure is often overlooked. In addition to using radioisotopes like carbon-14, DNA is also utilized in determining the age of fossils. Mitochondrial DNA (mtDNA) found in mitochondria (a cellular organelle which generates energy) provides valuable insights into the closest maternal ancestry of humans. This DNA, responsible for cellular mitochondrial functions, is composed of 16,500 nucleotides and includes 13 genes. It is solely inherited through the maternal line, which means it can also indicate non-sexual reproduction, thus not affected by gene swapping during sexual gene transfer. Mutations (variations) in mtDNA are observed up to ten times more frequently, making it a foundation for understanding our ancestors' timeline. Moreover, it can be extracted in larger quantities from bones. In the conducted studies, the birth of "Mitochondrial Eve" is estimated to have occurred approximately 140,000–200,000 years ago. In other words, our species formed at least 110,000 years ago and inhabited in Africa. However, for investigating the genealogy of "Adam," the most recent male ancestor, the "Y-chromosome Adam" is utilized. The examination of the male lineage pushes the timeline of intelligent humans back to 195,000 years ago. It is likely that the population of intelligent humans in Africa, around 150,000 years ago, consisted of approximately 20,000 individuals, which reached one million individuals during the exodus from Africa. Yet, there is still a missing link between humans and apes. This entity could have been somewhat bipedal and possibly had a larger brain than apes. In other words, walking on two legs may have occurred before significant brain enlargement.

This intermediary between humans and apes can be called "man-ape." The first trace of this creature was discovered by Mary Leakey in 1976 in Laetoli[20], northern Tanzania. She named it *Ardipithecus* and estimated its age to be 4.5 million years. However, the ideal body for walking on two legs includes short hands and long, strong legs. Ape feet work very well for hanging and swinging from tree branches. Meanwhile, in humans, the intestines are located inside a bowl-shaped pelvis and do not protrude outward. The hip joint in humans is also larger and, along with a strong thigh bone, supports the weight of the torso and head. The foramen magnum in apes is slightly further back compared to modern humans. In other words, the foramen magnum in humans is located in the middle[113] and aligns the head exactly with the spinal column.

The Aquatic Phase Theory was first proposed by the German pathologist Max Westenhöfer, titled "The Aquatic Ape Hypothesis" in his book "*Der Eigenweg des Menschen*" (*Man's own way*) in 1942. It was later revisited by the English biologist Sir Alister Hardy. However, it was ultimately Elaine Morgan, a Welsh scientist from England, who decided to examine this theory in her famous books "*The descent of woman*" in 1972[114] and later "*The scars of evolution*" in 1990;[115] and answer three important questions. The first question was why our brains are three times larger than those of apes? The second question was why we do not have hair, or fur? And the final question was why we can speak? This theory initially captured the attention of the scientific community, but later, several arguments were raised against it, although it was never completely rejected. According to Morgan, life in an aquatic and marshy environment was necessary for human evolution and the beginning of bipedalism. She believes that the Great Rift Valley near the Red Sea was the birthplace of early humans. Perhaps this region had been separated from the surrounding environment due to rifting forces and periodic floods. Additionally, unstable food sources could have aided in the formation of the aquatic phase. As a result, our ancestors lived in a region that was neither dry nor aquatic but rather relatively marshy. However, due to severe floods that occurred around 4–8 million years ago, we do not have fossils from that period.

She believes that starting to walk on two legs in open plains or forests is much less possible. In marshy areas and sometimes in the sea, living creatures need to keep their heads up to breathe. At the same time, having a downward-facing nose is better for swimming in water and prevents water from entering the nasal and respiratory passages. Additionally, due to the presence of a thick subcutaneous fat layer, it doesn't need hair to keep warm. For temperature regulation, hair and fur are beneficial for animals that have evolved in dry environments, but the presence of subcutaneous fat in aquatic animals such as whales, seals, manatees, and hippos is very helpful. The number of fat cells in the human body is ten times that of monkeys. Animals with abundant fat cells are either in hibernation or have an aquatic phase. In any case, humans are not part of the first group. However, unlike the hairless and slim monkey infant, human infants are mostly born without hair, with a round and bulbous appearance, and can float on water shortly after birth. Perhaps the aquatic phase has somehow been effective in starting to walk on two legs. The presence of hair is effective for maintaining heat, and of course, controlling body temperature. Its protective effect against blows, scratches, parasites, sunlight, and rain is undeniable. At the same time, its aesthetic importance has been proven in finding a mate, especially in birds. Considering that the ratio of body surface area to body volume is disproportionate in small mammals, and heat retention is very difficult, the presence of hair is necessary and helpful for small mammals. Conversely, for large mammals such as elephants, rhinos, and hippos that live in hot regions, having hair increases body heat. Perhaps in the transition from a life in the jungle on trees to a life in the plains, humans' early mobility led to increased heat production. Losing hair helps a lot in cooling the body and better functioning of the brain. Meanwhile, eccrine sweat glands help cool the body by excreting water as sweat with salt, but apocrine glands protect its fur skin from friction by producing oily secretions. However, humans have not lost all their hair. Scalp hair is probably effective in preventing direct sunlight, underarm hair, pubic hair, and thigh hair may have been effective in preventing surfaces from rubbing against each other. Additionally, eyebrows, beards, and mustaches may have had aesthetic importance. The frizzy hair of the African model acts as a radiator due to the movement of air among the hair, helping to cool the skin. Meanwhile, being without hair does not contribute significantly to reducing heat[114,116], especially when running in pursuit of prey or escaping predators.

The upright posture can be a result of evolutionary progression from wading and striving to move with two feet in water. The human interest in water and the ability to learn to swim may be justified by this assumption. Exiting the water after swimming is accompanied by a feeling of weight reduction, and standing afterward requires having a strong and effective sense of balance. Perhaps the good sensory balance state in humans is a result of the phenomenon of exiting water in their evolutionary path. About 9% of boys and 6% of girls have webbed toes, especially between the second and third toes. In addition, excessive webbing between the thumb and index finger hinders movement in 90% of cases. This condition can significantly aid swimming. Unlike monkeys, humans have the ability to excrete salt due to the presence of sweat glands. It is likely that these glands have evolved due to living in a salty environment, and in humans, the salt deficiency alert system is dysfunctional, unlike many other animals. Humans are the only primates that shed tears, and the taste is salty. This phenomenon bears a special resemblance to the tears of marine turtles[117].

The larynx in human infants is located higher than the trachea until three months of age[118] and gradually moves downward. Additionally, the soft palate can close or open the nasal respiratory pathway with upward and downward movements, which can be effective in producing sound and ultimately speaking. Moreover, humans are capable of voluntarily holding their breath for a limited time. These are tasks that the apes cannot perform. Unlike apes, humans do not have an estrous cycle. This cycle is an evolutionary tool that indicates the readiness of the female for mating and a desire to engage in sexual intercourse. This phenomenon is not observed in fish. However, the anatomical structure of the reproductive system, such as the hymen and major labia in the vagina, can prevent water from entering the female genitalia. Finally, it can be noted that the brains of aquatic animals, due to the consumption of seafood and the higher levels of omega-3 and omega-6 fatty acids, are larger than those of creatures living in dry environments.

As mentioned, the presence of severe floods 4–8 million years ago caused the extinction of fossils from this period, preventing human access to them. This phenomenon may have led to the separation of *Austra's* fossils from the apes. This indicates that this population may have emerged in a small and specific region of the Earth and gradually grown and developed. *Homo habilis* showed a great interest in living near water and lakes. Furthermore, the aquatic ape hypothesis may suggest that although modern humans are genetically very close to apes, they do not have much resemblance to them in terms of appearance and outward characteristics. Another theorist, Owen Lovejoy[67,73] believes that although the speed of movement, maneuverability, and static stability in walking on two feet are less than walking on all fours, the ability to reproduce may have been an influential factor that enabled humans to adapt well and comfortably to the evolutionary progression of walking on two feet.

There are two fundamental and effective strategies in the process of reproduction and/or reproductive behavior of living organisms. In the first method, a large number of sperm cells are simultaneously produced. This method works well in suitable growth environments, such as water for fish. However, in the second method, the number of offspring produced is low, but parents make efforts to care for them. Apes are positioned at the end of the second spectrum. In other words, the female does not enter the estrous cycle until up to five years after reaching maturity. With the disappearance of forests and the upsurge of cold weather during the Ice Age, some of them changed their behavior to increase their reproductive capacity. Therefore, during the estrous phase, all male animals can have sexual intercourse with the female, creating "paternity certainty"[119,120] for each male, which can be effective in their preservation, care, and nurturing of offspring. Similarly, in the evolutionary path of humans and monkeys and the change in resting place from trees to the ground, mutual love between male and female has developed, ultimately shaping habitual sexual behavior. Furthermore, having more than one child does not create any disruption in child care because male monkeys also intervene effectively in this matter. The simultaneous release of hands has also been very helpful. According to Finlayson[121], this intermediary creature between humans and monkeys, although it has adapted to the new environment due to the reduction of trees, still retains the ability to climb trees to escape dangers, find food for consumption, and sometimes spy and monitor the environment. Scientific studies have shown that chimpanzees in closed environments and cages have learned to wash their food roots and consume them later. According to Harvey's theory, they can have an episodic memory and wash their food while standing on two feet and then use it. At the same time, many macaque monkeys that live alongside coastal areas in Southeast Asia and Japan still exhibit such behavior.

SEXUAL BEHAVIOR

The sexual behavior of modern humans has been markedly shaped by civilization and culture. Nonetheless, it also exhibits notable distinctions from that of non-human primates and apes[122], particularly in terms of its intricacies and complexities. One of the first and most prominent differences is the estrous cycle. This cycle is hidden in humans and cannot be externally detected. In apes, the estrous cycle is characterized by swelling and reddening of the genitalia in females, indicating the onset of ovulation and readiness for mating and reproduction. Some non-human primates show only about two weeks of readiness for mating within a 20-year period. Furthermore, if chimpanzees do not conceive during this cycle, the endometrium is absorbed, while in humans, non-pregnancy is indicated by menstrual bleeding. Some animals have an annual cycle, such as bears, foxes, and wolves, while others have a semi-annual cycle, such as dogs. Some have multiple cycles, such as cats, cows, pigs, and horses. In chimpanzees, the estrous cycle typically lasts about ten days, and males compete with each other to mate, making it practically unclear which male is the father of the offspring. However, in humans, ovulation is hidden, so it was not necessary for men to queue up to mate with their female counterparts. Therefore, they engage in emotional and romantic relationships with each other, and with the identification of the father in such a relationship, childcare becomes obligatory for both parents.

The second prominent difference is the ability to engage in permanent mating without necessarily being tied to ovulation[123] in humans. Overall, these two factors have led to the formation of nuclear families. Thus, monogamy, alternating and regular mating, and ultimately the protection and care of the family become valuable. This phenomenon has been of fundamental importance in nomadic families and has become even more significant with the transition to agriculture for the preservation of pastures and land. At the same time, various degrees of dependency of children on their mothers can be observed. In monkeys, infants are dependent on their mothers in the first year, and this dependency decreases in subsequent years. In apes, this dependency is stronger, and they remain dependent for a minimum of two years and mostly up to four years. In fact, a combination of low reproductive rate, long birth intervals, and a relatively long childhood period that prolongs the time to sexual maturity causes tailless monkeys (apes) to face unfavorable conditions for reproduction.

Usually, it takes about 10 years for a female chimpanzee to reach sexual maturity. This lack of rapid reproduction has led parents in this species to be deeply concerned about the survival of their offspring[124]. The concern for the survival of their child requires the mother to have a longer lifespan in order to bring her valuable offspring to reproductive age. Meanwhile, males, in addition to producing sperms, take on the responsibility of protecting the group and preventing the killing of offspring by other males. However, they usually consume the most and best food, depriving females of food. Females are forced to eat the remnants of food. Although the female chimpanzee encourages the male for protection, she tries to have sexual contact with all males. In this way, the father of the child is not clear, and she creates a sense of paternity for all males. Although the female chimpanzee pursues a mating strategy with all males, she ultimately gets lost with one male in the forest during her estrous period. However, humans use a monogamous strategy that has evolved from and along with polygamy. This causes not only paternal care but also the relationship and dependence of parents to each other, which is a kind of biological tool for this work, to increase. In the course of evolution, paternal care, along with increased mating and sharing of more food, especially meat, has increased from about 2.5 million years ago. With the reduction in the time interval between pregnancies, gradually, other family members began to assist in childcare. This change in behavior is observed in gibbons, and even more so in owl monkeys and marmosets. In addition, in humans, male selection by females initially served to protect themselves against aggressive and dangerous males. In any case, initially, this behavior has been recognized as an important and effective factor, but in monkeys, frequent and multiple sexual contact serve to create confidence for all males in terms of paternity, causing a kind of mental confusion about fatherhood. Therefore, all males feel good about their offspring. In some cases, mate selection is also based on genetic characteristics. A beautiful, strong, and healthy male is preferred. However, except in humans, no case has been reported where mate selection is based on economic abilities or cultural superiority[125,126], which has been the result of cultural and economic evolution in the course of mental evolution.

In addition to male protection and prevention of infanticide by other males, obtaining food in exchange for sexual intercourse, stealing less cooked food in exchange for sexual services, and exchanging sexual relations for improving situations and socio-cultural correlations can be mentioned as other influential factors in mate selection. In the course of evolution, human descendants have gradually moved from polygamy to monogamy, reducing sexual competition, long-term monogamous relationships in order to have good social relationships and hidden, not transparent, and overt sexual relationships. The evolution of the mind and intelligence has played a major role in this matter. This new group of mothers during pregnancy and lactation require more energy. Therefore, they consumed more meat, fat, and other foods. This phenomenon caused other males (mates) in the group to share their food to some extent. This process has continued in the past two million years. This sharing also continued after childbirth, leading the mother to share information and help the child in hunting. This sharing effort has improved the hunting capabilities of the *Homo* species.

The physical structure of males and females in primates, especially their reproductive organs, differ significantly due to the secretion of male sex hormones (androgens), causing clear anatomical differences between male and female sexes, known as sexual dimorphism. The greater the difference in size between males and females, the less involvement of males in child-rearing. Although gorillas have large bodies, they have smaller testes. They try to transfer their genetic information to the next generation through their fewer sperm by establishing a harem and preventing females from mating with other males. On the other hand, chimpanzees, despite their relatively smaller size, have larger testes. Since they can mate with all females during their estrous cycle, they try to transmit their genetic information to their offspring by producing more sperm with their larger testes. In addition, their larger body size is used for competition with other males, and their large canine teeth are used for expressing aggression and occasionally bluffing rivals. Females, on the other hand, do not have much involvement in choosing mates and mostly invest their resources in the process of reproduction for the growth and upbringing of offspring. This phenomenon is more pronounced in birds. Male birds are usually more beautiful, sing pleasant songs, and build and decorate nests. Traditionally, mammals stimulate ovulation through sexual arousal, which is considered a natural trait. Therefore, prolonged sexual contact was deemed necessary for nocturnal mammals to be able to secrete the necessary hormones. Primates and other mammals that had to allocate less time to sexual contact due to environmental conditions[127] evolved into species capable of spontaneous ovulation. This process ensured both reproduction and survival.

The story of humans is somewhat different. Sexual intercourse in humans is not merely for procreation but also for sexual pleasure and the establishment of social and emotional connections. Humans have hairless bodies and rely on their linguistic abilities to attract the opposite sex. Men have stronger muscles and more prominent body features compared to women. Male genitalia in humans are thicker and longer than those of other primates and lack an osseous prominence inside. Unlike some primates where females play a role in selecting males due to their higher investment in the growth and nurturing of offspring, in the human model, both male and female genders strive to select better mates. It seems that for millions of years, women have paid attention not only to muscles and body features but also to the size of the male's evolving genitalia[128]. They have shown a greater inclination to mate with men who have longer and thicker genitalia, and perhaps from their perspective, better sexual satisfaction has not been overlooked in this regard. Therefore, in the process of evolution, the narrow genitalia gave way to thicker and more flexible genitalia. Strong men could have had aggressive and rough behaviors, so women tried to control their temperament by using sweet words. In the meantime, men also learned to conceal their rough temperament.

It seems that the most efficient biological change in human reproductive strategy primarily lies in concealed ovulation and constant readiness for mating outside the estrogenic period. Although the exact reason for this process is not properly identified, it enhances bonding with a mate. In non-human primates such as monkeys and apes where the period of sexual readiness is prominent, if the male can flirt or mate or have increased interaction with a female, he can still be confident in his paternity. In humans, since females do not advertise their reproductive readiness, the opportunity for specific sexual contact needs to be longer for fatherhood to be achieved. Most animals have behaviors similar to humans, except for chimpanzees and bonobos, who display sexually irresponsible, promiscuous, and uninhibited behaviors. In terms of sexual behavior, the last common ancestor of humans diverged from chimpanzees and bonobos and became more similar to gorillas and orangutans. Although both genders are covered in current societies, sexual attractiveness exists in various forms in both genders. All these factors contribute to the connection between the two genders, and of course, the influence of culture[129] has become fundamentally important in determining a good pair.

The last point is that the aquatic phase can justify the reduction of the human estrogenic cycle. Additionally, standing on two legs causes the vagina to be slightly repositioned forward. Although this phenomenon may simplify regular sexual intercourse for humans, it makes the process of orgasm in women more difficult and delayed. In many non-human primates, copulation happens very quickly, sometimes lasting only under a minute or even a few seconds. Therefore, females reach orgasm very rapidly. The time required for women to reach orgasm is, of course, longer and sometimes may not occur during sexual intercourse, but the necessary time for men to reach orgasm (ejaculation) during sexual intercourse is usually not more than a few minutes.

In humans, there is another dilemma. In addition to the concept of sex, gender identity plays a significant role[130] in the evolutionary structure of humans. The word "sex" encompasses a separate biological structure. Gender is genetically determined at the time of embryo formation and after fertilization based on the presence of XX or XY chromosomes. The presence of the *SRY* gene on the Y chromosome is the determining factor for the growth and development of testes. The development of ovaries is a predetermined phenomenon that occurs automatically in the absence of the mentioned gene. Following the formation of testes and the secretion of testosterone from them and its influence on the reproductive system, brain organization, and physiological aspects of the human body, the process moves toward masculinity. Other changes occur during puberty and ultimately remain the same, but the concept of gender identity is what shapes the social and relational structures of intelligent human beings. It includes norms, criteria, and power in different situations that can vary throughout life. Individuals have a personal sense of their "gender identity" and identify themselves in society accordingly, shaping their social interactions based on it. "Gender roles" can vary across a spectrum from femininity to masculinity, and eventually solidify as a form of "gender ideology" in individuals' minds and internalize as "gender stigma," which impacts human social connections and support. This collection represents the sexual laws governing society, and individuals usually conform to these norms. The characteristics and issues related to gender and gender identity are strongly intertwined with human biology, behavior, and health outcomes, including race, ethnicity, living conditions, socioeconomic status, food security, sexual preferences, and language.

The *SRY* gene, in the second trimester of pregnancy, triggers the secretion of testosterone (male sex hormone) and leads to the formation of a brain with male characteristics. In the absence of this gene, the sexual structure moves toward the formation of a female brain. However, it should be noted that intelligent humans are never simply a complete black-and-white sexual dimorph organism and basically (s)he operates as a spectrum. This spectrum includes a typical female individual on one end and typical male counterpart on the other end. Individuals can fall anywhere within this spectrum. Sex hormones are not only related to phenomena such as a person's knowledge of his own gender, male and female games and behaviors, and his sexual orientation, but also to a lesser extent influence on human cognitive traits such as empathy and sympathy, violence and aggression, social dominance, and mental rotation although it should be noted that these characteristics cannot be easily measured. However, complete and precise information in this regard is not available but in men, the volume of the temporal (auditory) and limbic regions, as well as the overall brain, is generally greater than in women. At the same time, women have larger cingulate and prefrontal lobes. The male nervous system is often superior to women's in performing motor tasks and sensory perceptions, while women demonstrate better social and verbal communication abilities[131,132] and have developed more elaborate instinctual analysis processes.

The interaction of environmental factors and nurturing with what is referred to as human nature is also very prominent. For this reason, anxiety and mood disorders; and mental trauma-related illnesses are more commonly seen in women or individuals who are perceived as female from the beginning. Sexual hormones affect neurotransmitters (chemical messengers in the brain), synapse formation during the prenatal and postnatal periods, and ultimately influence the period of adolescence. Changes in intracellular signaling contribute to the development of diseases such as attention-deficit hyperactivity disorder (ADHD). Epidemiological statistics indicate the occurrence of various disorders in childhood, including 5.9% for hyperactivity, 1% for autism, 7.7% for speech disorders (speech impediments, stuttering, and dyslexia), 9.7% for learning disabilities, and 2.8% for tic disorders. These studies have shown that these disorders or diseases are observed in males at rates 1.2–4 times higher than in females. The main hypothesis raised in this regard is the lack of balance between excitatory neurotransmitters (glutamate) and inhibitory neurotransmitters (GABA). These two are the most important chemical transmitters in the brain and likely play a crucial and effective role in regulating oxygen consumption and myelin production for specialized and efficient neural fiber function. Additionally, male steroid hormones can potentially disrupt the functional connectivity between the two hemispheres, which may partially account for the differences in functional abilities between men and women.

Neurobiological evolution has also played a significant and influential role in sexual behavior. With the onset of sexual activity or thoughts about it, increased activity in the hypothalamus, a region in the upper anterior part of the brainstem, the ventral tegmental area, and the anterior cingulate cortex can be observed. The intense pleasure and orgasm experienced during sexual activity have greatly contributed, not only as physiological bodily responses but also as deep feeling experiences within the brain, to the promotion of survival and preservation of the species. Sexual stimulation is accompanied particularly by reduced activity in the cortex of right amygdala and temporal lobe. The amygdala is a small part of the brain, located in the anterior part of the hippocampus and is considered the emotional center of the brain. Early studies by Heinrich Klüver (1897–1979) and Paul Clancy Bucy (1904–1992) members of Chicago neurology club prior to World War II demonstrated that by removing the amygdala complex in monkeys, they exhibited severe overeating, increased oral behaviors (hyperorality), a tendency to examine all objects orally (psychic blindness), increased autosexual, heterosexual and/or homosexual activity, and the absence of fear[133,134] toward previously feared stimuli. Subsequent research revealed that the primary function of the amygdala is fear management, and a healthy amygdala effectively controls sexual behavior and overeating, behaviors that have evolved over the course of many years. Unfiltered and raw sensory information is transmitted to the amygdala through the thalamus (literally means chamber), which acts as the gateway for sensory inputs of the brain. Fear behavior is a fundamental and ancient trait evaluated in mammals, and its presence has been highly influential in the evolutionary development of social cognition in human offspring. The amygdala has developed a high proficiency in interpreting visual information from facial expressions, which not only indicates age and gender but also processes subtle emotional signals, reassurance, and personal attraction. The reduced activity in the ventromedial prefrontal cortex is also prominent in this regard. It appears that the decreased activity in the cortex contributes to a lack of effective awareness of the surrounding environment during sexual intercourse. This issue likely aids in the complete establishment of sexual intercourse and the completion of the reproductive process in humans. Romantic love seems to have gradually emerged in the evolutionary journey of the human mind.

5 Tools and Skulls

TOOLMAKING

Crafting tools played a pivotal role in the advancement of early humans, benefiting them from two key perspectives. First and foremost, by making tools, humans were able to improve their daily functioning in life. They needed tools for cutting meat and animal carcasses, building shelters, and engaging in agriculture. However, the second function of toolmaking was the beginning of speech and language development. This phenomenon can be considered a revolutionary advancement in the lives of early humans. In fact, tools are essentially the functional scope of an object instead of the functional activity of the hands and mouth[135] to achieve an executive and performance goal. Many animals, such as birds, use simple forms of toolmaking. New World monkeys use their cupped hands to grasp and hold objects or food. Old World monkeys have the ability to grasp precisely with their fingers, using their extended fingertips. This way, they can manipulate small objects with their fingertips and thumbs. This process has greatly evolved in humans. Primates have a special inclination of using tools and are actually a type of user with an institutionalized habit of using tools. Primates have a specialty for using tools and are, in fact, a species of tool-using creatures with an established habit of using tools.

Having social relationships and individual cooperation has been a prerequisite for the toolmaking abilities of intelligent humans, a capacity we call "cumulative technological evolution." This technical innovation occurred during the Middle Stone Age, approximately from 80,000 to 30,000 years ago. This period of technical innovation was accompanied by the emergence of abstract thinking and symbolic behaviors, such as pigment processing and dyeing, personal adornment, the expansion of social networks, the creation of metal tools, and various forms of handicrafts. This collection of actions and behaviors signifies the beginning of modern behavior in intelligent humans. Scientific studies indicate that behavioral modernity was accompanied by changes in the cognitive capacity of intelligent humans resulting from the emergence of new genetic alleles. This novelty includes improved working memory, general intelligence, language-related capacities, abstract thinking, and greater flexibility in understanding the minds of others (theory of mind). Starting from approximately two million years ago, with the expansion of brain size, cognitive reserve in upright humans has also escalated, accompanied by longer lifespans and improved connectomic structures. With the increased volume, the brain can act as a cognitive buffer, and a diminution in its cognitive reserve could be associated with delayed clinical symptoms following various injuries or diseases. As it is well-known today, in educated individuals who develop Alzheimer's disease, their clinical symptoms are postponed and manifested with a greater delay[136].

In various monkey species, similar behaviors can be observed for extracting food using their fingers from a hole. This behavior is a kind of inherent and primitive characteristic that has been retained in African *hominids*. The second step was the use of thin sticks to make stone tools. This feature is almost unique[137] and specific to the human species. Essentially, the foundation of toolmaking was established by the skilled human (*H. habilis*). By having relatively broad and elongated fingers and the ability to grasp objects with relative precision between two fingers, they used simple tools for cutting and skinning meat from carcasses. The first discovered stone tools date back to around 2.5 million years ago in the Hadar region of Ethiopia. However, a significant number of these tools have also been found in the Olduvai Gorge in Tanzania, and it is called the Oldowan culture[138]. The use of tools in chimpanzees, especially among females, is very prominent. They use sticks to extract insects, termites, and ants from their nests, as well as to empty nutritious seeds from woody parts. Extracting insects and breaking open nutritious seeds are challenging tasks for young chimpanzees. The use of thin sticks to extract underground

DOI: 10.1201/9781003538400-6

food roots is primarily performed by male chimpanzees and is interpreted as a behavior similar to hunting. In some cases, chimpanzees also use small spears to hunt small mammals. Meat is a desirable food for plant-eating chimpanzees and is usually given as a gift or reward to chimpanzees that have shown more cooperation or to females during the mating period or other associated events.

This was the first stage of human evolution in toolmaking, which started in a region between northern Tanzania and southern Kenya, and essentially two types of basic tools were created. The first type of tools were core tools, that were sharpened on one side by removing different layers from the stone, and they were used as stone tools for cutting and butchering carcasses. The second group is called flake tools, which were initially sharpened on both faces to be used for skinning and cutting meat. Gradually, during the age of *H. erectus*, with the emergence of more complex social relationships and communication methods, the construction of these tools expanded and became more diverse. The second stage of evolution in toolmaking started from this point. Initially, simple observation, similar to what is still used today, was sufficient for making tools. Handyman (*H. habilis*) was able to carry their tools for long distances, indicating their cognitive ability to design tasks. They probably learned to collaborate and cooperate with their fellow beings. Furthermore, they learned to create new tools for themselves using simple tools. This meant they were compelled to engage in internal or collaborative problem-solving in their minds, which is one of the fundamental aspects related to the initiation of language and speech. The next point was the placement of an object inside another object, which itself influenced the emergence of generative grammar in language structure.

Bifacial toolmaking, with two working edges, is considered a characteristic of *H. erectus*, and it is referred to as the Acheulean culture. Further progress occurred during the era of *archaic H. sapiens*, with the creation of more complex tools for cutting, skinning, and breaking bones. This period of toolmaking was called the Levallois technique[139,140]. From the Acheulean period onwards, mere observation was not enough to enable toolmaking; the presence of necessary instructions was required. Instructions mean the existence of a system through which two individuals can establish verbal or non-verbal communication to transmit information. The development of spoken and written language played a significant role in this period.

The Oldowan cultural era primarily evolved during the time of *H. habilis*, and the Acheulean period was predominantly developed and crafted during the time of *H. erectus*. The second stage of toolmaking was undoubtedly more complex than the first stage. For example, using fire to shape a tool requires human intelligence and experience. This means that toolmaking had become a two-stage process, two distinct stages that required cognitive understanding. The second stage of toolmaking requires answering questions such as why one should heat a simple tool with fire. This phenomenon indicates cognitive flexibility or "creativity" and ultimately leads to the creation of cultural complexities.

The toolmaking culture created a specific personality and character for different individuals. This character was generally absent during the era of *H. habilis* and *H. erectus* due to the simplicity of toolmaking methods. At the end of the prehistoric period and the beginning of the era of *sapiens*, a new technology emerged, which ultimately led to modern humans. In fact, this stage, the Solutrean industry[141,142], is the last stage of early human toolmaking. The use of tools to shape a growing brain is highly significant and influences its evolutionary function. Henry Head and Gordon Holmes, in their famous article in 1911[143], pointed out that the use of tools helps in understanding body, posture, movement, and locality. The use of tools such as automobiles or motorcycles has amplified our perception of our own body image and created an enhanced form of it. Of course, it is only for the driver, not the passenger. In fact, even inanimate objects enter the realm of the individual's body image, which is important for the formation of the concept of "self" and differentiation from "the others." Furthermore, the use of tools also contributes to plasticity for body representation in the brain and the creation or expansion of the peri-personal space, which can be called a "body schema."

In the technological explosion era that began approximately 30,000–40,000 years ago, humans had invented about 150 types of tools for various tasks such as cutting, digging, and carving. They had the ability to connect these tools to each other for optimal use. Although the use of a piece of wood connected to a piece of stone is a simple example of toolmaking, it greatly aided human ability to initiate communication. This behavior indicates the ability to generate not only toolmaking but also effective speech.

TINKERED APE

The skulls of many mammals typically exhibit an elongated, narrow, and tubular shape, contrasting starkly with that of humans. Human skulls are characterized by their broad, short, and rounded structure, featuring a prominent forehead and lacking a snout. The neck is relatively short and vertical, connecting to the middle of the head rather than the back[144,145], unlike many other mammals. Furthermore, their tongue is round and muscular, recessed into the throat, and they have a protruding external nose with downward-facing nostrils. The whiteness (sclera) of the eyes surrounds the dark iris, and their focused gaze is evident from a distance. Hair is visible in certain areas of the head. Breathing is done through the mouth and nose, speaking is aided by the mouth, teeth, and tongue, and the ability to think is attributed to the presence of a relatively large brain. In other words, if someone asks us why humans are like this, what answer do we give them? In fact, the correct question is why the human head is shaped like this? Why does it differ significantly from other mammals? To answer this, humans should be compared to three large groups of primates: the first group is Egyptian chimpanzees, the second group is bonobos, also known as pygmy chimpanzees, and the third group is gorillas. Generally, animals with larger body sizes, like gorillas, have a lower metabolism, are mostly herbivorous, and have a greater tendency to consume high-fiber, low-energy food. Molecular studies in the 1950s showed that humans and chimpanzees share a last common ancestor, called *Pans*[146], approximately 8–10 million years ago when they separated from gorillas.

Human-like people and chimpanzees continued their evolutionary paths, and later, these two branches diverged with only up to 3% difference in genetic makeup. This means that their difference encompasses a total of 35 million base pairs in DNA structure, and of course, only three million pairs of these are functionally significant, while the rest are what is known as non-coding DNA. However, humans did not evolve solely due to genetic changes[39]. One of the most significant and prominent difference between humans and other primates is in the shape of the head. Initially, the neck assumed a vertical position. Front teeth and canines became smaller, and instead, the grinding surface and thickness of the molar teeth increased. A layer of enamel formed on the teeth, while the jaw muscles became stronger and larger, and the protruding nose remained on the surface of the face. As the brain grew larger and placed itself in the skull, it began to be round, the face moved back a little, and a chin, which was the lower jaw, was added to this set.

The structure of the skull is complex and integrated. Due to spatial constraints, any change in one part results in a change elsewhere. As a result, alterations in brain size have been accompanied by changes in skull size. This enlargement has led humans to only be able to hear specific sound frequencies, alter facial and eye sizes, and create significant changes in the biomechanics of the chewing muscles. However, it is not clear whether these changes have occurred as a result of adaptation to the environment and improving the survival abilities of living beings or if they are merely by-products of these changes. Is the entire process simply a random and accidental evolutionary change, or has it followed an entirely intelligent course? A prominent example is the shrinking of the face and the absence of a snout, which may be a form of adaptation for speech abilities, increased air turbulence in nasal cavities, trunk, and head balance, or perhaps all of these changes are behind-the-scenes reasons for its creation[147].

The next question is why the structure of the skull and the human head is so complex. There are multiple reasons for this. The first factor is the existence of various, different, and fundamental functions in the head. The skull is not only a home for the brain but also controls centers for respiration, blood pressure, temperature regulation, vocalization, vision, hearing, smell, taste, swallowing, balance, and movement. In the evolutionary process, an attempt has been made to improve the function of each of these centers in a way that does not disrupt the functioning or performance of other parts[148]. For example, the enlargement of the jaw to improve mouth closure and teeth and sound production, which is accompanied by proportional jaw growth, has not disrupted the functioning of other parts. These numerous and diverse components function as modular units which are independent processing units, specialized and frequently localized networks, separate from each other, and while they are functionally independent, they are relatively dependent on each other for serving a specific function. According to Georg Streidter, modularity could be some sort of by-product of pressure to reduce assembly and connection costs. In this system, about twenty bones for the skull, 32 teeth in the mouth, one brain, sensory organs, muscles, cranial nerves, arteries, and veins, despite their relative separation, contribute to the overall function of the skull. Ultimately, this collection works in perfect harmony, cohesion, and functional overlap. For example, the oropharynx or nasopharynx (oro: mouth; naso: nose) are common pathways for the entry of water, food, and air, as well as the exit of sound and mucus. During chewing, pressure is not only applied to the crown of the tooth but also to the root of the tooth and the surrounding tissues, gums, chewing muscles, and the temporomandibular (jaw) joint. Gene mutations can affect functional coherence. For example, cartilage may not form properly, causing the face to sag, and the function of the teeth for proper closure and chewing may be impaired. Premature closure of growth plates in the skull bones leads to a smaller head, changes in facial shape, and a reduction in cognitive capacity.

Unlike the common opinion, the greatest evolution in complex systems occurs in the cranial (skull) structures rather than the limbs or internal organs (viscera). The main difference between extinct human species and modern intelligent humans lies primarily in the shape and structure of the skull, rather than the hands and feet; it is essentially cranial changes rather than post-cranial changes. This evolutionary trajectory has led to the skull having both the greatest and optimal functionality while maintaining its best structural form. In fact, the skull and the human brain are complex structures that work in an intertwined yet fully integrated manner. The cranial bones have demonstrated the fundamental property of achieving integration through what we call biological tinkering. This ability is achieved through the presence of collagen tissue, which has the ability to stretch and accounts for about 60% of the bone tissue. This tensional effect can create either positive, outward, or negative, inward pulling forces[149]. This phenomenon allows the brain, cranial nerves, and sensory-motor organs to be well accommodated within it and function simultaneously and synchronously with each other.

SCOPE OF LIFE

Human life strategy is heavily influenced by the living environment, which, in turn, impacts life span. In unstable environments, where resources are scarce or unpredictable, living organisms tend to adopt a life strategy characterized by shorter life spans and a focus on rapid and frequent reproduction to ensure the survival of offspring. In this type of life, organisms will be highly opportunistic, infants will grow quickly, and little investment will be made in them. Naturally, their mortality rate due to natural factors and environmental predators will be higher than usual. Conversely, in more stable environments with abundant resources and lower risks, humans may adopt a strategy of longer life spans and slower reproduction, investing more time and resources in individual development and the care of offspring. When the living environment is stable, living organisms continue their lives at a slower pace. Birth and reproduction, and of course the number of offspring, will be lower, and more investment will be made in their growth. Dependence on parents increases, and ultimately their death occurs later. Furthermore, life span cannot operate outside the laws of thermodynamics. If more energy is consumed in one organ, less will be consumed elsewhere. Therefore, body size determines the amount of energy consumed. A larger body can better retain heat and avoid energy waste. It also reduces the risk of attack and damage by smaller organisms, thus increasing lifespan. Survival is closely tied to food consumption, especially meat consumption, and having functional teeth. Longevity depends a lot on the sensory information[150] that is transmitted to our brain from the surrounding environment by sensory fibers and the autonomic nervous system.

The age of the first molar tooth eruption is an important variable that can be used to estimate the lifespan of a living organism. Human molars grow at the ages of six, twelve, and eighteen. These ages correspond to 3.3, 6.6, and 10.5 years in apes. There is a direct and very close relationship between the eruption of the first molar and brain size, with a coefficient value of 0.98. The first molar eruption in chimpanzees occurs at around three years of age, which is very similar to *Austras*, whose brain volume is about 400–440 cc and is very close to that of chimpanzees. For skilled humans (*H. habilis*) with a brain volume of 600 cc, it occurs at around 3.8 years of age. In upright humans (*H. erectus*), it occurs at around 4.5–5 years of age (with a brain volume of 820–980 cc), and in intelligent humans (*sapiens*), that time will be at 5–6 years of age (with a brain volume of 1400–1500 cc). It seems that the evolutionary trajectory of the body has coincided with the evolution and growth of the brain.

The range of human life consists of five fundamental phases. The first phase is the embryonic period. Although it is longer compared to other non-human primates, it is relatively short overall. The second phase, according to Jean Piaget[151], is the infancy and childhood period, which appears to be the most prominent and influential stage of human life. In fact, a human infant is born premature. If the primates' growth pattern were to be followed, a human newborn would need to spend at least 1.5 years in the embryonic period within the mother's womb. Human childbirth is a type of mechanical birth. Due to the human ability to move on two legs, the pelvic outlet has become smaller. Therefore, the fetal head needs to exit the pelvis before it grows too large[152,153]. The average diameter of the fetal head during normal deliveries is about 10 cm, and if the fetal head is larger than usual, it would require a difficult birth and/or a cesarean section. After approximately 8–9 months, a human infant gradually acquires the abilities that are typically possessed by apes at birth, with some assistance. However, after passing 10–12 months, they completely outperform them. In fact, many animals have completed their brain growth at birth. For example, the chimpanzee brain is about 60% developed at birth. This percentage for a human newborn is approximately 24–30%, which is equivalent to 400 cc, the brain size of an adult chimpanzee. The extended period of childhood has a significant positive impact on human brain development.

The brain growth in chimpanzees reaches approximately 95% of their adult size at the age of 3–4 years, while in humans, it reaches for this stage around 6–7 years of age. However, the brain growth in chimpanzees is complete at the age of 10–12 years, whereas in humans, it is complete at around 14–18 years of age. Calculating brain growth based on changes in volume over time occurs much faster in humans compared to monkeys. This coincides with the myelination of nerve fibers, which increases the speed of message transmission. A ten-fold increase in conduction speed requires a hundred-fold increase in nerve diameter and, consequently, a thousand-fold increase in brain volume.

Overall, the primary strategy of humans is to acquire more energy resources and reduce mortality. They try to continue their overall growth simultaneously with brain development and evolution. This evolutionary process continues simultaneously in determining physical growth, sexual maturation, reproduction, child-rearing, and caregiving until complete independence and avoidance of death. In fact, the energy consumption strategy in the environment for a mammal is so programmed that it determines the time of birth, separation from parents, when its physical growth remains constant, when reproduction slows down, and when it will eventually die. This timing is longer for humans compared to many other mammals. English philosophers like John Locke (1632–1704) and Alexander Pope (1688–1744) believed that the long-term dependency of infants and children on parents and their relatively late maturity is the basis for the creation of social dependencies and family life. On the contrary, Dutch anatomists like Louis Bolk, Wilfred Clark, and many others believe that the slow growth trajectory of humans and their dependency on parents and nurturing by them are the by-products of the selection of larger brains[154,155], and ultimately having higher intelligence. Children start assisting other children at around 14–18 months of age, a behavior that is less observed in chimpanzees. Although children do well in sharing their belongings, chimpanzees only share worthless or less valuable items. In essence, humans have primarily invested in brain growth and higher intelligence to implement strategies for a longer lifespan and delayed mortality.

Severe infant helplessness, short birth intervals, and long-term dependency of children and adolescents on parents were factors that led to behavioral evolution in food sharing, living in a stable home, and gender-based division of labor. In simple terms, the most important strategy for humans has been to live slowly and die in old age[156,157]. The genes that have augmented cognitive capacity in humans during the reproductive period will be associated with reduced activity as age increases[158]. For example, the *ApoE* gene, which is effective in lipid metabolism, is very influential for the growth of the nervous system in childhood and adolescence, but with age and lipid metabolism disorders, it causes brain and heart strokes and ultimately death. However, it is still not properly understood whether late physical maturation and long-term lifespan have been the cause or effect of the development of a large brain, large body size, improved social behavior learning, and ultimately increased parental investment in children. Brain size plays a crucial role in determining energy consumption. The energy consumption for the human brain is about 20–25%, for chimpanzees 9%, and for many ungulates about 2%. However, a large brain has less tolerance for increased temperature, reduced oxygen, and blood pressure changes. Brain growth is costly, and this cost is expensive for long-term survival[159]. However, humans, with their long lifespan and slow pace of life, have at least one very fundamental characteristic, and that is the helplessness of their infants in self-protection. Infants who are completely dependent on their parents. Other primates, besides humans, cling to their mother's hair and hang onto it, but human infants need more care and always need to be looked after with both hands. Human hunters and gatherers did not place their infants on the ground except for short periods. Increased care limited the mobility and movement of parents and families.

This situation has led to an upsurge in mother–infant social relationships and is accompanied by increased social skills along with brain growth, speech ability, and mind reading. It takes approximately one year for a human infant to develop motor skills, similar to a baby chimpanzee. This delay is very prominent in humans[160]. The size of the human brain at birth is one-fourth of the adult brain. These numbers are half for monkeys and two-thirds for macaques. In this group of mammals, brain growth occurs slowly during the postnatal period. This period is about one year for humans and one to two months for chimpanzees and macaques. In fact, the inability of human brain growth during the first year is due to the lack of transferable energy for brain growth.

Another fundamental difference between humans and other primates is the existence of a childhood stage from three to eight years old. Other primates almost have a direct transition from infancy to adolescence. In this stage, the child's dependency on the mother falls, and they experience social relationships outside the family. It is likely that this phase emerged in part during the Pliocene epoch with the emergence of two new *Homo* species. Six years old is usually the time when the neural pathways for speech mature. Language acquisition occurs well at this stage and accents are formed. However, brain growth continues until the age of thirty, but this process did not exist in early *Homo* species. Scientific studies have shown that up to 20% of the nervous system is constructed and myelinated at birth, and this process continues until the age of 30. The last part of this process should be the prefrontal cortex. This evolutionary phenomenon led to the belief of politically protesting students in the 1960s that they should not trust anyone over the age of 30[161]. They probably believed that with the completion of neural myelination and the stabilization of the thinking process, these individuals would not engage in political and/or other high-risk behaviors.

This phenomenon is observed in birds in two forms. The first group is born after a short embryonic period and often without feathers. However, their ears and eyes are closed, and they are relatively endothermic, meaning they have the ability to regulate body temperature but need to be cared for in the nest for a while to grow. The second group undergoes a longer embryonic period and usually is born as single or twin offspring with acceptable motor development. They have some feathers, their eyes and ears are open, and they also have the ability to regulate body temperature. Another point to consider is that the narrowing of the pelvic canal and birth canal does not necessarily result in brain growth impairment and does not necessitate premature delivery. For example, infant apes are less capable than monkeys, and monkeys are less capable than prosimians, even though the pelvic canal and birth canal in all of them are sufficiently wide. We should look at it from a different perspective, and in fact, this phenomenon is a form of energy trade. How much energy and in what period should be spent to create and nurture a child? In other words, the more mature and developed the infant is, the less energy is consumed for brain growth. For example, the gestation period for chimpanzees is about 230 days, and for humans, it is 270 days. Although ape brains are smaller, their bodies are more mature and developed. Their motor system has better evolution and coordination compared to human infants, but in the human model, long-term investment is made for a larger and more capable brain so that the child can have a better life in the future. The ratio of the infant's brain weight to the adult human brain weight seems to be a good factor for evaluating whether it requires parental care or not. In the course of human evolution, *Austra* has had a wide pelvic opening, so the newborn had a skull size similar to that of chimpanzees. In the middle of this path, during the era of *H. erectus*, as the pelvic canal narrowed, degrees of prematurity and the need for maternal care (altriciality) became evident. The narrowing of the pelvic canal made the childbirth process more difficult for the mother, necessitating human assistance during childbirth, which gradually led to the development of a better social structure in humans. Generally, extending the duration of pregnancy beyond 40 weeks is very dangerous for the mother, and she needs to care for her infant outside the womb for the completion of physical and brain growth. This phenomenon is very evident in modern humans.

Giving birth to a baby is associated with a transition to motherhood. Changes in the nervous system due to environmental and physiological needs during life represent the process of plasticity in the brain. This process is achieved by regulating cellular activity, brain structure, and connectivity. Pregnancy in humans is a dynamic process that is associated with a significant fluctuation of hormones. These changes cause the maternal-caregiver circuit to develop, which includes behaviors with the help of which the mother tries to improve the offspring's survival. Factors such as the maturity of the baby, litter size, and access to food are involved in this matter. Finally, they cause mother–infant synchrony, sensitive mothering, positive communication, and affectionate contact[162].

The second characteristic of these creatures is slow and continuous growth, with a growth spurt during adolescence. Linear growth is not observed in many mammals, although weight increases with age. The growth spurt in Western boys occurs at the age of fourteen, while in girls it occurs at the age of 12. The main question is why growth does not occur continuously and uniformly. The weight of chimpanzees at the age of seven is equivalent to 80% of the weight of adults, but in humans, the weight of adolescents aged 11–12 is around 80% of that of adults. The exact reason for the growth spurt is not known, but it may be related to brain growth, which consumes a lot of energy until adolescence and inhibits physical and sexual growth. For a long time, the clear relationship between brain evolution and lifespan has occupied the minds of scientists, and it seems that lifespan is a form of adaptation in the evolutionary process. Thomas Kirkwood and Michael Rose[163,164] called it the "disposable somatic theory." It is assumed that the energy needed for brain growth is high until the period of youth, so sexual development is delayed, and menopause around the age of 50 in women is a form of energy allocation for raising their children. In other words, menopause is an investment in reproductive expenses, and thus the lifespan of primates depends on gender, with the lifespan of females being longer than that of males in most species. This path prepares the granddaughters to reach puberty, and they continue the cycle of birth and reproduction. Studies in Canada and Finland have shown that mothers with longer postmenopausal lives have raised more grandchildren due to the stability of their physical and mental abilities. Kristen Hawkes, professor of anthropology at Utah University, believes that with the emergence of the *Homo* species, young people have become more dependent on adults[165,166]. This phenomenon has made human childhood longer, and as a result, it has to go through a period of heavy learning and protection by parents to prepare the adolescent for adult life. The readiness of children for adulthood is also dependent on the cultural dominance of society and maybe a balance between paternal investment in child protection and finding a new sexual partner. This phenomenon can affect long-term parental relationships, both parents' care and protection of children, and sexual relationships between them (such as sexual pleasure or reproduction). This behavior has developed simultaneously with the growth of intelligence and the improvement of brain function and has been able to increase lifespan and reduce mortality. Technology is also a by-product of human cognitive functioning, which works well in preventing environmental hazards. The development of memory, along with intelligence and technology, helps us find and store new resources for the future. It can be concluded that lifespan and brain size, as variables representing intelligence, support each other. Scientific research indicates that the effort to find food and protect oneself from predators has been accompanied by an increase in brain size, but some animals, such as lemurs in Madagascar, have survived despite small skull size and have a long lifespan, up to 20 years, due to their habitat, protected from predation pressure.

Part II

Neurobiology

6 Sweet Dreams

The etymology or the root of the word "day" traces back to "di" or "dai," indicating a concept of division. Throughout centuries, humans have consistently divided time into days and further subdivided each day into hours[167]. However, the length of hours varied during different seasons, with longer hours in summer and shorter hours in winter, as the time between sunrise and sunset was divided into 12 hours. Gradually, starting from the 14th century, mechanical clocks and, in modern times, digital clocks mercilessly took control over human life. The circadian rhythm is likely the most significant mental source for understanding the concept of time. It is created by genetically encoded molecular clocks with a periodicity of about one day[168]. Day and night follow each other, and ancient humans counted uninterrupted days within themselves. We count all the changes. According to Aristotle, time can be defined as the measurement of changes related to movement. Even when we sit in absolute darkness, time passes because something inside us, in our minds, rotates, moves, and shifts. The passage of time has played a fundamental role in regulating the mechanisms of life from the beginning. Molecular, biochemical, and hormonal reactions have time-regulated processes. Among them, the human circadian rhythm of sleep and wakefulness is a 25-hour endogenous timed rhythm that is squeezed into 24 hours of the day.

Physiological periodic behavior and rhythms are seen not only in humans but also in other organisms (plants, animals, fungi, and bacteria) and are called circadian rhythms. This word is derived from "*circa* and *dies*" meaning "about a day" which is created as a result of the operation of an internal and automatic time-keeping system called circadian clock[169]. This system generally has a 24-hour time period that operates using environmental cues which are called *zeitgebers* in German. This kind of timing has a significant effect on controlling homeostasis and cell behavior[170] and is genetically controlled by *CLOCK* genes. The product of the activity of these genes are autonomous molecular oscillators that are constantly produced, then destroyed, and regulate the 24-hour cycle of sleep and wakefulness. But it is still not clear why this cycle is formed in a 24-hour cycle. It seems that the earth rotation around itself and the influence of sunlight is set the circadian rhythm in the course of evolution. This day and night cycle has been effective for the growth and reproduction of living organisms, and finally, with several changes during the evolutionary course, found its appropriate form. In fact, circadian clocks have evolved in organisms in such a way that they can effectively organize their temporal relationships with their environment and internal physiological processes. Disturbance in this system can be associated with many internal diseases. These diseases are seen in many shift workers. Lithium, as an inhibitor of an enzyme for "GSK clock modifier"[171] gene, can be used in the treatment of some diseases such as bipolar mood disorder[172] and cluster headache by prolonging the circadian rhythm.

In the scope of human life, energy consumption is closely related to sleep. Sleep is defined as a reversible behavioral state in which the organism has almost no physical activity and reacts less to external stimuli than in the waking state. In other words, sleep is a factor for resting and regaining energy, which plays an effective role in establishing or stabilizing homeostasis[173,174], and from a phenomenological point of view, it is considered a dynamic and complex behavior. Non-human primates typically go to sleep at sunset and wake up again at sunrise, following a nocturnal cycle that usually lasts 10–12 hours. Although it may seem that humans should follow a similar pattern, in practice, this is not the case. The duration of sleep in infants, ranges from 12 to 15 hours, in adolescents from 9 to 10 hours, and in adults, it is generally less than 8 hours. Studies have shown that sleeping less than seven hours is inadequate for proper growth, and sleeping more than seven hours in adults can lead to various diseases due to reduced physical activity. In a study conducted in two tribes in Africa (Tanzania and Namibia) and one tribe in South America (Bolivia), who lacked basic

DOI: 10.1201/9781003538400-8

modern amenities such as electricity, telephones, and lamps, the average time of sleep onset was about three hours after sunset. During this period, they often engaged in conversations, social interactions, and cooking using dim firelight, which had little impact on disrupting their sleep cycle. It seems that being awake after sunset is a characteristic of modern humans compared to non-human primates. The lengthening of nights in winter has caused the duration of sleep in these communities to be about one hour longer than in summer. Insomnia (sleeplessness), a pathological sleep disorder, which affects approximately 10–30% of modern societies, is only observed in 2% of these tribes, which resemble ancient hunter–gatherer societies[173].

Sleep is the fundamental factor for regaining energy and optimizing its consumption. Sleep, including hibernation, involves behaviors aimed at reducing energy expenditure. Some species of hummingbirds, which have a body temperature of 41°C when awake, experience a drop of four degrees during sleep[175]. In mammals, sleep leads to a decrease in energy consumption and requires waking up to obtain energy again. Sleep in mammals, such as humans, begins with a NREM (non-Rapid Eye Movement) phase and is typically followed by the addition of the REM (Rapid Eye Movement) phase after 1–1.5 hours. These stages continue in an alternating manner. The key point is that during the REM stage, the body's muscles become relaxed, while eye movements, which occur simultaneously with dreaming, significantly increase. Meanwhile, the evaluation of brainwave patterns (electroencephalography) during this sleep phase indicates brain activity similar to wakefulness[176] and that is why it is also called paradoxical sleep[177]. Studies show that the NREM phase is longer than the REM phase and physical–mental fatigue is reduced in NREM sleep.

Dreams are special mental phenomena. They are mental images with a narrative structure that are scattered over time, loaded into the mind virtually and in fragments[178]. They unfold and are less continuously observed. We live with them as if they are real and happening. Often, we ourselves are present in these dreams in an unsettling manner. Dreams usually relate to information from different parts of our memory that blend together in a tangled way and rarely replay as a complete episode. Impossible events are also seen in dreams, without a sense of timing or cause and effect. Although dreams are constructed based on what is stored in our memory, they often do not remain in memory for unknown reasons. Although they may not be recorded in memory due to the deactivation of the prefrontal cortex, they have adapted phenomena from an evolutionary perspective and coexist with life. Presumably, the brain does not like dreams to be mixed with realities and cause inappropriate reactions and behaviors in the organism. REM dreams are usually based on visual concepts, and about half of them also include auditory components. Less than 30% involve tactile or motor characteristics, and rarely do they involve olfactory or gustatory senses in the narrative. An interesting point is that despite paralysis, we have the ability to move, run, and escape from a fake attacker with great difficulty and weakness in our dreams and minds.

Sigmund Freud, an Austrian psychiatrist, did not believe that dreams were the result of the behavior of gods or demons; rather, he searched for them in the unconscious world of the mind. This unconsciousness encompasses disturbing thoughts with instinctual origins, such as sex, fear, anger, and aggression, which we tend to repress. The goal of psychoanalysis is to extract these thoughts from the minds of patients. According to Freud, dreams are a royal road to unconsciousness. These dreams require interpretation and often revolve around forbidden sexual relationships. While it is possible to interpret almost everything based on sex, it is not accurate to claim that all anxiety disorders (neurosis) stem from unsuccessful sexual experiences[179,180] from a psychological perspective. Dreams are primarily characterized by a sense of danger. Up to 40% of Finnish children's dreams and 80% of Kurdish children's dreams, who are living in Iraq and Turkey, are fearful or dangerous. Primitive humans were constantly faced with environmental hazards and predatory animals. Although this phenomenon has not entered the genome of intelligent humans, the fear of dangerous creatures in unfamiliar environments is embedded in the biological structure of humans. It seems that dreams are calibrated to indicate environmental hazards and early humans were more concerned about environmental dangers than internal illnesses. Dreams are generally not recorded in the brain and can be considered a form of mental wandering with a waste of time.

REM sleep begins during the fetal period and reaches its peak in the third trimester of pregnancy. Shortly before birth, REM sleep is accompanied by NREM sleep, followed by wakefulness. At the beginning of birth, the amount of REM and NREM sleep is roughly equal and gradually decreases to about an hour and a half during the night. In children, especially before starting school, dreams are very simple and consistent. Generally, children up to the age of four do not have the ability to mentally escape from the present moment and cannot engage with a continuous mental scenario. It is likely that from this age onwards, the ability for mental travel begins to develop.

REM sleep entered the sleep structure of land animals about 200 million years ago. This phenomenon has increased in humans after separating from marsupials (pouched mammals). Kangaroos, in terms of time, have about one-tenth of the REM sleep that humans have. The main point is why the brain during sleep, which is designed for energy conservation and reduction in its consumption in the course of evolution, periodically enters the REM phase and its activity rises while its energy and oxygen consumption are similar to wakefulness[181]. REM sleep was first reported by Aserinsky[182] in 1953 and later recorded in the electroencephalogram of a cat by Dement[183] in 1958. REM sleep is observed in many warm-blooded animals, including mammals and birds that can control their body temperature. Mammals are classified into three main subgroups. The first group is monotremes, which include egg-laying mammals such as the platypus and echidna. The second group includes marsupials such as kangaroos and North American opossums, whose newborns enter the mother's pouch after birth and remain inside until they fully develop. The third group is placental mammals, which receive nourishment from a placenta in the womb until full growth, such as whales, elephants, humans, and after birth, they are supported and cared for by parents depending on their brain growth and physical development.

It has been determined and shown in the studies that during NREM sleep, the brain temperature is lower compared to wakefulness, while during REM sleep, the brain temperature is similar to wakefulness. However, during REM sleep, the activity of REM sleep generators in the brainstem increases. It seems that monotremes are an intermediate form of cold-blooded (ectothermic) organisms, dependent on environmental temperature to regulate their body temperature like reptiles, while warm-blooded (endothermic) animals, which control their body temperature through heat production, belong to the group of birds, humans (placental mammals), and marsupials. It is done using three different mechanisms: using brown fat, shivering skeletal muscles, and constriction of skin vessels (cutaneous vasoconstriction)[184] for heat retention. Additionally, as the body temperature increases, the duration of sleep reduces. Studies have shown that the average body temperature of birds, humans (placental mammals), marsupials, and monotremes is approximately 41, 37, 35, and 31°C, respectively. Meanwhile, the average duration of REM sleep is 0.7, 2, 4.4, and 7.5 hours[185,186] for the mentioned groups, respectively.

REM sleep is not merely a passive process. Good REM sleep controls brain temperature and its metabolic process[187,188]. Preventing REM sleep in rats will result in their death due to metabolic disturbances. Scientists believe that at the beginning of NREM sleep, the body temperature, including the brain, decreases. When the temperature reduction reaches a critical point, the brainstem generators initiate electrical discharges to induce REM sleep, which, in turn, causes the brain temperature to return to normal through increased activity, manifested as dreaming, to maintain its function in controlling the body system and homeostasis. It appears that REM sleep may be a kind of by-product or epiphenomenon for temperature control.

Simultaneously, during REM sleep, by activation of cholinergic and glutamatergic nuclei in the midbrain, the body muscles become relaxed, and muscle tone reduces, which is associated with diminished energy consumption. In fact, noradrenergic neurons in the brainstem which are important for awakening do not fire. After a while, REM generators in the brainstem stop working, and the NREM phase resumes. This cycle usually continues until early morning, and eventually, the cold of dawn leads to awakening, increased activity, and body temperature. Overall, REM sleep seems to serve as a thermoregulatory mechanism for controlling brain temperature during sleep and controlling the autonomic system and homeostasis. Animal studies in cats have shown that about 15% of a healthy cat's daily sleep, which lasts 12–15 hours[189], is REM sleep.

If the brainstem is separated in a way that temperature control is lost, and the organism's temperature during sleep decreases to room temperature (23°C), the REM sleep percentage in the cat's total sleep cycle will reach 70%, and the REM phases will also become longer[190]. Ultimately, it appears that REM sleep has three important functions which achieved during the evolutionary trajectory. Firstly, it keeps the brain warm by maintaining constant brain activity. Secondly, it allows the organism to awaken and react to real dangers. Lastly, it reduces muscle tonus and body movements, leading to reduced energy consumption and lack of responsiveness and reaction to dreams during REM sleep.

7 Cognitive Organ

THE BRAIN

Basic human abilities, such as walking upright, using hands, creating useful tools, and speaking, primarily rely on the functioning of the brain. While most of these abilities are predominantly observed in humans, simpler forms of them can also be found in other animals. What characteristics does the human brain have and how does it work? Is the size and volume of the brain alone effective in enhancing its capabilities? Is it only the number of neurons (nerve cells) that is influential? Or is it a combination of those factors? How does energy consumption in the brain occur? If the brain is doing so much work, why doesn't it overheat like a car engine? Does it have a cooling system? The human brain is similar to other species in terms of size and function. To understand evolution in different time periods, in addition to fossils for anatomical evaluation, we also need to assess physiological and functional aspects, but unfortunately, fossils only help us evaluate anatomical changes. This method fundamentally changes our perspective on the basic events that have occurred in human development and/or evolution. It should be noted that the brain is a complex intertwined organ, and changes in it do not occur in a layered and segmented manner.

In simple terms, the human brain is composed of two nearly identical hemispheres that are stuck together with the help of corpus callosum (callus body). The cerebral cortex is made up of a collection of nerve cells (neurons) and is thicker in primates (including humans) compared to other mammals. It has a greater volume due to the presence of numerous convolutions (gyri) and grooves (sulci). The cortical extent increases with the size of the brain. Beneath it lies white matter, which is formed by the axons of neurons surrounded by fatty myelin sheaths, serving as insulation and facilitating the efficient and rapid transmission of neural impulses, which are essentially electrical waves. The circuits formed by complex interconnections of neurons with short and long axons, connected to each other at synapses, contribute to the functional properties that ultimately shape human behaviors. Electrical messages cannot jump from one nerve to another, so coded messages are transmitted to each other with the help of chemicals through very special connection points between two neurons called synapses; the first time was described by German pharmacologist, Otto Loewi[191] in 1921. In other words, the primary language of communication in the nervous system is electro-chemical. They transmit coded messages electrically through a neuron and chemically between neurons. The cortex, along with the underlying white matter, constitutes approximately 43% of the brain mass in a tree shrew, 66% in a marmoset, 76% in a macaque monkey, and 84% in a human. The cortical surface area can be considered as a representation of the number of neurons and the brain's capacity for information processing.

In 1791, a German anatomist, physician, and anthropologist named Samuel Thomas von Sommering thought that as the muscles become stronger by working and the skin of the hand is thicker, the brain also becomes bigger or stronger with activity[192,193] and its performance improves. Although this question was not answered correctly, phrenologists (craniologists) took great advantage of it at the beginning of the 19th century. At the same time, in 1830, a German doctor named Theodor Schwann (1810–1882) and a botanist, Matthias Schleiden, proposed the hypothesis of the existence of cells as the basic structure of the brain. After Camillo Golgi discovered a type of cell staining in 1873, called "black reaction"; a theory was ascertained under the title "neuronal hypothesis" by Ramon y Cajal in the 1890s. A little later, Ramon y Cajal, by using his prominent ability, visual language, expressed the concept of free arborization for neuronal projections, but it was Charles Sherrington, an English neurophysiologist, who before the end of the 19th century coined the term

DOI: 10.1201/9781003538400-9

"synapse"[194,195] (syn: together; haptin: to clasp) for the place where neurons connect to each other. However, Ramon y Cajal planted the wrong concept of the non-proliferation of neurons in the minds of scientists, so it is still taught in many medical schools.

Four fundamental theories have been proposed regarding the evolution and expansion of the brain[196]. Some studies attribute this phenomenon to the size of the body. In other words, the increase in body size has been accompanied by an increase in the survival capabilities of living organisms, thus requiring a larger brain for control[197]. In fact, brain growth is a by-product of body growth[198]. Another belief emphasizes opportunity focus, suggesting that the increase in brain size is related to specific traits within a species that facilitate better interaction with demanding environmental conditions. These new conditions likely arose due to changes in the ecosystem[199] related to the organism. Another group believes that multiple genes, from the outset through their specific mechanisms, have led to the expansion and development of brain volume, size, and function[200,201]. Ultimately, a set of constraints resulting from physiological limitations has led to a set of imperatives in brain size expansion, prompting the organism to demonstrate a process of overcoming difficulties. A prominent example is the inability of infants/young children to protect themselves, which is resolved simultaneously with physical and brain growth[202]. It seems that all of these mechanisms play a role in the development and expansion of the brain, and it is not sufficient to justify these changes relying solely on one mechanism.

What is certain is that initially, the increase in human brain size was a result of changes in their dietary behavior. This behavioral change can be traced back to the time when humans began walking upright, followed by their ability to create tools and an increased consumption of food, especially meat in its cooked form. However, after the emergence of Neanderthals and other types of intelligent humans, there is no clear evidence of a significant increase in brain size. Instead, cooperation among individuals for hunting and food procurement, particularly for obtaining meat, led to increased interpersonal communication and social growth. This phenomenon aided in cognitive improvement, increased social intelligence, and the development of complex toolmaking. In other words, in the initial stage, brain growth was primarily driven by the consumption of high-quality food, but in the second stage, it was solely the result of intellectual and cognitive development, which is a kind of cognitive-cultural evolution.

Can brain volume alone be the primary factor in this evolution? Before discussing that, it should be noted that we can only measure the volume inside the skull in fossils, which differs from the actual brain volume. Research by Rohrs and Ebinger[203] has shown that with an increase in body size, the intracranial volume increases more than brain volume, although not exceeding 10%. Francisco Aboitiz referred to this as the somatic determinant, which means the relationship between brain size and the bulk of body[204,205]. The average weight of the human brain in males is about 1440 g, and in females, it is 1230 g[206]. Table 7.1 compares intracranial volume and a possible number of neurons in different species.

TABLE 7.1
Brain Volume, Number of Neurons, and Life Span of Each Living Organism

	Intracranial Volume (cm^3)	Number of Neurons (Billion)	Million Years Ago (Mya)
Chimpanzee	375		Up to now
Australopithecus	450	30	2–4
H. habilis	610	40–50	2.5–3
H. erectus (early)	870		1–2
H. erectus (late)	950	50–60	0.5–1
H. heidelberg	1200		0.2–0.8
H. sapiens	1400	80–85	0.2 till now

The last common ancestor of humans and apes had a brain volume of less than 400 cc and lived in deciduous woodlands[207]. The brain volume of *H. habilis*, who lived 2.5–3 million years ago, was 1.5 times that of apes, and the brain volume of *H. erectus*, who appeared less than two million years ago, was about twice that of apes. For Neanderthals, this number was approximately three times that of apes. Genetic variations, in the form of various gene mutations (changes to your DNA sequence), have had an impact on the size, volume, anatomy, and physiological structure of the brain. According to Mary-Claire King and Allan Wilson[208,209], evolutionary genetic changes have shown more expression changes in their genes than in their nucleotide sequences. In fact, the nucleotide sequences for proteins in humans and monkeys are more than 98% identical, but their DNA sequences are about 95–96% similar. Microarray genetic studies have shown genetic differences between chimpanzees and humans, although these differences can be ambiguous. In other words, it is better to conduct more investigations on the specialization of phenotypes from previous phenotypes in the course of evolution. These changes have mostly been toward improving the function of the nervous system, plasticity, and reducing energy consumption. For example, the *MYH16* gene inhibits muscle growth in the cranial region but does not inhibit cranial growth. The deletion of one of the growth inhibitory factors by the *GADD45* gene has led to an increase in brain volume and expansion of the cortical surface. Another gene change, called *SLIT-ROBO*, settled around 2.4 million years ago, and resulted in increased processing and improved signals at the cortical level. Meanwhile, $FoxP_2$ has shown a difference of two amino acids between modern humans and Neanderthals/Denisovans. This simple difference has enabled modern humans to have the ability of learned vocalization, namely speaking[210] while Neanderthals did not have much more than simple verbal communication. It should be noted that the ability to speak is not necessarily and solely related to the activity of this gene and brain size. External stimulation is necessary for a child to start speaking and conversing. At the same time, the absence of this gene results in bonobos or other monkeys not learning more than a few words when exposed to a human environment. At least seven genes associated with microcephaly (reduced brain size) have been identified, and further studies have been conducted on two of them. The first one, *ASPM* (**A**bnormal **SP**indle-like **M**icrocephaly-associated gene), is located on chromosome 1 and is involved in the production and size of microtubules and the function of cilia in neurons. In 2002, this gene was reported in 56 individuals of a family in northern Pakistan[211] with a history of consanguineous marriage. The second gene called *Microcephalin (MCPH1)* which is highly pleiotropic (producing more than one effect shows itself in phenotype), is located on chromosome 8, and was reported in two families with seven affected individuals in the same year. This gene affects DNA repair and correction[212]. Mutation in these two genes has been highly influential in the increase of brain volume in evolutionary processes.

Various genetic patterns in humans have led to the growth of the brain in the prefrontal cortex, resulting in a three to five-fold increase in size compared to the chimpanzee brain during the first five years of life[213]. This increase in volume is accompanied by an increase in the number of synapses and their function. The highest increase in synaptic density is observed in humans between the ages of three and fifteen, whereas in chimpanzees and macaque monkeys, it occurs in the first few months of life. This phenomenon indicates that an increase in brain volume, associated with cognitive growth, takes at least 10 years in humans, whereas in chimpanzees and macaques, it occurs within a very short period after birth. Maturation of axonal tracts through myelination of nerve fibers and increased conduction velocity indicates the maturation of neuronal connections, allowing effective and efficient signal transmission at the cost of reduced flexibility in the brain network. This process primarily takes place during adolescence in chimpanzees, completed in male chimpanzees by the age of fifteen and in female counterparts by the age of ten. However, myelination of cortical layers in humans continues until the third decade of life. In other words, human infants undergo myelination of nerve fibers at a slower pace[214], about one-and-a-half to two times slower than macaques, and it should be noted that human sexual maturity occurs slightly later than in chimpanzees. This maturation and growth of the prefrontal cortex in humans continue for approximately five years longer compared to macaque monkeys and/or chimpanzees. This phenomenon, characterized by asynchronous growth in different structures, such as the completion of sexual maturation and the incomplete maturation of the nervous system, is known as "neoteny"[215]. The number of synapses per neuron in humans is approximately 30,000, which is more than two to three times that of mice and rats, which have around 9000–15,000 synapses per neuron. The cortical neurons, by creating multiple, stable synapses with relatively long fibers, establish specialized input–output connections in this system, which are highly efficient in information processing.

The human brain volume ranks fourteenth among two species of whales and eleven species of bipedal mammals. The weight of the brain in some species of sperm whales reaches up to 9 kg, which is approximately six times the weight of the human brain. The average weight of the brain in whales is 6800 g, elephants 4700 g, and dolphins 1700 g. The smallest brain among mammals belongs to bats (74 g), and the largest belongs to sperm whales (9 kg). In a phylogenetic (within a family) hierarchy, as the body size increases, the brain also grows but with a slight lag, indicating that the relative rate of brain growth is lower than that of the body[216] of an organism. Scientists refer to this relationship between body weight and brain weight as a negative allometric relationship. The relative brain weight in small mammals is around 10%, in whales 0.005%, and in humans approximately 2% of body weight, similar to dolphins and chimpanzees. However, it was reported that notable individuals such as Emmanuel Kant, a renowned German philosopher, had a brain weight of 1600 g, Johann von Schiller, a famous German writer, had a brain weight of 1785 g, Ivan Turgenev, a Russian writer, had a brain weight of up to 2010 g and Jonathan Swift, an English writer (with a brain of approximately 2000 cc), have been reported. It should be noted that famous writers like Anatole France, with a brain weight of around 1000 cc, also existed. It does not seem that brain weight and volume alone can justify the intellectual and practical evolution of modern humans. Another question is whether two brains of similar weight function in the same way. The brain weight of chimpanzees and cows is approximately 400 g each. However, chimpanzees' behavior and intelligence are clearly more evolved and complex compared to cows. Having a larger brain volume does not necessarily imply better functionality. Pagel and Harvey[217] reported that at least thirteen animal species have brains similar in weight to humans, which is around 1.5 kg.

Nerve fibers are made up of a plasma membrane that contains significant amounts of cholesterol and fat. This is why the brain contains a high amount of fatty tissue. Cholesterol and fat serve as the building blocks for the structure of the nervous system and its neurotransmitters. Changes in their serum levels, whether increased or decreased, can be associated with mental and neurological disorders, anti-social behaviors, and even suicide. Cholesterol and fat play a role in ketone production, which can be used by the nervous system to produce energy in the absence of glucose. It is likely that during the first two million years of human evolution in Rift Valley, Africa, and the primitive human life near the sea, mothers effectively aided the initial growth of the brain during the fetal and infant stages by using fish and shellfish. The volume of white matter in the brain is approximately three to four times that of gray matter and is a fundamental factor in brain development. Almost all parts of the human brain are more developed than the brain of the ape, except for the olfactory bulb, which indicates their better ability to smell. The greatest growth rate is observed in the frontal lobes[218], especially the prefrontal regions. An increase in the volume of each lobe can cause changes in the shape of the skull. For example, the growth of the temporal lobes contributes to the expansion of the middle cranial fossa and results in its rounding. The growth of the frontal and prefrontal lobes has contributed to the enlargement of the anterior cranial fossa and ultimately the prominence of the forehead.

In comparing the brain volumes of humans and apes across various parts, significant differences emerge. The frontal region, responsible for higher cognitive functions like reasoning and decision-making, is substantially larger in humans, measuring around 500 cc compared to the ape's 110 cc. This may suggest a considerable advancement in cognitive abilities in humans. Moving to the temporal lobe, associated with memory and language processing, humans again exhibit a significant increase, with a volume of approximately 250 cc compared to the ape's 60 cc. This expansion likely reflects the development of more sophisticated communication skills and memory capabilities in humans. The cerebellum, crucial for motor control and coordination, also shows a notable enlargement in humans, measuring around 175 cc compared to the ape's 50 cc. This suggests enhanced motor skills and coordination abilities, especially in human hands compared to their ape counterparts. Finally, considering the overall brain volume, humans significantly surpass apes, with a brain volume of approximately 1350 cc compared to the ape's 400 cc. This comprehensive increase in brain size likely correlates with the complexity of human cognitive abilities, including language, problem-solving, and social interaction[206,219,220].

Accurate estimates of the number of brain neurons were made since 2004. Before that, it was speculated that the number of brain neurons was around 100 billion, ten times the number of supporting glial cells. Therefore, the common assumption among students was that humans typically use only 10% of their brain capacity. The number of brain neurons remains constant at the beginning of birth, and the increase in volume is mainly related to the growth of other parts within the brain tissue (parenchyma). Overall, recent studies have shown that the human brain consists of approximately 86 billion neurons, and there are approximately the same number of glial support cells in the brain. Contrary to the belief, around 69 billion neurons are located only in the cerebellum, with an average weight of about 154 g[221]. Sixteen billion neurons are found in the cerebral cortex, and the remaining one billion fills the spinal cord, brainstem, and diencephalon. The number of neurons in the cerebral cortex is approximately six billion in chimpanzees and 1.7 billion in macaque monkeys[14]. These numbers are 14 million neurons in mice and 31 million neurons in rats. In primates, unlike other animals, the number of neurons increases with weight and brain volume. For primates, the chimps exhibited the highest number of neurons, a total of six billion, with a corresponding brain weight of 286 g. This was followed by the Rhesus monkeys, with 1.7 billion neurons and a brain weight of 69 g. It was shown that marmosets had 245 million neurons and a brain weight of 5.6 g, while the lemur mouse had 22 million neurons with a brain weight of 0.9 g. On the other hand, nonprimates showed distinct patterns. For example, the giraffe displayed the highest number of neurons among nonprimates, with 1.7 billion neurons and a brain weight of 398 g. The mean number of neurons in pigs with a brain weight of 42 g is 303 million neurons and it is 71 million neurons in rabbits with a brain weight of 4.4 g; while rats had 26 million neurons and a brain weight of 0.9 g. These findings shed light on the variations in brain weight and neuron count between primates and non-primates, highlighting the diverse evolutionary adaptations among different species[206].

In terms of the number of neurons, the human brain is considered an evolved and scaled-up brain. If a rodent like a human were to have 100 billion neurons, their brain would weigh 30 kg, and their body weight would be 80 tons. If humans were to follow the model of rodents or were forced to live like a whale in the sea, they would never be able to stand on two feet. It was once believed that the glial cell/neuron ratio in the brain indicated progressive cognitive growth. However, it is now simply understood that this phenomenon represents the average number of neurons[14].

Does increasing the number of neurons correlate with increased intelligence? In the conducted studies, the brain of an Asian elephant weighs approximately 2.5–3 kg and contains around 257 billion neurons, but 98% of them are located in the cerebellum. In other words, the ratio of cerebellar neurons to cerebral cortex in the brain is 44 times higher, and in humans, this ratio is approximately four times. Comparative anatomy of different mammalian brains revealed intriguing variations in their brain weight and neuronal composition. At the apex, it was the human brain, renowned for its complexity, has a substantial weight of 1400 g, accompanied by a remarkable 16 billion neurons within the cerebral cortex, signifying the intricacies of human cognition. Moving on to our primate counterparts, chimpanzee brains weigh in at 400 g, with six billion neurons in their cerebral cortex. In contrast, the elephant, a mammal of vast size, possessed a brain weighing 4000 g, albeit with a surprisingly modest 5.6 billion neurons in the cerebral cortex. This observation suggests that brain mass does not necessarily correlate linearly with neuronal density across species. Delving further into the primate spectrum, the monkey exhibited a brain weight of 87 g, accompanied by 1.7 billion neurons in the cerebral cortex.

While smaller in size compared to humans and chimpanzees, monkeys display significant neuronal density relative to their brain mass. Lastly, the brain of the giraffe, an emblem of the savannah, with its towering stature, the giraffe's brain weighed 700 g, with 1.7 billion neurons in the cerebral cortex. This interesting juxtaposition of brain size and neuronal count hints at the unique evolutionary adaptations of giraffes to their environment. Those studies underscore the intricate relationship between brain weight and neuronal composition in mammals, highlighting the diverse evolutionary trajectories that have shaped the cognitive capacities of different species[206,222,223].

The human nervous system is considered a highly metabolic demanding and expensive tissue, consumes up to 25% of the body's energy for less than 2% of its weight. Overall, the nervous system, along with the digestive system, kidneys, and heart, have the highest energy consumption. Our nervous system as well as other parts are entirely aerobic organs; hence, they cannot reserve any glucose[198] and are dependent on a continuous flow of saturated blood with oxygen and glucose. Therefore, brain growth and expansion require constant increase in blood flow[224] and vascular capacity. In the evolutionary process, leaf-eating monkeys, which exclusively consume tree leaves, have larger intestines and smaller brains compared to fruit-eating monkeys. Leaf-eating monkeys require less mental-cognitive capacity due to the abundance of food in the jungle, resulting in smaller brains. However, depending on the type of consumed food, the digestive system should be more efficient and larger. It can be inferred that the size of the brain reflects the amount and type of food intake by the digestive system and hunting strategies. Among humans, a group has managed to find more energy resources for themselves to aid in the production of energy for a high-consumption system.

The brain primarily depends on glucose as its nutrient and energy source. For the human brain cortex, the glucose consumption rate is approximately 1.32×10^{-8} micromoles per minute, slightly higher for the mouse brain at around 1.39×10^{-8} micromoles/min. However, mice are not more intelligent than modern humans. Overall, the energy consumption for every billion neurons per day is equivalent to six kilocalories, so the human brain consumes about 516 kcal/d. It should be noted that the brain constitutes approximately 1–2% of body weight but consumes around 18–24% of the body's total energy. This energy consumption is not only due to the presence of a large number of neurons but also the numerous synapses they form. Each neuron forms between 10,000 and a 100,000 synapses with surrounding neurons. Perhaps due to this reason, the rate (slope) of increased glucose and oxygen consumption, or in other words, brain metabolism, is higher in the brain compared to the body mass as the brain volume increases. About 60% of generated energy is for maintenance of ionic balances; mainly for Na^+/K^+ pump[225] which transports three sodium ions outward against two potassium ions inward the cell membrane.

In fact, the function of the brain depends on energy metabolism, synthesis, and degradation of neurotransmitters (chemical substances responsible for transmitting nerve signals). In comparison to humans, who use over 20% of the body's energy, other primates such as rhesus monkeys and apes use approximately 11–13%, other primates 8–10%, most mammals 3–5%, and others consume between 0.5% and 8% of their energy related to the nervous system[226,227]. The reason for the increased energy consumption in humans compared to macaques/chimpanzees may be is the higher metabolic rate in human brains[198,228]. During the last three million years of evolution, cerebral perfusion increased 1.7-folds and brain size also increased 3.5-folds. Totally, it shows that cerebral perfusion only by using carotid arteries was increased by almost six folds[224]. This phenomenon is likely due to the occurrence of more genes involved in metabolism. One of these genes is *GLUD2*, which activates a specific enzyme called glutamate dehydrogenase, leading to increased metabolism of glutamate. Glutamate is the most abundant and active excitatory neurotransmitter in the brain, activated in the last common ancestor of humans. However, studies have shown that lactate metabolism, which is related to obtaining energy through anaerobic means, is higher in chimpanzees than in humans. It is not accurately known to what extent this metabolic variation is related to the environment and dietary habits or which portion of it is associated with genetic patterns.

Various reports have shown that the occurrence of multiple genes and a wide range of different metabolites in the human brain, in contrast to the brain of other primates is different. Three main groups of genes, including regulatory genes, protein-producing genes, and new human-specific genes, are involved in this process. Today, it is widely believed that evolutionary changes are more dependent on gene-related changes rather than different protein sequences. Simply put, environmental factors can influence gene regulators to some extent, which in turn leads to changes in the molecular phenotype (external appearance) of the gene, causing changes in its functional phenotype, such as the production of amino acids and proteins. These changes ultimately enhance human cognitive abilities. A prominent example of such a change is observed in two amino acids that have altered the regulatory activity of a transcription factor called $FoxP_2$. In humans, if there is only one copy of this gene, it leads to speech impairment and cognitive disorders, while naturally, this factor is not present in mice. However, if both copies are introduced into the mouse genome, it increases synaptic activity and dendritic branching of neurons, resulting in enhanced learning abilities[229] in mice. Another example is the existence of the *CLOCK* gene, which acts as a regulator and controller of the circadian rhythm[230,231] in humans. Actually, it was introduced in 1984 by Jeffrey Hall and Michael Rosbash, both working at Brandeis University in Boston. They with Michael Young at New York University succeeded in isolating the *Period (Per)* locus which was later called *CLOCK* gene[232]. Dysfunction and impairment of this gene not only disrupt the sleep–wake cycle in humans but can also contribute to the development of diseases and psychiatric disorders such as manic behaviors. The sleep–wake cycle has evolved under the influence of the human genetic structure, and it will exist in the absence of signs and symptoms of the circadian cycle, although its accuracy will decrease[233].

Another important question is how many genetic events need to occur for the formation of the modern human genome. The key point is that mutations need to reach a vital frequency in order to become fixed in the population. The influential factors include the mutation rate (m), the selective value of mutations in the population (S), and the population size (n). Mathematical and statistical studies have shown that the expected time for each mutation is approximately 300 years. Therefore, in the past 20,000 years, about 700 mutations have occurred. However, considering the low selective value of mutations in the population (S = 0.01), only about 14 valuable genes have emerged[234,235]. According to available evidence, the evolution of the human brain has undergone two major phases. The first stage was long and slow, accompanied by an increase in brain size and cortical expansion. The second stage, which has recently occurred, involves shape changes in specific regions of the brain and has led to the re-creation of new cognitive traits. This process ultimately caused a cultural explosion around 200,000 years ago, primarily driven by several genetic mutations. This cognitive novelty has facilitated the accumulation and transmission of cultural information, but the high speed and short time for cognitive emergence have created drawbacks due to the lack of precise regulation in the processes, resulting in crude and unrefined changes. As a consequence, there are numerous psychological and cognitive disorders resulting from impaired brain growth and development. Many of these mental disorders are unique to humans and have not been observed in nonhuman primates. Molecular genetic studies may potentially contribute to their treatment[236]. It should be noted that for survival, all intelligent and sentient species need to possess the ability of error monitoring;[237] otherwise, they are prone to extinction. Evaluation of errors is due to sensory experiences, which are not necessarily complete because we generally do not have a comprehensive and true sensory perception of the surrounding world. The brain works as a probability calculator that updates its inference from the surrounding environment by receiving information from the outside world. The brain can fill in our sensory gaps through the knowledge we already have. In addition, correcting our mistakes is also a necessity to learn new things[238].

Regardless, what matters is the production of energy for the brain and other bodily components, If 2% of the weight of a gorilla, which has an average weight of about 150 kg, is attributed to its brain, it would require the gorilla's brain to have nearly 200 billion neurons, which is actually more than double the current number of neurons in the human brain. Gorillas and orangutans spend approximately 7–8 hours each day eating fruits and plants to meet their daily energy needs. This is roughly equivalent to the time when *H. habilis* and/or *Austras* were engaged in eating all through the day. With an increase in brain size to the mentioned extent, monkeys must spend many hours daily eating raw food and fruits, which exceeds more than half of the day. This is almost impossible, especially during periods of food scarcity or heavy rainfall. Currently, the ratio of a 150-kg gorilla's weight to the weight of its brain (450 g) is more than 330 times. For a 75-kg human, this ratio is only about 57 times[216,239]. For the addition of one billion neurons in a small animal, the increase in eating time is less than an hour. To constitute 2% of the weight of a gorilla's body as the brain, it would require adding more than 120 billion neurons to the existing number of neurons. These two sentences imply adding 730 kcal to their daily energy consumption, which increases their eating time by approximately more than 2 hours and sometimes up to 12 hours in total. Studies have shown that the size of orangutan brains in the forests of Borneo, Indonesia, where they consume less or lower-quality food, is smaller compared to those living in the forests of Sumatra.

The skilled and tool-maker human (*H. habilis*) was engaged in eating for approximately 7.5 hours according to this calculation. With the increase in brain size during human evolution, upright posture, and then the development of the intelligent human, the continuation of this trend was not possible. The most significant discovery of humans during this period was fire, and their most important invention can be considered cooking[240], especially cooking of meat. Cooking not only improved the nutritional value of food by up to twice but also significantly reduced the time required for eating, eliminating the need for the *H. habilis* to spend eight or more hours a day on eating and sometimes spending a similar period of time looking for hunting or appropriate food. Early humans, especially during the *H. erectus* period, attempted to utilize energy-intensive foods such as fruits, seeds, and meat, in addition to rhizomes and tubers as edible stems and roots, for brain growth. According to Sonia Ragir[241], many roots and stems contain enzymes that inhibit food digestion. Humans gradually learned to inactivate these enzymes through processes such as cutting, pounding, grinding, fermenting, or drying food. Processing and detoxification of food were generally performed by women, and children learned it alongside their mothers. Richard Wrangham[242] believes that the habit of using fire began around 1.5 million years ago in East Africa, but other researchers believe that the use of fire started during the Ice Age. However, early humans had already found ways to improve food digestion, such as cutting, pounding, grinding, and tenderizing food, especially meat, which increased calorie production. The slicing of meat using teeth, using the contraction of the bilateral muscles of the lower jaw, and separating the food into multiple pieces. The movements of the lower jaw are performed vertically, forwards, backward, and sideways with the help of five masticatory muscles; and the opposite side is responsible for the opposite movements, maintaining the balance of this cutting-milling system. The chewing force for humans is approximately 200–400 Newtons per square meter (N/m^2) and is up to five times greater for orangutans. Repetitive chewing movements exert pressure on the facial and mouth muscles and cause changes in facial shape. These precise movements on small morsels allow sensory nerves to transfer sensory messages to the brain through the tongue, jaw muscles, and ligaments around the teeth, and motor fibers provide appropriate force feedback to the corresponding muscles. Overall, this process, with increased energy absorption in a short period, enabled early humans to have more time for other activities such as using their hands, organizing group hunting, and ultimately becoming social. The use of fire, along with changes in the living environment, led to changes in social dynamics, romantic and sexual relationships between the two sexes, and ultimately the sexual dimorphism of humans. This, in turn, improved memory structure, the ability to plan tasks before performing them, and self-control and logical thinking[243]. In other words, the most significant achievement of cooking was companionship and synchronization with cultural changes, and the current human is the product of the agricultural revolution that followed. In fact, consuming raw food alone could not and still cannot produce a significant amount of energy. That is why large primates, unlike modern humans, have small brain despite having a robust structure. In other words, the time limitation for consuming raw plant-based food during the day restricts brain growth, a limitation that disappears in humans due to meat consumption and cooking. Aiello and Wheeler[244] demonstrated that all high-energy-consuming organs (heart and intestines) except the gut have grown in proportion to body size in primates. This means that as the brain size increased in evolutionary history, the intestines became smaller (a large brain at the expense of a small gut) to achieve energy balance which we can call it "brain-gut trade-off." Cooking meat, especially in its cooked form, has significantly contributed to this matter.

It can be considered that the limitation of cerebral metabolism was the main restricting factor in the survival process. In other words, there is a very close relationship between finding new methods of nutrition and the cognitive growth of the brain. A small number of mammals, such as lemurs, undergo hibernation to conserve energy for survival. According to approximate calculations, the average weight of an *Austra* and *H. habilis* has been approximately 33 and 38 kg, while the brains of each contain around 34 and 40 billion neurons, respectively. This indicates that both required 7.5 hours of food consumption to maintain energy throughout the day, a method that needed to be changed. Another essential point is examining the efficiency of energy consumption by the brain compared to the body. Does the cerebral system operate optimally in terms of energy consumption? Mathematical and statistical studies have shown that with a ten-fold increase in body weight, brain energy consumption increases ten-fold, and with a hundred-fold increase in weight, energy consumption increases by 31.6 times[245].

The role of the cerebral cortex is highly important in this context. The surface area of the cerebral cortex in small mammals such as shrews is about 0.8, in rats 6, in cats 83, and in humans approximately 240 cm^2. The thickness of the cortex in shrews is about 0.5 mm, while in humans it reaches about 2.4 mm. However, the largest area of the cortex, which has developed in humans, is the "association cortex." It is speculated that in terms of classification compared to the primary sensory-motor areas, the association cortex is superior and can connect different functional regions of the brain, not only in the cortex but also in the brainstem and thalamus[246]. Although the human sensory-motor cortex is roughly equivalent to that of chimpanzees and apes, the association cortex in humans has expanded up to three times more. In other words, in humans, the association cortex predominates clearly and cannot be compared to other primates, and it has a significant impact on the internal structure of the human brain. Another important point is the possibility of the existence of specific areas and centers due to the diversity of brain functions. Studies have shown that similar areas to the speech motor (Broca's area) and sensory (Wernicke's area) centers exist in the brains of monkeys, but in intelligent humans, their functions may have emerged. Therefore, it can be concluded that there is no specific special function that indicates the existence of special centers in the cortex. However, it is possible that new areas and regions such as the middle temporal gyrus for word comprehension and the insula gyrus as a region for self-awareness have emerged in the evolutionary process[247].

Additionally, the human prefrontal cortex is approximately three times larger than that of monkeys. Interestingly, although the cerebral cortex accounts for about 82% of the brain's weight, only about 19% of neurons are located there. In other words, it comprises approximately 80% of the volume but contains only 20% of the neurons. In guinea pig, the cerebral cortex forms slightly over half of the brain's weight but still possesses about 20% of the neurons. Various studies have shown that, in mammals, the cerebral cortex constitutes approximately 50–80% of the brain's weight and contains 15–25% of the total neurons. These numbers are almost entirely reversed in the cerebellum. The cerebellum comprises about 10–20% of the brain's weight but contains 70–85% of all neurons. The exact reason for this is not known, but it may be a by-product of evolutionary processes[248]. The large size of the cerebellum relative to the rest of the brain, which has different structures, the greater number of cerebellar neurons (which are four times more than the cortex on a fixed basis), and the somatotopic topography (mapping) in the cerebellum, which is very similar to the cerebral cortex, can all be reasons why the structure of the cerebellum has developed in coordination with the association cortex and has a joint collaboration with it[249]. However, it is possible that the smaller number of neurons located in a broader region enables them to create neuronal networks with optimal and efficient functions. These small networks can continue their work with low energy consumption and, if necessary, establish communication through dedicated nuclei and centers to transmit neural messages to other networks. This approach allows neural networks to continue their work at a faster speed and with lower energy consumption, preventing excessive growth, slowing down, and excessive heating of the brain due to energy consumption.

Another important point is the heat generated from energy consumption and how the brain cools down. If the body temperature is below 35 or above 41°C, it may cause damage to the brain. Its temperature may fluctuate between ±2 to 3°C following exposure to variant physiological challenges such as eating. Any type of brain activity[250,251] such as sensory perception, motor activity, and thinking excluding muscle contraction use ATP and increased temperature. There are temperature-sensitive receptors in the hypothalamus in primates and mammals. Blood at different centers in the brain is almost cooler than those parts of brain tissue and will reduce the temperature. Dogs cool themselves by panting, but humans use their five million sweat glands instead. Panting is impractical for humans, especially during running. Chimpanzees, on the other hand, use the sweat glands in their palms, soles of their feet, and scrotum (covered with hair) to sweat. However, sweat glands are present throughout the body in humans, particularly in the head and face, due to the extensive connection between superficial veins and the skin. The scalp hair covers only a small part of the head and can act as a protector against sunlight.

The cerebral cortex in humans is generally composed of six layers of cells. The function of the outer two layers (layers two and three) is mainly related to connecting different surface areas of the brain, while the deeper layers, five and six, connect to the lower regions of the brain and the spinal cord through their long axons. Sensory fibers typically synapse with the fourth layer[252]. In the cerebral cortex, the size of the speech area has expanded regardless of the size of the brain, but the primary sensory and motor areas have not undergone significant growth and expansion in evolutionary terms. They can function well and perform specialized tasks without the need for assistance from other parts. This means that independent working centers are created, and the ability to perceive conceptual phenomena increases. Often, sensory stimuli excite more than one sensory modality. The area and cortex related to speech are among the most important. Similar speech areas with less evolution can also be seen in other primates. This indicates that entirely new neural circuits have emerged in humans. These similar centers have mirror neurons that can discharge electrical activity to execute the same action by observing it. The speech areas in the human cerebral cortex are up to six times larger on the left side and up to five times larger on the right side compared to chimpanzees. This phenomenon indicates disproportionate growth of the cortical region related to speech in the evolutionary process of humans. The same has happened for the temporal cortex of the brain. The temporal cortex in humans, primarily related to hearing, is larger than similar regions in bonobos, and its main parts are associated with the hippocampus for episodic memory control and the amygdala for emotional understanding[253,254]. These two parts fundamentally contribute to self-awareness and verbal communication. Although the visual cortex does not have extensive coverage, it is specialized in shape, color, position, and motion perception.

The relative sizes of the brain and the reconstruction of neural connections in the brain have been highly effective in the process of human behavioral evolution. For example, the olfactory sense presentation in mammals such as dogs is much greater than that of raccoons, which rely on their hands, or the larger size of the temporal lobe (auditory cortex) in bats for sound perception or monkeys that use their tails to grasp tree branches, which are longer and larger compared to ordinary tails[14]. In other words, brains regulate themselves with their bodies through more precise methods, and it's not solely the size and volume of the brain that is influential. This regulatory process is not necessarily formed through genetic transmission but rather dependent on ontogenetic (related to the origin of each organism) growth and adapts to the body's needs. The neocortex of mammals often possesses the property of plasticity, meaning that one region can gradually take on the function of a damaged section, thereby preserving certain predetermined human behaviors without the need for training, education, or environmental influences. Babbling in infants is an example of this phenomenon that does not require instruction.

The cerebral cortex of an elephant is approximately twice the size of a human's, but the number of neurons in the gray matter of an elephant's brain is estimated to be about one-third of that in humans. By this calculation, the estimated number of neurons in the prefrontal cortex of modern humans is around 4.3 billion. For a baboon, this number would be 230 million, for a macaque 137 million, and for a marmoset, it would be approximately 20 million. This number of neurons allows for human behavior, morals, policy, and planning to take place. The number of neurons present, which is indicative of human cognitive capacity, is determined by the genetic structure of humans. This number of neurons has remained relatively constant for about a 100,000 years, but their cognitive abilities have significantly augmented during this time, which reflects the cultural evolution[12,206] following initial genetic evolution.

Another point is that the size of the brain is relative to the size of the living organism. The size of the brain in vertebrates is proportional to their body size. As the size of animals increases, the size of the heart and liver, which are directly related to metabolism, also increases. In a study, it has been shown that the brain weight of a tree shrew is about 0.1 g, while their body weight is no more than 1–2 g. In comparison, the brain weight of a blue whale is 9 kg, and its body weight is 100 tons. In other words, with a 100,000-fold increase in brain weight, the body weight has increased by approximately one hundred million times. In primates, the increase in brain weight is accompanied by a coefficient (slope) of 0.7 with the increase in body weight. This slope has been calculated to be between 0.5 and 0.9 for other mammals.

It was uncovered that the body weight relative to brain weight across various mammalian species revealed fascinating disparities that shed light on the intricacies of neural evolution and adaptation. Beginning with the humble guinea pig, a body weight of 500 g against a brain weight of 3.6 g, underscoring the relative proportionality of its neural mass compared to its overall physique. Similarly, in the case of the marmoset, scientists encountered a matching brain weight of 6–7 g, yet with a slightly lower body weight of 0.3–0.4 kg. This subtle variance in body mass suggests potential differences in metabolic rates and ecological niches among these small primates. Investigation on the rabbit, an animal known for its reproductive prowess, agility, and efficiency revealed despite its petite little body weight of 4.6 kg, has a brain weight of 9.1 g. Transitioning to larger primates, the capuchin monkey exhibited a considerable body weight of 3.3 kg alongside a robust brain weight of 52 g. This notable brain–body ratio hints at the advanced cognitive capabilities of these primates, which may be linked to their complex social structures and problem-solving skills. In contrast to the terrestrial lifestyle of most mammals, we also explored the intriguing case of the common and European moles. Despite their diminutive body weights of about 42.7 g and 95.3 g respectively, their brain weights of one g each; probably signify the importance of sensory processing and navigational prowess in subterranean environments. Despite different opinions[255], these findings may underscore the diverse adaptations and evolutionary trade-offs observed in mammalian neurobiological evolution, reflecting the intricate interplay between brain size, body mass, and ecological pressures across species[206,222].

However, scientific studies have shown that when body weight increases a 1000-fold, the number of spinal cord neurons increases tenfold. In fact, this is based on the fact that the number of neurons in the spinal cord does not depend on the weight and surface area of the body but has a direct correlation with the length of the spinal cord. Furthermore, as the body size increases, the density of neurons in the brain reduces, but the size of neurons does not increase. The size and volume of the brain are not solely determined by the number of neurons. The presence of glial support cells is also important in this matter. The term "glia" (means glue) was first used by Rudolf Virchow[256,257] in 1856 to refer to nerve glue. Several years later, Franz Nissl[258,259] introduced the concept of glial cells for their key role in mental functions. In the 1950s and 1960s, the ratio of glial cells to neurons was examined in various studies. It was finally determined that the number of glial cells is related to the size of neurons. The primary function of glial cells is metabolic support for neurons, so the larger the size of neurons, the greater the number of glial cells. The ratio of glial cells to neurons in the modern human brain is 1.05.

Perhaps one of the most fundamental questions is how valuable a neuron is in evolutionary terms and how much energy investment should be made in it. Neurons generally do not consume a significant amount of energy to generate electrical discharge; instead, they require about 90% of the energy to return to their initial state and prepare for the next discharge. Approximately 13% of the energy is needed to maintain this state. The larger the neuron, the higher the energy consumption. Studies by Polish physicist Jan Karbowski[260] in various mammalian species have shown that glucose consumption increases with an increase in brain tissue mass, with a slope of +0.15. In other words, if the weight of the brain is increased ten-fold, glucose consumption increases by approximately 70%. At the same time, the rise in energy consumption with an increase in brain tissue volume has a slope of +0.85, and with an increase in the body size of an organism, it has a slope of +0.75. In other words, the growth rate in glucose and energy consumption with an increase in brain mass is greater than the increase in body mass of the organism. Other studies indicate that glucose consumption by each neuron in modern humans is approximately 5.44×10^{-4} micromoles/min, which means each neuron requires about 3.3 billion glucose molecules per minute. In other words, for the entire brain with 86 billion neurons, five grams of sugar is sufficient for an hour of function. This amount is equivalent to 129 g of glucose in 24 hours, which is about 516 kcal/day. It should be noted that while the number of neurons in the cerebellum is about four times that of the cerebral cortex, the energy consumption in the cerebellum and cerebral cortex is approximately equal, and the reduction in synaptic activity or the lessening in the number of synapses is likely a significant factor in the reduction of energy/glucose consumption by the cerebellum. Overall, the energy consumption of the brain is approximately 500 kcal/day, equivalent to 60 watts of electrical energy. Brain energy consumption remains constant during rest and activity, unlike muscles whose glucose consumption increases significantly with activity. In fact, the brain is a highly efficient machine in terms of energy consumption.

Obtaining energy from the natural environment is very challenging. It is not only the availability of resources that matters but also factors such as the size of an organism's mouth and the type of food consumed (raw or cooked) and their effects. In non-human primates, the duration of daily feeding is about 7–8 hours, and as body weight increases, the number of neurons reduces. Beginning with a body weight of 25 kg, our estimations revealed a substantial neuronal count of 53 billion neurons. This suggests a dense neural network relative to body mass. As we progressed to larger body weights, we observed a gradual reduction in the estimated number of neurons. At 50 kg, the neuronal count decreased to 45 billion, highlighting a proportional decline in neural density as body size increases. Further analysis at 75 kg unveiled a notable reduction in the estimated number of neurons to 30 billion. Finally, at 100 kg, estimations yielded a significantly reduced neuronal count of 12 billion. This observation suggests a further decline in neural density, possibly indicating trade-offs between brain size, metabolic demands, and ecological factors in larger-bodied organisms[206].

Therefore, for a 75-kg primate, there would be more than 30 billion neurons, and it is not possible to generate more energy than this through the consumption of fruits and raw plants. Nutritional support for a larger number of neurons is not feasible. It seems that this group of animals has shown a greater emphasis on increasing muscle mass rather than brain size. The early human, *Austra*, chose a different path. Initially, *Austras* had a small stature, but gradually their body and brain mass escalated. They could only obtain a total of 200 kcal of energy per hour of eating. Over a period of 1.5 million years, the number of neurons increased from 30 to 50 billion. In the following similar period, until reaching the form of modern humans with 86 billion neurons, they had to manage energy consumption in some way. There were different ways to control consumption. The first way was to reduce body size, then to reduce brain energy consumption, and the third way was to increase eating time. All three methods were challenging and sometimes impossible to overcome food shortages. It seems that initially, they chose to increase the consumption of high-energy foods in the short term. Early humans were initially among the gatherer species. They would leave their homes in search of food and roam in the surrounding environment. Walking on two legs reduced energy consumption. At the same time, early humans tried to take longer strides with their longer legs. Their gluteal muscles (buttocks) grew larger, and by strengthening the muscles and ligaments in the neck, their heads became upright. The balance system related to the semicircular canals in the inner ear evolved in connection with the visual, vestibular, and cerebellar systems of the brain. The shortening of toes and the arrangement of the foot and other toes next to each other also aided movement on two legs. This set of changes improved humans' endurance running ability. In addition to collecting grains and edible plants, early humans gradually became hunters. The consumption of meat was accompanied by rise in the number of neurons around 1.5–2 million years ago. The ability of brain neurons for processing memory, planning, personal control, awareness, communication, and language learning was effective. A new event also occurred around 1–1.5 million years ago. Still not intelligent, humans accidentally learned to use fire for cooking food. Cooking not only improved the taste but also increased the speed of chewing, swallowing, and digestion. This accidental discovery had a significant impact on their lives. The breakdown of collagen fibers due to heat weakened the tough cell walls of plants and facilitated the absorption of fat and carbohydrates. The result was the shrinking of teeth, certain skull bones, and the shortening of the intestine. Gradually, human beings who were constantly on the move became farmer and befitted a sedentary life. Years later, industrial agriculture became the primary source of food production for modern humans. But what was the reason for the enlargement of the brain in the evolutionary process? Was it simply the ability to walk on two legs, or is it considered a kind of followed or tailed event? There are two main theories on this matter. The first theory relates to ecological intelligence, which means the struggle in the environment for a better life, including obtaining food, water, and air, improving nutrition, and living conditions, has been instrumental in brain development. The second theory is related to social intelligence and it recognizes the importance of living with other beings and the necessity of cooperation with peers.

For now, it is assumed that brain growth for problem-solving through intelligence and the cognitive outputs derived from it, such as ideas, beliefs, and cultural values, has likely reached its perfection due to energy consumption limitations and neural processing constraints. Generally, the brain with a volume of 1500 cc produces approximately fifteen watts of thermal energy. Therefore, we need a kind of radiator to cool it down, and this radiator is the blood circulation in the brain. The growth and escalation in brain weight require faster and greater blood circulation to solve the issue of brain heat production. Approximately 14% of blood circulation is dedicated to the brain[261]. If we only consider the heat and the heart's ability to effectively circulate blood, the brain volume could increase up to ten times, but there are other factors as well. The biophysical constraints of the brain are mainly related to the processing time of neuronal signals and synaptic transmission. It appears that the brain has reached its maximum capacity for processing information based on neuronal structure. The factors influencing this capacity include nerve conduction velocity, processing time in synapses, pulse length (width), and neuronal density. In fact, nearly ninety billion neurons cause approximately 10^{15} synapses to process or act per second. Studies indicate that the brain operates at around 20–30% of its desired capacity. Therefore, increased and improved brain activity requires structural upgrading and organizes the nervous system to control thermodynamic and electrical signals processes. Taking all these conditions into account, the human body can reach its maximum processing capacity with a brain volume of at most 3500 cc, not more. It should also be noted that the functional evolution of the brain is separate from its structural and anatomical evolution.

ENCEPHALIZATION QUOTIENT

Yet, the fundamental question persists. What precisely determines the superior intelligence of modern humans? Although brain size and volume, the number of neurons in the brain and cerebellum is influential factors, they still do not provide a good index to prove the better functioning of the human brain. For years, it has been implicitly assumed that a larger brain is a sign of higher intelligence but the brain merely acts as raw material in natural selection. Studies by Alfred Binet, a French psychologist in the 20th century, and subsequent research have shown that the heritability estimate of intelligence based on brain size is at most no more than 0.4[262], given that this estimate has been determined between zero (minimum) and one (maximum). The brain cannot be considered as a standalone functional unit because multiple genetic and environmental factors play a role in its performance.

Scientists have long learned that to evaluate a phenomenon, they should measure another factor or phenomenon. It was calculated the ratio of the cross-sectional area of the corpus callosum (the white matter connecting the two hemispheres) to the medulla oblongata. The former is related to brain size and volume, while the latter is related to body size and weight. The basic problem was that this ratio was only significant in placental mammals[263]. However, Eugene Dubois, a Dutch anthropologist, introduced the concept of cephalization[264] in the nineteenth century. Its fundamental concept was that by having the brain weight and body weight, an index could be proposed to determine brain growth. In the 1970s, Jerison[265] used the same concept and introduced an index called "Encephalization Quotient" (EQ) which is a dimensionless index. This index is the result of the ratio between the actual size of the brain and the expected size. In other words, this index represents the size of the brain as it is and what it should be. Just to reiterate, this index indicates the brain size that should be based on the available weight and body size, calculated according to Dubois' formula. Essentially, this index shows how much the brain weight should be in grams based on the body weight in kilograms. To put it in another way, it can be concluded that according to the brain weight of an organism, how much its body weight should be. Has this ratio been observed in all animals, and does it vary with cognitive intelligence? Jerison considered this quotient to be related to intelligence, the capacity to construct a perceptual world, in complete correlation. Since these calculations are performed on logarithmic axes, the results show the ratio of the actual size of the brain to the predicted size. EQ = 1 indicates the normal proper ratio of brain and body growth. If EQ is more than normal, it means that brain growth is more than predicted body growth; and vice versa.

In this computational model, humans are superior regardless of the group they are compared to (vertebrates, mammals, primates, etc.). As weight increases, the body shape changes. In mathematical models, doubling the surface area corresponds to a three-fold increase in volume. However, in animals, this increase in volume does not occur uniformly in all body parts. The changes in shape with body size are referred to as "allometry"[266]. The first step in allometric analysis is to linearize the curves. This is achieved by taking logarithms. Harry Jerison believed that EQ could be a measure of the number of brain neurons. According to him, it was the number of neurons that contributed to cognitive adaptability or, in other words, biological intelligence. He believed that every organism requires a specific number of neurons to function, and therefore, a certain volume of the brain or number of neurons is necessary for maintaining the motor, sensory, and cognitive systems. However, EQ is always a relative unit, a ratio, and is heavily dependent on the reference value[267] in calculations.

Using this calculation method, the modern human brain ranks first by a significant margin from other animals. For example, a squirrel weighing 100 g with a brain weight of 24 g has an EQ of 2. This number can vary to some extent in humans depending on body weight and brain size. Overall, modern humans are placed in the highest category with an EQ of 7–7.5. In simpler terms, the human brain should be seven times smaller in relation to body size for getting EQ = 1. Tolerating sevenfold increase in EQ indicates a growth in brain synapses without generating excess heat in the brain. In the study by Ralph Holloway and David Post[268,269] in 1982, an EQ of 9.6 was calculated for the intelligent human. This number was considered to be 4.8 for the *Austras*, 7.0 for the skilled human (*H. habilis*), and 7.3 for the erect human (*H. erectus*). The EQ for dolphins is slightly higher than 5, and for chimpanzees, it ranges from 2.2 to 2.5. These figures are 1.2 and 1.0 for dogs and cats, respectively, and less than one for horses, sheep, and mice. The brain volume in mammals, according to Jerison's theory, increases with a coefficient (slope) of 0.67. In the study by Martin[228] and colleagues, this number was considered to be 0.76. These values are 0.75 for primates and 0.54 for different animal species. It should be noted that the significant increase in EQ in human-like creatures did not coincide with the emergence of intelligent humans but occurred later. Secondly,

EQ does not necessarily represent the concept of intelligence. Intelligence is a broader and more complex concept. EQ is mostly used as a coefficient to measure the cognitive abilities of organisms, and in other words, the correlation between EQ and intelligence does not imply their equivalence.

Overall, the growth in size and volume of the brain alone is not sufficient to justify the increase in cognitive power and capacity of intelligent humans. Ralph Holloway[270,271] divided the evolutionary growth of the brain into three stages. The first stage is called the "*Australopithecus* phase," which lasted from six to two million years ago and was gradually accompanied by the growth of brain size, or in other words, cranial volume. The next stage, which lasted up to 200,000 years ago, belongs to *H. habilis* and *H. erectus*. Although it was accompanied by an increase in brain volume, the functional abilities, toolmaking, mobility, and change of habitat indicate an increase in their mental and intellectual capabilities. However, in the third stage, which includes the period of *sapiens* and *Neanderthals*, cognitive capacity significantly increases, but *Neanderthals* fall behind in this competition compared to *sapiens*. Despite the lack of volumetric brain growth, functional and social abilities evolve, and among them, the ability of speech and language plays a fundamental role in this victory[272].

8 Plastic Brain

NEUROPLASTICITY

Nerve cells communicate with each other through synapses, employing an electrochemical language. For the first time, Canadian psychologist Donald Hebb in his famous book "*The organization of behavior*" (1949)[273] proposed the meaning of the of concept of memory is deepening and strengthening of synaptic connections. A quarter of a century later, two neuroscientists, Timothy Bliss from Great Britain and Norwegian Terje Lomo, showed that continuous nerve stimulation with a frequency of 10–20 Hz, known as long-term potentiation (LTP), causes significant improvement in neurochemical signaling that can last for weeks. This phenomenon is achieved through the release of glutamate, its attachment to cell surface receptors called NMDA, and the facilitation of sodium and potassium permeability. This phenomenon is accompanied by other biochemical changes in the post-synaptic neurons and finally by releasing chemical substances known as neurotransmitters, they affect the presynaptic neuronal membrane in a retrograde manner. The publication of their ideas in 1973 led to the understanding of the concept of learning and memory in the hippocampus. The current opinion of scientists is that by strengthening a certain group of neurons in a network in the hippocampus, memory is strengthened[274]. Remembering something (retrieval), in fact, is a type of reactivation of the same neuronal network in the hippocampus.

Plasticity in the nervous system refers to the improvement and expansion of neuronal and synaptic connections through the influence of neurotrophins (chemical compounds effective in the growth of basically neurons and other parts of the nervous system; trophe means to nourish) during growth and development. Plasticity is the capacity for change, which is caused by environmental influences. It is the malleability capacity of the brain. Plasticity is the ability to change and respond to a specific or novel stimulus in the environment, new experiences, and unexpected events. These changes are caused by the brain's ability to build or prune synapses, and it is called neuroplasticity. In fact, the unit of memory is the number of neuronal branches and their spines that are formed on the surface of dendritic and/or axonal synapses[275]. That is why the ability to store information in memory is very large. Continuous nerve stimulation and LTP is a type of functional plasticity, but memory can be considered its structural type. Pruning dendritic spines can greatly help fine-tuning neural circuits. However, there is a critical period for plasticity. There are certain time intervals to learn a language, use the dominant hand; and improve vision and hearing senses.

This phenomenon simultaneously contributes to anatomical changes such as sensory field expansion, enlargement of cortical areas, increased connectivity between cortical and subcortical regions, and increased gray matter in the cortex related to the storage of complex visual movements. Additionally, the development of physical skills, such as music training and playing instruments, leads to changes in the shape and structure of the sensory and motor cortices. These types of neuroplastic changes occur more rapidly, completely, and effectively in children. Ultimately, the process of learning to take steps and walk improves the brain's structure and function[276]. Crawling and using all four limbs for walking also increases synaptic connections and creates more synaptic branches in the cortex. Language, music, and writing education represent culturally complex behaviors that involve transforming sensory inputs into motor abilities, accompanied by changes in the organization and size of different brain regions[277].

Plasticity and genetics are two sides of the same coin, where the expansion of one restricts the other. Neuroplasticity, contrary to the old belief, is a rule rather than an exception[278], allowing the nervous system to evolve and continually adjust itself in response to various experiences. One social custom in ancient China was binding the feet of young girls to have what was considered more beautiful feet. Studies suggest that long-term foot binding[279] along with reducing the somatosensory

DOI: 10.1201/9781003538400-10

representation of the feet in the brain cortex, expands the cortical sensory area for genitalia[280] and strengthens the sensory perception of the nearby genitalia, leading to an increase in genital sensation, making it more excitable. This phenomenon may potentially influence the increased genital representation in the somatosensory cortex, impacting the sexual tendencies and behavior of Chinese girls, and creating more desirable partners for Chinese men.

But plasticity is not always considered a positive process. Addiction and chronic pain are obviously maladaptive forms of plasticity. Addiction bypasses the reward system in the brain[281] and eventually hijacks it. Even after stopping opioid drugs due to the formation of multiple synapses in the neural pathways related to pleasure, it causes cravings and drug-seeking behavior which remains for a long time. Nerve growth factor is a substance that is necessary for the development of sensory nerves at the beginning of embryo formation, but gradually its behavior changes and becomes an important factor for causing pain due to inflammatory processes in the body. Diseases that are associated with constant pain cause the pain sensation to be transmitted to the thalamus through spinal cord fibers, and increasing the number of synapses, like addiction, it causes the pain memory to remain after treatment. Except for opioid drugs, the use of recreational medications, pathological gambling, hyper-sexuality, and shopping[282] can be among the maladaptive forms of plasticity in the brain.

The primary factor contributing to increased energy consumption in the brain is the transmission of electrical signals by neurons. There are two mechanisms for reducing energy consumption in the brain. First, throughout the course of evolution, the *Homo* species has shown a tendency to shrink the wiring system that is nerves and their connections, inside the brain. Second, there is an attempt to optimize energy consumption by increasing efficiency and improving network performance. As a result, instead of working in a random manner, neurons form small networks referred to as nodes[283]. The connection of these nodes creates clusters and ultimately modules within this structure. Each of these modules is connected to other regions and networks through a central hub. In this state, modules with different functions are formed for behavior, general intelligence, working memory, and other specific cognitive traits. The inter-network connections cause different functional networks to be in close proximity to each other and form a central core. This working method, which has emerged in the human evolutionary structure, not only exhibits high efficiency but also has low costs and energy consumption. The anatomical and functional diversity, structural connectivity, and information processing centers result in variations in the shape and cognitive structure across different individuals. Disruption in this structure can be associated with diseases such as Alzheimer, Parkinson, Huntington, multiple sclerosis, seizures, ALS, and epilepsy[284]. Mental disorders such as depression and mood disorders are also not exempt from these effects. The modular system not only allows for increased internal connectivity within the network but also significantly increases the speed of message transmission, reduces message delays, and ultimately facilitates learning and cognitive processes.

We need to pay attention to the fact that in larger brains, the size of neurons does not increase; instead, their number increases. Therefore, establishing connections between them becomes important and their organization will be completely different from smaller brains for better functionality. In primates, the expansion of white matter volume in the brain is associated with the length of myelinated axons, not the number of neurons. Hence, the brain must grow in a way that doesn't negatively impact internal communication by elongating axons and reducing the speed of neural transmission. To accomplish this, the neural structure has evolved to create a complex and comprehensive map of neural fiber connections. This collection consists of small islands of neurons (nodes), each with specific functionality, which can connect with each other when necessary, ultimately reducing the specialization intervals between these functional components of the cerebral cortex. This phenomenon causes the number of information processing centers to proliferate in the form of networks. The findings based on this set of network communications, referred to as "connectomics" by Olaf Sporns[285–287] from Indiana University, finally lead to asymmetry and lateralization of brain function and consequently influencing the volume and size of the brain. According to

Jon Kaas[288,289], distance is equivalent to time, which reduces the speed of neural message transmission and energy consumption, ultimately resulting in functional asymmetry as a solution for establishing neural connections. In other words, evolutionary pressure has led to highly specialized tasks, such as speech processing in a specific region of one hemisphere, enabling consecutive and sequential control. Overall, structural or functional connectomes create their own topography of a small world, saving energy by shortening pathways, ensuring stability and resilience, and ultimately providing flexibility for changes in different animal species or different conditions. If we consider the increase in brain volume as a type of hardware change, the formation of connectomes can be seen as a type of software substitution. Connectomes on the multimodal cortex promote convergence between networks with different functionalities[290]. This evolutionary process likely began in humans around seven to eight million years ago, acquiring greater abilities and complexity and improving cost-effectiveness[291]. For example, in bonobos, the connection between the amygdala and the anterior cingulate cortex is much stronger than in chimpanzees, which results in better anger control in bonobos compared to chimpanzees. This type of multimodal connection is more pronounced in humans than in other primates through white matter. Although the presence of connectomes has been effective in preventing excessive brain growth, there is still not sufficient connectivity between the two hemispheres. In fact, the corpus callosum connects a maximum of two-thirds of the axons between two hemispheres[292], finally causing the functional asymmetry between them. Studies have shown an inverse relationship between the size of the corpus callosum and the functional asymmetry of the hemispheres[293].

PREFRONTAL CORTEX

While the prefrontal cortex (PFC) is associated with the largest expansion of the brain cortex in evolutionary history, the greatest overall brain expansion is attributed to the frontal cortex[294], which is actually the forehead region. Throughout history, the evolutionary manifestation in intelligent humans has included speech ability, cognitive knowledge, intellectual supremacy, abstract thinking, map design capacity, prediction, self-awareness, empathy, and emotional responses. With its growth, the parietal cortex behind it has become smaller, and the occipital (visual) cortex has been completely pushed back. In humans, the frontal cortex is composed of two parts: the motor cortex and the PFC, which forms its anterior part. Although they are not well distinguishable anatomically, they have been recognized as separate entities histologically for years. We humans, and chimpanzees, with about 98.8% genetic similarity, are very close to each other behaviorally and cognitively. Abilities such as toolmaking, engaging in complex social interactions, and self-awareness stem from this genetic commonality. This cognitive set not only enhances our understanding of ourselves but also completes our ability to envision the future. However, all of these cognitive functions require the integrity of the PFC[295,296].

Research on the anatomical structure of the brain in patients with brain injuries has shown that the PFC has the most connection with the thalamic nuclei and is a good place for information interaction between different parts of the brain. The PFC accounts for over 29% of the total brain cortex in humans, 17% in chimpanzees, and 11% in macaque monkeys. For the dog and cat, the figures are 7% and 3.5%, respectively. It is mainly involved in complex social behaviors and executive function tasks. The available information indicates that the main characteristic in the human brain cortex is related to the structure of connectomes, the diversity and expansion of superficial layers, and the PFC, which are considered relatively recent in terms of classification[297]. Recent studies estimate that the number of neurons in the human PFC accounts for about 8% of the total, and it seems to have little difference from other primates. The PFC receives its information through the internal and posterior nuclei of the thalamus. Although the growth of the PFC is slow, it has shown noticeable development in humans[298] due to evolutionary processes. At birth, there are more synapses and connections in the PFC, and gradually, with age, the number of synapses with incorrect neural connections reduces, a process known as pruning, and the thickness of PFC decreases too, which usually takes place until the age of thirty. It should be noted that the maturation of the brain is completed by the age of sixteen. Essentially, the PFC is a part of the frontal cortex that makes up less than 30% of the brain cortex. Its main importance and functional ability are not only due to its size but also because of the extensive connections and the presence of chemical neurotransmitters that affect its activity. In addition, these changes are accompanied by progressive expansion in volume, brain rotation, receptor modulation, and the reconstruction of neurotransmitter functions. The multi-billion-dollar pharmaceutical industry is based on these substances that alter the functional performance of the brain without affecting its size and volume and give rise to various neurological and psychiatric disorders[213,299]. Damage to the PFC is associated with behavioral maladies, inability to engage in abstract thinking, being compelled to perform tasks automatically, quickly, and without thinking in response to sensory inputs or enjoying primal instinctual needs without effort in controlling them.

The frontal cortex is dedicated to eye and limb movements, the expression of emotions, speech ability, and visceral control, but the posterior cerebral cortex is dedicated to sensory reception, memory (mnemonic), and recall[300]. In other words, the frontal lobe cortex is the "doer" and the posterior part of the cerebral cortex acts as a "sensor"[301]. But it should be kept in mind that although the frontal cortex is highly specialized, it is almost unable to do anything without the sensory cortex; it is practically nothing. In addition, the motor system of the frontal cortex is matched with the cognitive and emotional systems, so it can plan movements and set goals for different times and places. Finally, humans can take certain actions with different motivations or reasons. In this way, in the course of evolution, the concept of intentional and/or voluntary actions has emerged, and perhaps this is the reason for the increase in the size and volume of the frontal cortex in the course of evolution.

Although anatomically, PFC has different sides, for a simple explanation of its physiological function, intelligent behavior, cognitive functions, and speech ability can be attributed to the dorsolateral level, while the orbitomedial remains dependent on emotional behaviors, responses, and reactions. In other words, cognitive behaviors in the external (lateral) part are considered the same as emotional behaviors in the internal (medial) part of the PFC. Temporal organization is the cognitive function related to the external side of the PFC, including planning, decision-making, and attention. Every intellectual plan has a goal and a time dimension that must be implemented in a timed manner. Planning and decision-making are inseparable components of PFC performance. The occurrence of these two requires attention, which is completely dependent on working memory. Working memory is a type of temporary activation of long-term memory networks that are updated for the organization of actions, and of course, it occupies a small amount of memory. In other words, working memory is a type of sustained attention to sensory perceptions or movement actions or a combination of the two, which represents the cognitive or behavioral goals of humans[302] that have gradually developed in the course of evolution.

Their temporal formations are all aimed at the future. Understanding the concept of the passage of time and the emergence of concepts called future and past, in the evolutionary process of *sapiens*, and the creation of language as one of the most prominent human characteristics, have been the results of the evolutionary process of PFC in the past few million years. Structural damage to the PFC not only slows voluntary movements but also causes confusion, ambiguity, and uncertainty in doing things due to a lack of understanding of time. With time organization, the fluency of the word gets meaning, novelty finds its place, presentation of reason and logic gets the application, and creative activities are formed. This work is done with the help of cognitive networks or "cognits" to represent knowledge which includes our semantic memory, remote memory, and culture established in our mind. Cognition is a complex and intertwined structure that is connected through connectomic structures (abstract and wireless types of communication), manages time, and redefines the information in memory distributed in the cerebral cortex. Cognits[303,304] overlap, intersect, and interact a lot with each other.

Neural networks form the substrate for cognits, which contain a permanent but modifiable repository of long-term memory and knowledge. Any neuron or group of neurons can be present anywhere in the association cortex and form parts of memory or items of consciousness. This structure extends to the hippocampus, which is part of the ancient cerebral cortex. The hippocampus works to consolidate explicit memory. The continuous electrical activity of cognits in the cerebral cortex is associated with the process of producing an immaterial, impalpable, and insubstantial phenomenon called consciousness, to distinguish self from the others. Consciousness is the mechanism by which the brain prioritizes information related to what we want to do and allows us to access that information. For this reason, consciousness shows itself as a dynamic stream, with a limited capacity. We do many daily tasks such as driving without the intervention of consciousness, but for tasks such as reading, writing, speaking, etc., we have to use working memory, which itself requires attention and the existence of consciousness[305,306] is mandatory.

The cognitive function of PFC is different due to the presence of sensory inputs from three parts. Internal sensory information known as interoceptive, such as visceral pain, is transmitted through the spinal cord and vagus nerve to the brainstem, thalamus, and limbic system (evolutionary primitive part of the brain and referred to as the "reptilian brain," involved in emotion), and finally to the medial part of the PFC. The information of external senses known as exteroceptive such as vision and hearing are transferred to the layers in the posterior part of the cerebral cortex. In addition, information related to memory, previous knowledge, work experiences, cultural categories, and moral values are placed in the association cortex. Those signals permanently affect the PFC in different ways. As a result, decisions are made based on risk-benefit. In fact, it is a type of occurrence of possibilities based on the input information to the PFC, which is necessary to stabilize homeostasis. In other words, "free will" or more precisely "purposeful intention to do something" or "experience of volition" does not exist in the way that is generally formed in people's minds, and in fact, it is formed based on the determinism that governs the system of PFC sensory receptions. It means freedom of action is relative and probabilistic to the sum of all inputs to the PFC. In other words, in many cases, our brain prepares actions and movements, before we choose them. Diseases of the frontal cortex take away the freedom of action from a wise person. A patient with obsessive-compulsive thoughts and actions is a prisoner of his obsessive behavior, and a patient with brain damage in that region of the cerebral cortex shows his/her abnormal behavior with the inability to make decisions (impulsivity). A famous example in the history of medicine is the left frontal lobe injury of Phineas Gage, a railroad worker in South Cavendish, Vermont in 1848. After the incident, he was no longer a normal person; his desire to have sex with any woman he saw, his irritability, and irrationality made him a different person. Remember that all these phenomena are for maintaining homeostasis[301,307].

Dopamine is the main neurotransmitter in the brain, and along with acetylcholine, has a higher predominance in the left hemisphere. Norepinephrine and serotonin (hydroxy-tryptamine) exhibit a greater function in the right hemisphere[308]. After events that occurred in East Africa million years ago, early humans were forced to hunt in very hot plains. This process increased the likelihood of heat stress and the resulting damage, so the need for the ability to control and regulate body temperature became crucial for humans. Dopamine has been effective in regulating the body temperature of early humans and has increased their ability to move and hunt. At the same time, the increase in calcium metabolism in muscles due to aerobic activity and the increase in the consumption of the amino acid tyrosine through meat intake led to further expansion of dopamine in the brain cortex. Gradually, the function of dopamine, which was mainly for controlling brain temperature, changed and it helped the growth and function of the brain, especially in the frontal lobe, to acquire positive emotions and pleasurable experiences.

Many scientists believe that in the evolution of simple mammals to modern humans, the human lifestyle has been changed to a warm-blooded animal with the capacity to stabilize the temperature which we call "endothermic" or "homeothermal" life. A warm environment makes neurons, neurotransmitters, and enzymes function properly[309] and their activity is not related to the ambient temperature. This makes them able to live easily in their niche. The process of temperature regulation is clearly dependent on the brain. Temperature-sensitive neurons convey sensory information to the hypothalamus in the brain through the spinal cord. The autonomic system will raise temperature (warming) through shivering and reduce it (cooling) through vasodilation and sweating[310,311]. Impaired temperature control is observed in many brain diseases such as Alzheimer's, Parkinson's, multiple sclerosis, and other peripheral nerve diseases.

The location of dopamine secretion is a ventral tegmental tract in the midbrain, at the top of the brainstem. About 400,000 neurons project from this center to the nucleus accumbens. This nucleus, which is a part of the basal ganglia in the brain, is involved in the process of learning, controlling voluntary movements, and organizing habits (habit formation) by referring neural messages to the cerebral cortex and amygdala (almond nucleus) which controls human feelings, emotions, fear and anxiety. In other words, dopamine is released following motivation, the nucleus accumbens determines the goal action to achieve the reward[312], and the reward is expressed in the form of enjoying food, sex, and quenching thirst presented in the amygdala and cerebral cortex. Therefore, the reward center is probably the nucleus accumbens, which determines how and/or how much the reward should be; dopamine is the chemical substance that forms the understanding of the reward in the mind; hence, it is called the "pleasure molecule"[313]. This process caused different parts of the PFC to develop different functions. The dorsolateral part (back and the external sides) became important for movement planning and execution and working memory, while the ventromedial part (front and internal sides) played a role in generating mental vigor and effective speech. The orbitofrontal part (eye and forehead region) is mainly involved in emotional, behavioral, sexual, and food-related functions. Finally, the frontopolar part which means the pole of the forehead, located in the most anterior part of the frontal cortex, contributed to abstract thinking, metacognition, and theory of mind. Therefore, structural impairments and functional disorders in these areas can be accompanied by various cognitive and behavioral symptoms. Damage to the dorsolateral part is associated with difficulties in action planning, working memory impairment, and difficulties in organizing thoughts. Orbitofrontal impairments are manifested by impaired behavioral control, personality changes, lack of empathy, irritability, and impairments in socially based judgment. Other impairments in these areas result in apathy, speech disorders, reduced motor and speech motivation, cognitive impairments, and spatial perception. It is evident that the function of dopamine is effective in controlling memory, cognitive flexibility, movement planning, abstract thinking, perception of temporal order, and the generation of new ideas. Excessive dopamine in the brain is manifested in diseases such as schizophrenia, autism, and attention deficit hyperactivity disorder[314] (ADHD). This structural and biochemical complex in the frontal and prefrontal cortices led early humans to learn to improve and enhance their behavior through imitation. Imitative behavior had a fundamental influence on his cultural evolution, and with the formation of imitative circuits in the brain, different brain regions evolved independently of each other, resulting in what we now call "mosaic cognitive evolution," which was more prominent in the PFC, a specialized area in the human brain. All three, executive functions, emotional responses, and social behaviors are the products of a complex social system, and their evolutionary importance in humans compared to other primates is evident. This process likely emerged around 1.9 million years ago when the frontal lobes took on a square shape in *Austras*. Gradually, neurons and their surrounding extensions became denser, their size increased compared to the chimpanzee type, and by creating more connections, they contributed to the improvement of cognitive abilities, intelligence, and attention. This complex is effective in the development of the theory of mind, which represents the maturity of cortical brain circuits and has the ability to gather and analyze sensory information[315]. The narcotic effect of drugs on the mind and the commitment to their constant use is also due to the reward that the brain receives. This prize is awarded for facilitating the release of dopamine and its impact on the amygdala and tegmental area in the brainstem, which is carried out by the insula. The release of dopamine prize in an addicted individual after drug use is accompanied by feelings of pleasure and euphoria.

9 Lopsided Ape*

ASYMMETRIC ORGANISM

In 1861, Edwin Smith, an American Egyptologist and antiquarian, purchased a collection of hieroglyphic papyrus from a broker in the city of Luxor at a low price[316]. Fifty years later, James Breasted, a professor of Oriental Studies at the University of Chicago, succeeded in translating them. This collection was about 48 patients with brain injuries that probably date back to 3000 years ago. So far, this is the oldest human manuscript written about the unilateral functioning of the brain[317,318].

In the first half of the 19th century, phrenology (craniology), which is a kind of pseudoscience, dominated the minds of many doctors. In fact, it was suggested a local place in the brain for any work. To diagnose which place is for what kind of job, it was referred to the bumps and depressions on the skull. Around the same period, Franz Joseph Gall, an English physician and anatomist, claimed since he was young, those changes in the cerebral cortex caused the development of mammals and humans compared to other animals[319]. He considered the different abilities of his classmates in different academic fields to be related to the changes in the shape of their skulls. During his career, he collected about 400 skulls from different people and tried to examine them based on their natural appearance and physiological function[320]. He considered nineteen out of 27 human characters is in common with animals. Although its unscientific nature was proved later, the concept of localization had opened its place in neuroscience. He tried to examine them based on their natural appearance and physiological function. Therefore, Plato and Galen's idea that the soul exists in the brain, Descartes' theory of duality and the theory based on the states and feelings of Baron Etienne Condillac (1714–1780), a French priest, and John Locke, an English philosopher, all were questioned. According to him, human beings have 27 inherent capabilities (talents) such as maternal instinct, verbal memory, sexual behavior, the joy of fighting, the ability to understand words, understand time, etc. For each of these characteristics, he considered a region in the cerebral cortex, but since it was not possible to evaluate it directly in the human brain, he chose the changes and projections on the surface of the skull for this task. He carefully examined the skulls of a large number of imprisoned criminals and a group of mentally retarded people. Although he can be called the founder of craniology, his topographic map was nothing but an illusion[321].

After facing significant opposition, Gall[322] was labeled a materialist, deprived of teaching at the University of Vienna, threatened by church believers, and exiled from Paris. Phrenology, however, was merely a slight modification or rearrangement of Nemesius' theory, who was the fourth-century Christian bishop of Emesa (present day Syria). Nemesius, in his treatise "*On Human Nature*," believed that the human soul resided in the center of the brain, within the ventricles, while Gall attributed it to the cerebral cortex; a theory called "functional modularity." His disciple, Jean-Baptiste Bouillaud[323,324], attempted to validate the professor's theories through practical experiences. By examining patients with brain injuries, he concluded that the speech area was located in the frontal part of the brain. However, the publication of his findings in his book entitled "*Traité clinique et physiologique de l'encéphalite" (Clinical and Physiological Treatment of Encephalitis)* in 1825 did not receive much attention.

Although most living organisms, including primates and humans, belong to the bilaterian group (symmetrically two-sided), two main characteristics of humans are entirely asymmetrical. The first is the use of hands, and the second is the ability to speak[325]. Generally, in humans, the left

* I borrowed this term from Michael Corballis: The lopsided ape (1993, Oxford University Press).

DOI: 10.1201/9781003538400-11

hemisphere predominates for language and hand use, while the right hemisphere predominates for visual–spatial function and attention. Baboons and monkeys do not exhibit this degree of hand dominance. The specialization of the hemispheres is a human phenotype that is not seen in other animals[326]. Anatomically, the anterior part of the right hemisphere is slightly rotated inward, resulting in the entire brain undergoing a torque toward the left. Essentially, the human brain is characterized as bilateral. Individuals are classified as left-handed or right-handed, and the lack of hemispheric symmetry is largely accepted as a human characteristic. However, this most significant characteristic of hemispheric specialization has demonstrated itself as a form of global right-handed superiority. At least, until a 100 years ago, left-handedness was increasingly condemned and stigmatized, while right-handedness was highly valued.

According to Diocles[327] right brain perceives and the left brain understands. In the writings of Gilles Li Muisis (1272–1352), a French Benedictine monk, chronicler, and poet, in the 13th century, it can be observed that he claimed that good priests can be found among right-handed individuals. Aristotle depicted a table in his book "*Metaphysics*" where he placed good, bright, and masculine phenomena on the right side and bad, dark, and feminine phenomena on the left side. He derived this interpretation from Pythagorean philosophy. The first treatise on iconography explicitly examined asymmetry. This treatise, written by a monk named Dionysius, in the monastery of Mount Athos in the booklet called "*Painters' manual*," around the years 730–734 AD, specified the patterns of placing images of religious figures. In images related to the crucifixion, the bad individuals are almost always depicted on the left side of Jesus Christ, while the good individuals are portrayed on his right side. A study has revealed that in painted images of Adam and Eve in the 15th and 16th centuries, Eve was placed on the left side in 82% of cases[328]. This number lessened to 61% in the 20th century. Concepts such as Eve being created from Adam's left rib are also constructed and addressed based on this premise. This indicates that the placement of Eve on the left side was not accidental. This nonbilateral, asymmetric, and right-dominant thinking is also manifested in literature and painting in many body parts. James Joyce, in his book "*Ulysses*," refers to the sensitivity of the left breast, the proximity of the left hand to the heart, the asymmetry of the scrotum, and the elevation of the right testicle[329]. In Italian sculpture, in more than half of the cases, the right testicle is carved higher than the left. In Aristotelian thinking, the right testicle is the place of sperm production for male infants, and the left side is for females, while the right half of the uterus is the place where a male infant is positioned. This method is also recommended in the book "*The principles of conception*" written by Ida Ellis, published in 1891, where it is suggested that tying the left testicle with a plastic band can result in having a male child and tying the right testicle is necessary for producing a female child. The male-dominant and asymmetric thinking is likely the main reason why the right testicle has been placed in a higher position than the left in many sculptures in the Middle Ages and afterwards[330].

In the 12th century, sacred texts considered left-handedness as a conflict between sacred and earthly phenomena. This dualistic thinking also permeated the medical community. In the Gospel, at least 25 negative qualities for left-handedness are mentioned, contrasting with a hundred valuable qualities associated with right-handedness. This rationale can be observed in many religions and cultures[331]. Left-handedness was considered related to a society with less development (especially non-Europeans), more tendency to crime, a higher likelihood of mental retardation, and various diseases. Gradually, in the 19th century, the functional asymmetry of the two sides was justified by claiming that each side fulfills a different function. They generally regarded the left hemisphere as associated with humanity, will, masculinity, reason, and logic, while the right hemisphere was seen as prominent in animalistic behavior, instincts, femininity, and madness. They assumed that the left brain was linear, logical, and susceptible to reason, while the right brain was holistic, creative, intuitive, and emotional. In the 1970s, this concept entered the realm of political discourse, labeling the left brain as representative of militarism–industrialism and the right brain as representative of peace-seeking and love. The left brain was seen as representing the West, while the right brain represented the East. However, the concept of dominant and nondominant hemispheres originated in the 19th century. Dualism simplifies facts. In the left hemisphere, there are three main one-way circuits. Language, dominant hand movements (usually the right hand), and the residual old mirror system, which is the main platform for creating all three circuits. Although left-handedness was more observed in men, it was labeled as a sexually transmitted disease and described as a feminine and primitive condition. The prevailing belief was that left-handers were more likely to be homosexual men (gays), while homosexual women (lesbians) were more likely to be right-handers. Left-handedness was considered abnormal and more likely to be associated with mental illnesses such as schizophrenia, mental retardation, autism, dyslexia, and language impairment.

The first time a Frenchman named Marc Dax (1770–1837)[332,333] introduced three patients with speech disorders associated with left hemisphere dominance in 1836. However, it was the introduction of two patients with aphasia (speech disorder) by Pierre Paul Broca on April 18, 1861, just one day after dissecting his patient's brain and presenting the results at the Paris Anthropological Society that solidified the concept of hemisphere dominance[334]. His first patient, Louis Leborgne, who had been admitted to the Bicêtre asylum 21 years earlier due to an inability to speak. Whenever he asked a question, he could only utter the word "tan tan," and became known as "Tan" patient. Contrary to phrenologists, he believed that brain lesions indicated its function, and the precise location of each lesion needed to be identified. However, it was Korbinian Brodmann (1868–1918), a German neurologist, who in the early 20th century, through numerous dissections of the brains of monkeys and humans, was able to name 52 areas on the cortical surface of the brain based on its cellular structure[335–337] that are still used by physicians. The initial belief at that time was that the left hemisphere was responsible for humanity, willpower, and masculinity while the right hemisphere controlled instincts and femininity. At that time, Broca proposed a law stating that the left hemisphere is dominant in 95% of right-handers and more than 70% of left-handers. Some studies even evaluated the genetic correlation of left-handedness, and William Brandler at the University of Oxford proposed the existence of the *PCSK6* (pro-protein convertase subtilisin/kexin type 6) gene, which could strengthen left-handedness[338]. However, this study was not later replicated by other researchers and did not gain acceptance in the scientific community. Nonetheless, some studies have also shown that left-handed individuals tend to be more intelligent, although this has not been conclusively proven. Numerous studies have shown that over 95% of right-handers and less than 70% of left-handers have their speech center in the left hemisphere of the brain[339]. At the same time, for every four men, three women are left-handed. In addition, based on epidemiological research, it has been determined that the lowest rate (nadir) of left-handedness was in the years 1890–1895, and it has reached its current steady state (asymptote) since the early 1950s[340]. Probably due to the industrial revolution in the 19th century, left-handed people were seen more and their inability to work with industrial tools that were necessarily made for right-handed people was clearly visible. Therefore, the social pressure to change the behavior of left-handed children and teach them to use their right hands had increased clearly[341].

Cesare Lombroso (1835–1909), an Italian police officer and specialist in criminal affairs[342–344], along with French anthropologist and sociologist, Charles-Marie Gustave Le Bon (1841–1931), believed that left-handedness was associated with mental weakness and asymmetry[345]. Lombroso considered asymmetry equivalent to savagery and sexual deviance. However, drawing on the theories of French physician Benedict-Augustin Morel (1809–1873)[346,347] and French psychologist and philosopher Théodule-Armand Ribot (1839–1916), he added the concept of degenerative mental and motor disorders to left-handedness[348]. Despite being Jewish himself and strongly detesting anti-semitic ideas, Lombroso did not attribute criminal behavior in Jews to race, bodily anatomy, or left-handedness. Instead, he interpreted it as resistance against the improper behavior of other members of society, a completely dualistic and contradictory interpretation. Ultimately, he believed that right-handedness represented masculinity and power, while left-handedness represented femininity and weakness. Additionally, he believed that left-handedness was an inherited trait that could not be treated, and there should be no attempt to change handedness preference. The most famous case of left-handedness was that of King George VI, also known from the movie "*The King's Speech.*"

Official royal historian Sir John Wheeler-Bennett wrote that when they forced the king to write and eat with his right hand[349], his speech disorder began. In psychology, this phenomenon is referred to as "misplaced sinister."

Contrary to the assumption of bilateral symmetry in evolutionary processes, the value of being bilateral is considered significant. Many human activities such as walking, running, and swimming require bilateral coordination and have been a crucial factor in the survival of the human species. However, two primary human behaviors, the use of the dominant hand for many daily tasks and the ability to speak, are completely asymmetrical. It is not only the preference for hand use but also the preference for other body parts. Generally, 88% of the population is right-handed, 81% are right-footed, 71% have a dominant right eye, and 59% have a dominant right ear. In other words, right-handers are mostly right in other parts of the body, and left-handers are left. Other studies have shown that 95% of right-handers are right-footed, while only one-third of left-handers are right-footed. Moreover, two-thirds of right-handers consider their right eye to be dominant, compared to only half of left-handers. In other words, left-handers are more mixed-handed. However, this pattern is not absolute. In Antaisaka, Madagascar, there is a tribe with a population of less than 100,000 inhabitant and almost all of them are left-handed. Both Hebrew and Persian are written from right to left, but the number of left-handed Hebrew speakers is greater than the number of left-handed Persian speakers[350,351].

Unfortunately, left-handedness has always been stigmatized. Positive values such as goodness, light, brightness, endurance, and masculinity are associated with right-handedness in the Bible, while characteristics such as femininity, darkness, and wickedness are attributed to left-handedness. This mindset led to the mandatory use and writing with the right hand for children in countries like Spain, Italy, and many Eastern European countries in the previous century, except for the Czech Republic and Slovakia. Structural and societal changes have resulted in an increase in the number of left-handed individuals over the past century. The left-handed population in Australia increased from 2% at the beginning of the century to 13% in the 1960s. Greater tolerance for left-handedness in America and Western society led to a clear increase in their statistics in the late century, reaching around 11–12%. However, the negative cultural perception of this behavior in the Middle East caused the majority of the population in those societies to remain right-handed, around 96%. In many traditional societies, such as the Zulu tribe in Africa and some Middle Eastern countries, using the left hand would result in severe punishment, including scalding the hand with boiling water. In English schools in the 1920s, methods such as sitting on the left hand, tying the left hand behind the back, constant humiliation, standing in a corner of the class, and prohibition from participating in collective activities were used as ways to treat left-handedness. King George VI (1895–1952) was repeatedly forced to write with his right hand.

Subsequent studies have shown that forced hand switching has a negative impact on children's mental growth and can lead to speech impediments. Laura Smith from Clark University and Harvey Ernest, a professor of histology and embryology at the University of Virginia, have shown that a higher percentage of speech impediments are observed in retrained left-handers. Similar studies by Lewis Terman[352,353], a psychology professor at Stanford University, estimated a four-fold higher risk of speech impediments in these individuals compared to the general population. At the same time, the likelihood of speech impediment improvement is also much higher with a return to left-handedness. In 1966, Michael Corballis and Ivan Beale[354,355] introduced the theory of conflicting hemisphere functions resulting from forced right-handedness and attributed it as a cause of speech impediments.

In Alsace, France, being left-handed is considered a form of insult, and Russians use it as a metaphor for being deceitful and unreliable. In Mandarin Chinese, it is used to mean strange and unusual. In Malawi, this concept is strongly gendered and signifies femininity, weakness, and being inferior. In the Buddhist tradition, Buddha shows the path to Nirvana with his right hand, and a saying attributed to the Prophet of Islam states that if the sun were placed in his right hand, he would not deviate from his path. Some famous left-handed individuals include Julius Caesar, Alexander the Great, King George VI, Charles Loman, Leonardo da Vinci[356], Prophet David, Gerald Ford, Ronald Reagan, and Martina Navratilova. Among these individuals, da Vinci had the ability to write backward and in mirror writing. Additionally, Albert Einstein was right-handed but a slow learner and had a slower speech acquisition, and King George VI was originally left-handed but was forced to write with his right hand, resulting in a degree of speech impediment.

The fundamental problem is that in many societies, almost all tools, instruments, and devices are designed for right-handed people. Zippers, ordinary scissors, doorknobs, can openers, and school chairs are among the most common items that are typically designed and made for right-handed individuals. Perhaps for this reason, left-handed individuals are more prone to accidents, and the mortality rate for individuals over 33 years old is higher among left-handed individuals compared to right-handed individuals. However, after the invention of writing, the act of writing exerted more pressure on left-handed individuals to change their dominant hand. Nevertheless, it seems that in some sports such as tennis, baseball, table tennis, and American and European football (soccer), left-handed individuals have performed better and been more successful.

Different studies provide different statistics regarding the number of left-handed individuals. This figure is estimated to be between 15% and 20% in America and Western Europe and between 0.06% and 2.8% in East Asia (China and Taiwan). Overall, approximately 10–12% of the population is left-handed, and this number has remained consistent through archaeological studies related to the nonstandard use of stones for separating rock layers over the past 5000 years. It appears that left-handedness is more common in men, with a ratio of around 2:1 up to 5:4 compared to women, and many of them are ambidextrous. Some individuals use only their right hand for writing and prefer to use their left hand for other daily tasks such as using scissors and knives. Nowadays, the term "non-right handedness" has gained more popularity.

Genetic factors have a predominant influence on handedness. In a study, it has been shown that if both parents are right-handed, over 92% of their children will also be right-handed, and if both parents are left-handed, this number will be slightly over 45%. However, if one parent is right-handed, slightly over 80% of the children will be right-handed. Similar figures have been reported in various studies. In 1981, Norman Geschwind, along with Peter Behan and Albert Galaburda[357] at Harvard University, proposed the theory of fetal testosterone stress and its impact on the left planum temporale. They believed that this impairment leads to the shift of the speech area toward the right hemisphere and ultimately results in left-handedness. This phenomenon can be associated with learning disabilities, and since testosterone levels are generally higher in male fetuses, left-handedness is more commonly observed in males, although strong reasons for this have not been presented. Meanwhile, Stanley Coren[358,359], a Canadian psychologist, considered left-handedness as a result of the inability to achieve right-handedness. He interpreted left-handedness as a type of mental imbalance and insanity.

Marian Annett presented the "right shift" theory based on years of study and data collection. Her model has been debated and examined for many years in at least two books and numerous articles since the 1970s. Her model suggests a genetic mechanism that predominantly drives individuals in the population toward right-handedness. In other words, humans do not have a specific genome for becoming right-handed or left-handed; instead, there is a genetic influence that determines right-handedness. If individuals inherit a tendency toward right-handedness, they will most likely be right-handed, and if they do not inherit such a tendency, right-handedness or left-handedness will be based on chance. In other words, right-handedness is influenced by genetics, but left-handedness does not have any genetic compulsion, and Annett introduces a single-gene model with two alleles: RS+ and RS−. This model can create three genotypes: RS++, RS+−, and RS−−. The "right shift" allele (RS+) has a tendency to localize the speech language to the left hemisphere and leads to minor impairment in the left upper limb. Individuals with the RS+ allele have the dominant "right shift" gene, and if the RS− allele is completely neutralized. The RS+ allele has a higher prevalence in women compared to men; therefore, according to this interpretation, all individuals with RS+ RS+ or RS+ RS− are right-handed, except for around 3–4% who will be left-handed pathologically.

The remaining 25% of individuals are RS− RS−, and since this allele has no functional effect, half of them, i.e., 12.5%, will be right-handed normally. Thus, overall, approximately 10–12% of the population becomes left-handed, which is also supported by epidemiological studies[204,205,360–363].

This method demonstrates why there is a form of collective bias toward the right. The process of evolution has driven the human species toward right-handedness. At the same time, speech language is localized in the left hemisphere. It seems that this form of right-handedness inheritance and the preference for language in the left hemisphere have been the best genetic probability for human evolution. This theory can still explain how right-handed children can exist in a family where both parents are left-handed. In other words, the absence of a right-handedness factor leaves the probability of becoming left-handed or right-handed solely to chance. Furthermore, RS+ individuals have more weaknesses in mathematics, while RS individuals show more difficulties in speech and reading.

Thomas Carlyle, a 19th-century historian, and Philippe Henri de la Creste[364] at the University of London believed that right-handedness was due to defensive behavior in the past. They insisted that the primitive humans held a sword or spear in the right hand and a shield in the left hand to protect their heart, which led to the superiority of right-handed warriors. However, Stanley Coren considers this theory as a historical one and states that initially the primitive human did not have a shield and did not use it, but they were right-handed. French biologists Charlotte Faurie and Michel Raymond have proposed the "negative frequency-dependent advantage" theory[365], suggesting that the scarcity of a trait in a community, such as left-handedness, can cause that group to have an advantage in certain specific matters, and as their number increases, their advantage diminishes. Overall, all of these theories suggest the influence of environmental pressure on the survival of genetic models.

One of the most important questions about the asymmetry of humans and the preference for right-handedness is how it can be justified by the theory of natural selection. The first point to consider is that natural selection is not the cause of phenomena but rather the result and the effect of the organism's ability to adapt to the environment. Therefore, species that have adapted themselves expand their genomes by reproducing their generations, although this is not always the case. It is possible for a transferred characteristic to contribute to the development and growth of an organism's work and not affect its reproductive ability and proliferation, but it ensures survival or even improves it, such as hoofed mammals that have learned to swim in the sea. This effective occurrence in growth can expand evolutionary processes and is referred to as the "evolutionary-growth hypothesis." The adaptation ability of a living being to the environment and establishing social communication has great importance in the evolutionary growth of the organism. This situation indicates the inheritance of the comprehensive whole derived from the functioning of the genome, the epi-genome, the organism, and the surrounding environment during growth. This complex and intricate structure continuously corrects itself through feedback loops and, of course, has no partial preference over another.

Prior to Annett's theory, the most fundamental theory regarding evolution was the postural origins hypothesis. This theory is essentially based on the logic that early humans used their right hand to grasp onto trees and their left hand to crush edible kernels and occasionally fruits. With the transformation of their ecosystem and their transition from arboreal beings to creatures with a terrestrial lifestyle, the preference for using the right hand became dominant. The growth of an organism in a niche where parents use their right hand leads the offspring to behave in the same way. In other words, the lack of hand asymmetry and right-handedness in offspring is intensified through their interaction with right-handed mothers. The potential environmental influence on the development of such a trait depends on the sensory, biomechanical, and motor characteristics of the organism during its growth period.

Jakob von Uexkull in 1909 coined the term "Umwelt" (a Deutch word that means environment) to describe the individual's whole subjective perception and sensory experience of the surrounding world[366], which is related to the environment. This quality shows how individuals interact and experience the environment through their sensory-perceptual processes and biomechanical functions during that period of time. For example, each of us may have a slightly different perception of the scent of a flower. Therefore, both the individual and their Umwelt can change over time due to environmental variations and differences in the organism's experiences. However, some qualitative properties of an organism's Umwelt, such as gravity, atmospheric pressure, the amount of oxygen in the air, solar heat, and social participation, remain constant and uniform. However, for some phenomena, such as underwater atmospheric pressure in the oceans or at the top of mountains, the education of children may not necessarily be equal due to the presence of more knowledgeable companions and parents, and there may not be any limitations on that.

However, in the evolutionary-developmental model, we need to consider the importance of using one hand and the lack of symmetry in limb usage. It seems that the preference for using one hand is primarily for performing various manual tasks, which sometimes in society are commonly performed with one hand in particular. In a society where some individuals perform tasks with one hand, it is better for apprentices who want to learn that task to use the same hand. This helps them learn the task more quickly and have no doubt about using the dominant hand for that task. Therefore, right-handedness in the population is the result of an incidental consequence of a growth package and does not incur any cost for being left-handed or ambidextrous. However, according to some researchers, right-handedness may have started during the fetal and infant stages. In many cases, the fetus's head is inclined downwards and toward the right side due to gravity. Additionally, the infant is highly dependent on its mother in the first two years of life and is often in her arms. Consequently, it spends a significant amount of time in an upright position during the fetal and breastfeeding periods. This causes the vestibular system to be more stimulated on one side than the other, and the breastfeeding child, due to the preference for head orientation, looks toward the right and gradually starts using the sensory-motor system of its right side. However, proving the role of prenatal vestibular experience remains a key challenge that has not been fully understood.

The main question of why humans are mostly right-handed still persists. Structurally, humans are bilateral, using both hands. They walk on two legs. However, functionally, the preference for one hand becomes important. It seems that humans use both hands simultaneously for reflexive actions[89], such as grabbing a ball during various games. But for intentionally performed tasks with deliberate manipulation but not reactive movements, such as writing, brushing teeth, handling a knife, using scissors, picking up a matchstick to light it, and holding a spoon while eating, the preference is associated with the prominent hand.

From an evolutionary perspective, many animals exhibit a preference for one limb over the other. Parrots, for example, are usually left-footed and pick up food with their left foot, but they lean to the right when perching. Cats are typically left-handed, while mice use both upper limbs equally. Large monkeys are right-handed in 65–70% of cases, orangutans in up to 90% of cases, and sea lions in 77% of cases. However, the preference for using hands in mice is not related to parental preference. Other studies show that monkeys, lemurs, and orangutans demonstrate a mixed preference in about half of the cases. Chimpanzees and nonhuman primates generally prefer the left hand. These creatures use the left hand preferentially for grabbing insects, food, and objects, as they rely on the right hand for hanging and balancing. However, humans, after evolving to walk on two legs, started using the right hand for specific and specialized tasks that played a dominant role.

Gabriel de Mortillet (1821–1898), the French historian and archeologist, demonstrated in the Somme Gravels region of France that our ancestors were mostly left-handed during the Paleolithic period but gradually became right-handed during the Neolithic period. *Neanderthals* also used their teeth to grasp meat and usually used the right hand to manipulate sharp stone tools for butchering. The difference in pressure on the teeth led to the development of angled lines on the front teeth, indicating how they grasped and used their hands. It is possible that mothers used their left hand to better understand the heartbeat of their infants while holding them and used the right hand for throwing objects and self-defense. The asymmetry is not limited to hands and speech[367]. In modern humans, asymmetry is also observed with a preference for the right side for speaking and the left side for emotional expressions. Additionally, the left hemisphere is generally better suited for understanding social matters.

The most significant evolutionary changes in human asymmetry occurred approximately 40–60 million years ago (during the Eocene geological period). Initially, this process manifested in the development of opposable thumbs. Subsequently, movements became more precise and delicate. Skills in performing intricate tasks and the independence of fingers improved. The next prominent and dramatic change occurred when insectivores (insect-eaters) evolved into primates. During this period, the connection between primary cortical neurons and secondary neurons in the forelimb spinal cord became closer. This phenomenon led to faster neural transmission and reduced synaptic connections. The extension of primary neuron fibers toward the lower spinal cord and reaching the lumbar region resulted in the separation of hand movement from leg movement, allowing each to function independently and more comfortably. Just as legs became free for bipedal locomotion, hands became free for manual activities. From the middle to the late Eocene period, with the appearance of early primates, the speed of neural transmission increased, and the muscles associated with limbs became larger and stronger. This, in turn, affected the expansion of the cerebral cortex for motor control. The thumb gradually positioned itself against other fingers, and the ability to grasp objects precisely between two fingers emerged. Furthermore, anatomical studies have shown that the cross-sectional area of neural fibers, extending from the cerebral cortex to the spinal cord and their extent toward the legs, is directly related to right-handedness.

Presumably, with the beginning of walking on two feet and the freedom of hands, there was not much inclination toward bilateral symmetrical behavior, and the free hands could play a special role in other matters. Perhaps the most important of these is human speech. When asymmetry has more benefits, it naturally becomes dominant in evolutionary processes. The beak shape of birds in finding food seeds and insects, as well as the shape of the noses of many other organisms, supports this idea. However, asymmetry did not occur suddenly as a predetermined event; rather, it gradually emerged and followed Darwinian laws. At the same time, cerebral hemisphere asymmetry is not exclusive to humans, but it is highly symbolic and specific in humans. It seems that laterality in evolutionary processes is a step forward and a kind of genetic specialization. It should be noted that currently, by default, the genetic status of humans is bilateral asymmetry, which in some cases, such as hunting and sports, can be helpful.

The freedom of hands, along with walking on two feet, led the early humans to use both hands complementarily. Usually, they would hold something with one hand and perform a specific task with the other hand. These behaviors are asymmetric. There are numerous examples in this regard, for example, they would hold stones with the left hand and throw them with the right hand, effectively using the left hand as a magazine. In building and using other tools, they also employed this method. It is likely that women also held their infants with the left hand and protected themselves and their infants from enemies with the right hand. In animal examples, such as crabs, they have lightweight claw for catching prey and use heavier claw to kill it. In other words, early humans were initially generalists and then gradually became specialists. The preference for using one limb in evolutionary processes opened a new window for early humans. It seems that cerebral asymmetry is a kind of adaptation in human evolution. Humans belong to the group of bilaterians, but about 1.5 million years ago, they gradually becamc unilateral. This indicates the efficient operation of a system on one side and, ultimately, a reduction in the time it takes to perform tasks, an increase in speed, and a decrease in energy consumption. The absence of asymmetry may potentially give rise to some conceptual thoughts such as delusion and illusion.

The pattern of laterality likely has a genetic basis in humans. Different forms of asymmetry in the human brain include internal thinking, vision, attention to the surroundings, language, and speech. The phenomenon of asymmetry has also manifested itself in other organs of the body such as the heart and liver. Despite this, it has emerged later in the brain but has been more influential, intense, and prominent. It seems that in the evolutionary process of the brain, the ability to perform tasks and the independence from spatial environment, along with bilateral symmetry in the structure and function of the brain, have been accompanied by a kind of nonessential constraint that has had a significant impact on brain function. Therefore, the nervous system has evolved toward asymmetry. Gradually, hand movements have also come into play to aid human speech ability, which will be discussed in subsequent sections.

In conclusion, it is necessary to mention a few points. First, laterality is not a sudden and random process but has gradually emerged and followed Darwinian laws. Second, as language becomes more complex, asymmetry increases and intensifies. This phenomenon may be due to limitations in message transfer between the two prominent hemispheres. Third, although cerebral asymmetry is not exclusive to humans, it is highly symbolic and specific in humans. It seems that laterality in the evolutionary process is a forward step resulting from genetic predisposition. It should be noted that bilateral symmetry is the genetic default state in humans, but in specific conditions where asymmetry is favored, it becomes selectively advantageous over symmetry.

CROSSED FIBERS

Since ancient times, and possibly predating the era of the esteemed physician Hippocrates, it has been understood that the motor control center of any organ we utilize resides in the opposite hemisphere of the brain. Almost always, the hemisphere controlling the movements of the right hand is located on the left side. Nerve fibers (axons), originating from the giant cortical neurons called Betz cells, after Ukrainian anatomist and histologist Vladimir Betz (1834–1894); controlling the movements of both upper and lower right limbs start from the left hemisphere and form a dense bundle in the lower parts of the cerebrum; and in the lower part of the brainstem (medulla), before entering the spinal cord, up to 90% of them cross over[368] and intersect to the opposite side. These fibers continue their path downward in the outer part of the spinal cord and when they need to branch out to the corresponding limbs and muscles, they connect to the second motor neuron through a synapse. The axons of the first neuron which is located in the cerebral cortex are called corticospinal fibers, and the second neuron, which carries the nerve impulse to the muscle, is called the motor branch[369,370] of the spinal cord.

The fundamental point is why this crossing over happened in the course of evolution?[371] What has been the significance and impact of this complex structure in the anatomy and, of course, the physiology of modern humans? The preference for using one-sided limbs is actually a qualitative change that appeared with the emergence of prosimians during the Eocene geological period, about 50 million years ago, and with the approximate formation of opposable thumbs and the development of dexterous hands for grasping objects[372]. The gradual changes have led to improvements in movements and skills, and increased independence of the fingers in performing precise tasks. Until the late Eocene period, the speed of nerve message transmission increased in such a way that, with a greater influence on the muscles and their enlargement, their functional abilities improved significantly, along with the expansion and development of the cerebral cortex[373,374].

The next dramatic change occurred during the transition and evolution from insectivores to primates. During this time, the connection point of corticospinal fibers to the anterior surface of the spinal cord became closer, allowing messages to be transmitted at a higher speed and with fewer synapses. Simultaneously, the development of these fibers downward led to an increase in the ability to walk on two legs and an improvement in hands for manual tasks. Studies have shown that the preference for using a limb is directly related to the cross-sectional area of nerve fibers that come from the cortex to the spinal cord and their development toward the lower parts of the spinal cord and the lower limbs. Gradually, the hands, which were suitable for hanging on branches and had seemingly parallel fingers but were actually in the opposite position, transformed into their current form. As a result, the ability to grasp objects with the hand increased significantly. Ultimately, the precise grip of small objects became possible between the index finger and thumb. In other words, the placement of the thumb in opposition to the other fingers played a crucial role in the precise grasping of objects. However, it should be noted that in many other animals, hands are used for movement and locomotion, and since they have specialized for this purpose, their fingers are fused together.

The first point that comes to mind regarding the intersection of nerve fibers is the mechanical stability of the spherical structure of the brain. Human brain tissue is made from fluid-saturated, porous, and nonlinear solid material. The stretch response of white matter tissue reveals a peculiar nonlinear mechanical behavior, with different stiffness in tension and compression, during loading and unloading. Perhaps the most common theory about intersection is the collection or diversion of a reflexive living being against the likelihood or occurrence of a dangerous threat. When a dangerous agent threatens a living being from one side, the organism bends itself toward the other side to escape the danger. For this action, it needs to activate the motor system in the opposite direction, which emphasizes the importance of the passage of motor nerve fibers from the midline[373,375]. This can be seen in the simple model of the Braitenberg vehicle. According to Valentino Braitenberg[376] (1926–2011), every intelligent motor system has a sensory system. This sensory system maintains two properties, taxis (orientation) and kinesis (movement) within itself in an established manner. This motor system uses an oar model, just like when an oar is used for movement and the boat arcs and turns in the direction the oar is pushed. In other words, when the sensory system of the organism is attacked or threatened from one side, it transfers the sensory information to the motor system of the opposite side. Stimulating the motor system in the presence of an intersection in the corresponding fibers leads to avoidance of danger. It has been demonstrated in mechanical–electronic models made by Braitenberg in 1965 it was shown that if sensory stimulation is accompanied by motor stimulation in the opposite direction, the toy moves further away from the stimulus. Simultaneous stimulation on both sides causes the toy to move forward. If there is no intersection, the toy moves toward the positive sensory stimulus. In the evolutionary process, the dangerous agent can be considered as a positive stimulus. This phenomenon is exactly like the movement of an oar in water and the movement of a boat following it. The movement of the oar to the right in water moves the boat forward and to the right. That is completely like dangerous stimulation of a fish on one side and contraction of all her muscles on the other side, causes to deviate to the opposite side of the stimulus. In scientific terms, taxis (oar direction) determines kinesis (movement direction).

In organisms with separate limbs such as a frog, unlike limbless organisms such as fish, it is necessary to combine the bending of the flat limbs on one side with the straightening of the underlying limbs on the opposite side to escape from a dangerous situation. The sensory system also needs to intersect simultaneously. The first sense is the sense of balance, followed by the superficial senses of the limbs including touch, pain, and temperature. This form of intersection of sensory fibers is also necessary for three-dimensional perception. Failure to intersect sensory-motor fibers can cause dysfunction and impairment for individuals. This phenomenon is also observed in patients with spinal cord injuries. There are specific neurological disorders in which fiber intersection did not occur during the embryonic period. This phenomenon leads to difficulties in motor-spatial learning. For example, in Joubert syndrome, due to the lack of fiber intersection, the patient experiences severe slowness and clumsiness in movements.

The interpretation of mathematical models of motor systems demonstrates that the number of neurons that can connect two points in a nonintersecting system and simultaneously transmit accurate and reliable information ranges from 100 to 500. To ensure better message transmission, a greater number of neurons with intersecting fibers are required. In the presence of nonintersecting fibers, not only spatial perception in three dimensions become problematic, but the accuracy of transmitted information also decreases[377]. Impairment in spatial perception with nonintersecting fibers leads to difficulties in movement, displacement, and body orientation. This intersection can be created symmetrically from left to right or vice versa. However, it should be professed that this significant amount of intersection is not possible without a suitable genetic structure that can manage it from the early embryonic stage. Proteins created from these genes are observed for nerve fiber orientation and error reduction in finding paths in vertebrates. In general, it seems that the motor system, which involves gross and coarse movements that do not require extensive spatial perception, does not require intersections. However, for delicate movements of limbs in which spatial perception is crucial, the presence of intersections is highly necessary and mandatory.

In addition to the intersection in sensory fibers, intersections are observed in other external senses such as vision and hearing. Visual sense intersection appears to be the most important[378,379]. Intersection in the visual cortex creates three-dimensional and depth perception. In animals with eyes located on both sides of the head, three-dimensional vision is absent, but panoramic vision allows awareness of potential hazards in the surroundings. In animals with forward-facing eyes, the visual field overlaps, leading to three-dimensional and depth perception[380]. This event improves motor skills. It should be mentioned that the projection of images from the nasal (medial) half and not the temporal (lateral) half of the retina requires intersection, and this incomplete intersection is necessary for image fusion in the brain and the creation of a consistent and continuous image. The incomplete intersection of these fibers results in confusion of external images in the mind. This phenomenon is observed in albinism[381,382] and some other diseases. Additionally, the intersection of sensory fibers is essential for the fusion of superficial and visual senses. This fusion allows a living organism to react more accurately, correctly, and quickly to the environment and surrounding predators. In some anomalies and diseases that are associated with motor fiber crossing disorders, such as Klippel-Feil[383] and Kallman syndrome[384], they are associated with relatively severe problems in horizontal eye movements and progressive curvature (scoliosis) in the spine.

The sense of smell is an exception. Almost all the nerve fibers related to the sense of smell go to the same hemisphere of the brain and do not intersect. It appears that the sense of smell does not have a direct impact on the motor system, so it does not require intersections like many other senses. However, the sense of smell without intersection in its fibers can create its own topographic structure in the brain. Presumably, for this reason, it was the first sense to evolve and works very well and strongly in many animals. However, in humans, with the development of a strong and effective sense of vision, the sense of smell has lost its function and importance, and its anatomical structure in the brain's cortex has become smaller compared to many smaller animals.

In addition, lateralization and asymmetry have predominantly occurred in humans. However, simultaneous brain growth in primates, especially humans, has shown more development compared to other monkeys. The expansion of brain volume has been accompanied by a significant increase in neurons, a substantial growth in neuronal synapses, and ultimately the formation of multiple modules with numerous interconnections. Modules are functional columns of the brain cortex that are created to prevent excessive brain growth. They also prevent excessive connectivity, mixing, and inefficiency of the brain. The fundamental importance of these modules lies in creating regional connections in the brain cortex, which allows the brain size to remain relatively constant in proportion to its increased functionality, ultimately contributing to brain lateralization.

10 Feelings and Sensations

PERCEPTUAL MATURATION

The survival of living beings relies on the sensory system before the necessity for a motor system emerges. Senses serve as conduits, transmitting internal or external information, manifested as a form of physical energy, to the brain through nerve fibers in the form of electrical impulses. This process enables living beings to develop a comprehensive and coherent understanding of their surrounding environment. Perception is a skill with which the sensory organ measures the received information to make a decision. In the course of evolution, sensory perception has been associated with learning and significantly changed human cognitive ability. Sensory perception through peripheral nerves has a subconscious effect on our daily life. We usually don't pay attention to it and only when abnormal signals like pain and/or muscle weakness are sent to the brain, it draws our attention[150]. Sensory perception is a necessary factor for initiating motor actions. Small mammals underwent further expansion approximately 65 million years ago because they initially had secure and low-risk lives. Hair or fur provided sufficient heat preservation during nocturnal activities and metabolic control, and since they were nocturnal, they did not require vision and sensory-motor function through the superior colliculus in the midbrain. The superior colliculus, as a sensory-motor relay structure in the evolutionary course, has specialized in detecting, marking, and orienting toward environmental events and happenings, playing a significant role in motor responses and attention to the surrounding environment in the visual system[289,385]. The new brain cortex (neocortex), had not undergone extensive evolution yet. During the same period, the tectum (roof) of the midbrain gradually evolved to control vision and circadian rhythm realization (day–night cycle) in birds and reptiles. The presence of sensory protective hairs on the skin and face of mammals allowed for the reliable detection of objects and other living beings from a relatively long distance and enabled the processing of incoming information in the primary and secondary sensory-motor cortex of the brain. The auditory system gradually developed. The lower jaw became smaller, the cochlea of the ear became larger and more efficient, and the perception of sound frequencies for auditory sensation was increased. This led to the expansion of communication between offspring and parents. Subsequently, the taste sensation and olfactory system developed[386] simultaneously. Olfactory receptors and the corresponding genes evolved. The size of the brain has enlarged relative to the body. By means of an upsurge of sensory input, the structure of the visual organ expanded. Furthermore, its corresponding part in the brain also improved significantly with the enlargement of the eyes, and the expansion of the surface area of the visual cortex and temporal lobes, resulting in better and flawless sensory perception.

The sense of smell has undergone significant evolution in primates, excluding humans, and the survival of many mammals is indebted to the use of the sense of smell. In humans, the olfactory bulb, compatible with the retina in the eye, has a volume of about 100 mm^3. This bulb is 60 times larger in dogs and even two to three times larger in chimpanzees and gorillas. In humans, the escalation in brain size has been accompanied by reorganization in the brain, leading to a reduced emphasis on the sense of smell and a greater investment in the development of visual senses, including binocular and three-dimensional vision. Actually, the sense of smell depends on what we see[387] or what we imagine. It seems that our vision impressed other exteroceptive senses such as hearing, touch, and taste too. This is how our brain is trained. Furthermore, the sense of smell, while it is known as an exteroceptive sense and provides us with the external and objective characteristics of material also has an interoceptive character and shows our desire and inner feeling toward the external object. In the course of evolution, primates changed from a macrosmatic organism to a microsmatic organism. Macrosmatic animals such as dogs and/or pigs have a bony transverse blade behind

DOI: 10.1201/9781003538400-12

the nose, separating the retronasal (behind the nose) olfactory space from the nasopharynx and cuting off its connection with the respiratory system. In *sapiens*, at the same time as the ability to perceive aromatic molecules has dwindled, the sense of sight has been strengthened, the transverse bony blade has been removed and the mouth has become a second nose[388]. The retronasal sense of smell makes us feel the taste of food well and the pleasure of eating is doubled. That is why bitter coffee is easy to drink with a pleasant smell perceived in the second nose. Stinky French cheese is no exception to this rule and eating it is more pleasant than smelling it. In fact, the responses of the nervous system in the course of evolution to perceive aromatic substances are completely different in the posterior (retronasal) and anterior (orthonasal) parts of the olfactory system[389]. The evolution of the senses, sight and hearing, has caused them to gain a superior position in our conscious awareness, but the sense of smell is still in our subconscious awareness and the perceptions of the sense of smell are still in the background of our mind; significantly changes our understanding of the surrounding environment. This evolutionary regression has caused the mental depiction, in other words, the olfactory memory of primates, to show fundamental weakness[390].

In monotremes (egg-laying mammals) such as echidnas and platypuses, the presence of motor cortex is questionable, and the greatest development is observed only in the sensory cortex. Marsupials such as kangaroos and koalas are similar in this regard, although the visual cortex shows greater growth and development. The motor cortex gradually emerges in placental mammals such as armadillos. In insectivores such as hedgehogs, shrews, and moles, further evolution is observed in the motor cortex. In prosimians such as lemurs, (common) mice, and shrews (tree mice), the primary and secondary sensory cortex develops. The brain volume in early mammals is relatively small compared to their body size, and their neocortex was not considerably expanded[391] regarding their little prefrontal cortex. Studies have shown that after the sense of smell, the sense of hearing develops, followed by vision, and then the ability to perceive three-dimensional (binocular) vision. The sense of taste, the motor system, and eye movements gradually developed, and ultimately, the sense of touch and the modulations related to lateralization in the cerebral cortex expanded. Many primates have a strong reliance on their sense of smell, but in some, like squirrels, the primary and secondary visual cortex show greater growth and intensify their visual abilities. Simultaneously with the growth of visual senses in other primates such as rabbits, in addition to the visual cortex, the cortex of temporal lobes also developed[1].

Over time, in lemurs, especially in flying lemurs, and then in tree-dwelling rodents (shrews) with a circadian rhythm, three parts in the brain; thalamus, superior colliculus, and finally lateral geniculate body developed. Simultaneously with the development and improvement of visual acuity (sharp vision), eye movements also evolved[392]. Meanwhile, with further evolution of the sensory-motor cortex, the presentation of motor organs such as limbs in the brain cortex increased. The improvement in visual status compared to the sense of smell in primates, especially humans, has been very important and valuable[393]. The early primates were small, arboreal, and many of them were nocturnal. As their body size enlarged relatively, so did their brains, and their eyes also grew larger and moved forward on their heads. Simultaneously with these changes, the enlargement of the temporal cortex can be observed in animals such as lemurs and galagos. The main importance of visual structure development is in creating coordination between the hand and the eye. This phenomenon has led to an increase in access to food while hanging from tree branches. The expansion of the brain's visual cortex gradually accompanied the expansion of the temporal cortex for hearing frequencies, and the parietal cortex for tactile perception, as well as the motor cortex. This process has resulted in fine finger movements and improved eye control in tree-dwelling animals such as moles, prosimians, and galagos.

Following this process, the development of the sense of touch was placed on the agenda in the insula and parietal cortex. The growing number of superficial skin receptors in monkey tails initially contributed significantly to the development of a specific form of touch sense. Gradually, in apes, the brain showed more growth compared to tail-dwelling monkeys, and areas for language, face recognition, and object perception emerged. The sense of touch has grown in primates in a similar way, but in humans, the connection between different parts gradually improved with the expansion of the association cortex, resulting in the human brain's cortex being three to four times larger[394] than that of chimpanzees. The sensory cortex has become so entrenched that by losing or amputating limbs, in many cases, patients feel pain in the severed limb in various forms. George Dedlow's life story is that of a soldier who suffered severe limb injuries during the American Civil War in the 1860s, caused by hollow-based, soft lead bullets known as "Minié balls." He lost all four limbs and was left with only a torso, resembling a strange larva-like living creature. Silas Weir Mitchell[395], an American surgeon who amputated over a thousand limbs during the American Civil War, delves into the concept of phantom limb[396–398] in the story that he wrote "*The Case of George Dedlow.*" Limbs that do not exist, but the pain is felt[399,400].

The sense of touch establishes social communication and plays an effective role in establishing bonding, intimacy, learning, and memory during childhood. In other words, it has the main and central role in the emotional development of the child[401]. The term autism, literary means self, was first expressed by Swiss psychiatrist Eugene Bleuler in 1907, but in 1943 Leo Kanner, an Austrian-American psychiatrist, entered this term into pediatrics. At the same time, Hans Friedrich Asperger (1906–1980), an Austrian psychiatrist, wrote his treatise on a special form of these patients, later called Asperger syndrome, in Germany. Although Kanner had a complete command of the German language, he did not mention Asperger's name in his article. Later, Edith Sheffer in her book "*Asperger's children*" accused him of collaborating with the Nazi during the Third Reich[402]. Simon Baron-Cohen, an English psychologist, believes that an abnormal sense of touch is the main cause of autism. In his opinion, tactile perception and increased skin sensitivity play an effective role in socialization, development, and repetitive behaviors. According to India Morrison, the skin is a "social organ" due to the role it plays in social perception and communication[403]. Numerous receptors on the surface of the skin provide various types of perception. Sensations such as understanding form and texture using Meissner's corpuscles, stretch and movement with the help of Ruffini's corpuscles, pain and temperature receptors (cold and heat); and perception of location, and balance with the help of proprioceptive fibers could be understood.

SENSE OF SURVIVAL

The structure of every living organism is fundamentally rooted in the principle of survival. The process of maintaining the body's structures within an optimal physiological range for efficient function, known as "homeostasis," serves as the cornerstone of survival. The foundation of homeostasis is the intelligent and rigorous management of energy, enabling it to be used for essential tasks, repair, defense, growth, participation in reproduction, and the safeguarding of future generations. This process involves neural autonomic, endocrine, and behavioral mechanisms and is often accompanied by rapid recognition and understanding of physiological problems and changes, leading to both automatic physiological reactions and a type of mental experience known as feelings[404]. The sensory afferent presentation of these feelings reaches the insula, while the motor efferent presentation reaches the cingulate cortex, which is responsible for controlling mood, cognition, behavior, and environmental attention. Bilateral damage to this system in humans is accompanied by a lack of voluntary inclination to speak, silence, and immobility known as akinetic mutism. A classic example is the control of body temperature, which, contrary to popular belief, is not simply a mechanical and simple control but heavily relies on our internal sensations and the amount of pleasure we derive from the ambient temperature. The advertising of drinking Coca-Cola with abundant ice in hot summers follows the same rule. Various pieces of evidence indicate that bodily states are represented in various regions of the brain, including the cerebral cortex and other higher cortical regions. The sensory afferent system in other primates is not as well evolved as in humans, and thus, due to a lack of direct sensory presentation in the prefrontal cortex, their perception of bodily sensations is not as well developed as in humans, which can create a fundamentally different concept of consciousness and self-awareness compared to humans[405].

Feelings have undergone changes over time, becoming transformed, acquired, and then relocated. Subsequently, they have been placed onto a more complex system for understanding the nature and structure of physiological phenomena, in order to understand the good from the bad, the correct from the incorrect, and the dangerous from the safe. In other words, sensory experience leads to the creation of a new level of behavioral regulation in humans. By accepting the fact that feelings have emerged and evolved in the course of evolution, their functional role in the systemic levels of the nervous system has become evident. Some feelings are related to the control of internal homeostasis, such as thirst, hunger, and pain, and are referred to as homeostatic feelings. Others are a result of an emotional response triggered by external stimuli and are referred to as emotional feelings or simply emotions. Acute pain is a type of sensory experience that alerts the organism to environmental threats and has a significant impact on homeostatic control. This phenomenon does not apply to chronic pains with complex and unusual structures that involve both the peripheral and central nervous systems. Even though the initial damage has ceased and disappeared, the pain persists. Perhaps the set point of this system in the homeostatic system has malfunctioned[406].

William James, American philosopher and psychologist[407] in 1884, correctly believed that some mental perceptions of everyday facts stimulate affection. He referred to them as emotions, which lead to the co-occurrence of such mental states with bodily manifestations. The brain typically attempts to predict bodily conditions based on the likelihood of emotional reactions. Its fundamental importance and primary function are optimal energy consumption. Studies have shown that the more accurate and better the interoceptive sensory information received from within a person, the more accurate the prediction will be. Emotions are not occasional phenomena but rather continuous and enduring behaviors based on goals and for the satisfaction of homeostatic and social needs, often occurring without individual feelings. Craig[408] refers to them as "homeostatic emotions," and Damasio[409,410] refers to them as "background emotions" that manage behavior.

According to Damasio, three types of sensory systems are important in shaping our awareness of "self": exteroceptive, interoceptive, and proprioceptive. The concept of interoceptive sensation entered the field of medicine over a 100 years ago[411] by Nobel Prize-winning scientist, Charles Sherrington. It has been of fundamental importance in distinguishing it from the five external senses. Initially, this term was only used to refer to the innervation of visceral organs, but subsequent research expanded its definition[412]. In essence, interoceptive sensations refer the individual to their own physiological baseline. In various literatures throughout different periods of history, interoceptive sensations have been attributed to different bodily states. These sensations have been categorized into a spectrum and given the concepts of good and bad, positive and negative. Good or positive sensations are effective in survival and evolution, while bad sensations can hinder survival and evolution by presenting numerous challenges or acting as general obstacle. Thirst, hunger, shortness of breath, pain, stress, discomfort, pleasure, happiness, sexual pleasure, and even more complex sensory-social responses such as blushing, empathy, encouragement, humiliation, worthlessness, maternal or romantic love, and other emotional states, encompass a range of these senses.

Changes in the body's internal environment, such as contraction of the abdominal muscles, heart rate, concentration and serum level of various metabolites and electrolytes, and many other pieces of our bodily internal information, are conveyed to the brain through thousands of receptors that are mechanical and/or chemical in nature within the interoceptive nervous system (INS) with the help of the vagus nerve or peripheral nerve branches that go to the first layer of the spinal cord[413]. Those nerve branches extend from the spinal cord to the thalamus. This system of transmission in the spinal cord called "spinothalamic tract" (means that those nerves begin in the spinal cord and terminate in the thalamus). The transmission of sensations of pain, temperature, osmotic changes in body fluids especially blood using osmoreceptors, and changes in body metabolism using metaboreceptors is done through thin, lightly myelinated fibers, while the transmission of vibration and position senses uses thick, myelinated nerve fibers. This sensory system neurologically enables the body's general sensory capacity for homeostasis, including sensations of pain, temperature, itchiness, smooth muscle distension, sensual touch, hunger, thirst, taste (gustation), cardiac and respiratory activity, and air hunger. In other words, human interoceptive sensations reflect the individual's homeostatic system and conditions, which are transferred to the relevant centers in the brain through thin nerve fibers with low myelination and cause appropriate behavioral response[414].

According to Damasio's belief, if stimulating factors have previously elicited an emotional response, they can also initiate an emotional response in subsequent instances without exerting a noticeable effect on the body. Perhaps there is a form of interoceptive memory or the ability for mental travel within the insula which means island. It is a small part of the cortical region deeply located in the brain and perceive our interoceptive memories as our true inner sensations in the past. The sent information gradually reaches the insular cortex[415] and the sensory-motor cortex of the brain. Sensory information, before reaching the sensory cortex of the brain, passes through the anterior insula for the examination and correction of sensory errors, the amygdaloid complex (almond-shaped nucleus) for understanding the significance of sensory stimuli, the nucleus accumbens (lying or resting nucleus) for evaluating the rewarding aspects of stimuli, and the orbitofrontal (means eye-forehead) cortex for evaluating tissue and stimulus-dependent attributes. Additionally, the main information related to pain sensation is also transmitted to the sensory cortex and insula through relatively thin and unmyelinated type C nerve fibers, from the skin surface, which is the site of intersection between internal and external senses. Almost every form of interoceptive sensation, before reaching the insula, travels to a region called the thalamus[416]. It was previously believed that the thalamus was only a center for relaying sensory messages, but it is now believed to play a role in creating a direct cortical representation for all homeostatic senses, which can be considered as a mirror-like image of the functioning of the sympathetic and parasympathetic autonomic system. The insula is a small and distinct lobe located within the depth of the Sylvian fissure of the brain, surrounded by the frontal, parietal, and temporal lobes. It significantly contributes to the perception of internal (visceral) sensations and body homeostasis. Damage to the insula, which typically occurs in cerebrovascular accidents, can have a clear impact on the perception of taste intensity, smell-related taste perception, temporal segregation of sounds, superficial sensation and awareness of the body, neglect of visual and auditory disturbances, disruption of voluntary actions, fatigue, feelings of disgust, speech disorders, decreased verbal memory, and a tendency to use recreational drugs. Damage to the right insula, associated with impairment of the sympathetic system, can lead to weakness, inattention, and fatigue, while damage to the left side, mainly associated with parasympathetic system dysfunction, can result in auditory and speech impairments. Damage to the anterior insula is accompanied by disturbances in intrapersonal sensations and inner-personal space, pain, mood disorders, and anxiety[417]. Destruction of the posterior portion affects extrapersonal interaction, pain perception, and other internal senses to a greater extent. In summary, considering the centrality of the insula in coordinating external and internal environmental functions, damage to it will be accompanied by disturbances in sensory perception from the outside, internal senses, awareness, emotions, language, and the regulated functioning of the sympathetic and parasympathetic systems for controlling homeostasis[363].

The feeling of pain is not merely a distinct sensation from other senses, but rather a distressing experience that represents prominent elements of dissatisfaction, suffering, and sorrow[418]. Pain acts as a signal that attempts to preserve the integrity and unity of the body. The insula is not only a part of the cerebral cortex for processing internal sensory information, but it also integrates and regulates two-way signals of cardiovascular, respiratory, and emotional aspects to generate and modulate our emotional experiences of internal and external events. Scientific studies have shown that slow breathing accompanied by rhythmic speech or Zen meditation can induce synchronized cardiovascular and pulmonary systems with optimal oxygen consumption, ultimately improving mood.

The skin serves as the site of contact and perception of internal and external senses, and the boundary between these two is not properly separated. This phenomenon is referred to as the sense of touch. Changes in the surrounding environment transmitted by external senses, including vision, hearing, smell, taste, touch, and balance, primarily contribute to emotional reactions. Together, they create a neural topographic map for the external and internal worlds through sensory inputs from the external and internal body. These continuous and real-time changes shape the human mind, constantly altering this map through daily experiences. In addition, unused synapses are constantly pruned, and this process, which usually continues until the age of thirty, occurs after improving sensory perception. The human brain has more than ten times as many synapses as cells. Pruning of synapses is a kind of genetic process using complement cascade with the help of microglial cells in the brain, however, environmental factors and individual sensory and motor experiences[419] are also effective in it. Failure to do so correctly leads to the occurrence of synesthesia (shared perception) which refers to the stimulation of one sense to produce sensation in another organ, such as experiencing specific tastes for words (lexical-gustatory synesthesia), perceiving specific colors for letters and numbers (grapheme-color synesthesia), or feeling textures for sounds (auditory tactile synesthesia). This is probably due to remained connections between the two sensory systems. Richard Feynman was a famous physicist diagnosed with grapheme-color synesthesia, and Wassily Kandinsky Russian painter, could see Wagner's opera, called Lohengrin (son of Percival) in color because of his sound (tone)-to-color synesthesia[420]. Another famous example is Solomon Shereshevsky, a Russian journalist, who was born in 1896 in Torzhok, 145 miles north of Moscow. He had a very strong memory so that he did not use paper and pen to take notes and remembers all the sayings and writings even from years ago in his mind. As a mnemonist, he interviewed eminent neuro-psychiatrist Alexander Luria for a long time at the Moscow Psychiatric Center; and also a film based on his life called "*Memory Mysteries,*" directed by Lyudmila Malkozova, was made after the 1991 Soviet revolution. Shereshevsky described the words of the director of the psychiatric center, Lev Vigotsky, as a "yellow crumbling speech" and the voice of the famous Russian director Sergei Eisenstein "like the flames of fire that are drawn toward him." This description caused Eisenstein to bring him on the stage in a meeting of the film festival and to mention him by the name of "his old friend"[421,422]. According to Craig[423,424], the right insula is involved in a wide range of processes, including pain, internal senses, and emotion-based, social, and cognitive functions, and it holds particular importance for emotional states. However, its most important function is subjective awareness.

The integration of information in the form of topographic images in the right insular cortex plays a fundamental role in the structural basis of subjective awareness regarding the "sentient self" or the "me." In other words, self-awareness and differentiation from "the others" (self from nonself) are shaped based on the perception of sensory feelings. Scientists believe that the presence of large von Economo nerve cells in the anatomical structure of the insula is essential for efficient and rapid communication between the insula and the prefrontal cortex[425]. The number of these cells is directly related to brain volume and plays a significant role in brain's emotional development, ultimately contributing to sensory information processing for decision-making. This phenomenon highlights the role of the emotional brain in cognitive function. From a classification perspective, the number of these neurons decreases in the order of human children, gorillas, bonobos, and chimpanzees, and they are absent in macaque monkeys. These neurons are also found in dolphins and elephants and generally hold higher existential value in mammals[424] that place greater importance on social interactions and cooperation.

Information is transmitted from the "non-neural inside" (the body) to the "neural inside or core" (the brain). There is a very special and specific connection between the "recipient" (as an object) which is the body itself and the "perceiver" (as the subject) which is the central nervous system (CNS). It should be noted that the perceiver, which is the CNS, is itself part of the recipient, which is the actual body that constitutes us and is inherent within it. This connection is permanent. In terms of external sensory reception, there is no order between the perceiver and the recipient. For example, the visual system cannot directly modify the received images. In fact, external objects are only sensed by receptors located at the distal ends of the corresponding nerves, and they do not have multiple entry points. In contrast to external senses, internal senses have multiple entry points and receptors even at the level of a single nerve branch. The interoceptive nervous system (INS), unlike the external sensory system, can be influenced centrally or peripherally in various locations[426].

Bodily awareness consists of two components: body ownership and agency which means the ability to control the body or be the cause of events. Ownership is an experience that people feel in their own body and consider themselves the owner of it. In other words, the brain differentiates its body from the other. Some diseases are associated with not recognizing or not accepting ownership. For example, in alien hand syndrome, which occurs following a stroke, the patient lacks control over limb movements that they acknowledge as belonging to them. They know that the affected limb is part of their own body; however, they may feel a sense of disconnection or lack of ownership over their actions. However, agency is a feeling in our existence that shows that we have authority and control over our bodies and are responsible for our actions and behavior. The agency depends on forward modeling in the brain. A brain that predicts the results of its future actions and compares them with what happens. This comparison shows us how much control we have over our behavior and actions. The feeling of agency causes a conscious experience of free will and a better understanding of body ownership. The schizophrenic patient is associated with an obvious problem in agency and considers his actions and thoughts to be in the hands of others[427].

Sensations can be distorted. Lewis Carroll writes about sensory distortions in "*Alice's Adventures in Wonderland*," distortions that are sometimes seen in migraine patients[428] too, or certain types of seizures. In anorexia nervosa, a type of simple sensory distortion leads to a feeling of fullness, causing the patient to develop delusions about their self-concept and body image, ultimately resulting in severe thinness and disruption of menstrual cycles, causing lasting harm to the patient. Fear of ugliness is seen daily in all societies and manifests itself as a morbid fascination with undergoing cosmetic surgeries. The high statistics of cosmetic surgeries, accompanied by exorbitant costs, support this point.

Variations in the neural map due to the reception of different sensations from within and outside the body lead to the emergence of emotional responses. In fact, emotional responses are a type of practical program that is pre-planned as a physiological instinctive response to correct abnormal conditions resulting from the reception and understanding of internal and external sensations by the brain, ultimately reaching homeostasis. These responses include fluctuations in organ function and internal environments such as heart rate, respiration, hormone secretion, and smooth muscle contractions, changes in facial muscles and limbs such as readiness for running and escaping, and alterations in cognitive processes such as behavioral responses, attention, and thinking. Practical programs can be divided into two groups: drives and emotions. The former encompasses general instinctual needs such as hunger, thirst, sexual desire, exploration, recreation, protection of relatives, and marital connection and intimacy. At the same time, emotions include feelings such as sadness, fear, anger, discomfort, happiness, embarrassment, pride, contempt, empathy, and encouragement, which are mainly triggered by external stimulation or by recalling it, and they accompany a corresponding reaction, which can be pleasant or unpleasant, after a conscious evaluation of bodily information, usually based on learning. These reactions, in some cases, are so severe that they can cause fluctuations in physiological body functions such as temperature or cardiovascular activity. Neurons in the insula have learned to calculate their own margin of error, but sometimes this estimation of internal feelings is not accurate, which can serve as a basis for anxiety development or consolidation[429]. This depends on monitoring and determining the feasibility and level of risk.

The mechanisms that calculate the level of stimulation and response are learned simultaneously and participate in regulating the body's homeostasis, and making decisions in complex and uncertain situations easier. Emotions are a kind of feeling that is accompanied by the action of the autonomic (involuntary) nervous system simultaneously[430]. Emotions are not occasional phenomena; rather, they usually occur continuously, optimize energy consumption, and, according to Darwin, their main purpose is to respond to the homeostatic and social needs of living organisms. Our awareness of emotional behaviors has evolved over time and has strengthened our emotional communication. For example, seeing a dangerous grizzly bear creates fear, which in turn initiates an internal physiological activity (increased heart rate, adrenaline secretion, and muscular readiness for escape), and this set of simultaneous reactions is manifested as fear in facial muscles.

In this example grisly bear is the stimulus, physiologic changes of the practical program are the emotional response; and fear as a conscious experience shows itself as a kind of feeling[431]. There are other classic examples in this regard. For instance, receiving a strong punch (stimulus) results in changes in blood pressure, heart rate, respiration, increased tear secretion, and changes in facial muscles (practical program), accompanied by the experience of discomfort as a sensory experience. In another example, the stimulation and pressure of a sharp object on the sole of the foot (stimulus) leads to pulling the foot backward, local blood vessel dilation, increased attention to the site of injury (practical program), accompanied by a sensory experience called pain. In other words, emotions are mental experiences that arise due to changes in the body's state, and their classical pathway involves external stimulation, perception by the sensory organs, changes in the neural map in the brain, initiation, and occurrence of action plans, changes in bodily state, and ultimately the experience of feeling. Mental translation of internal changes within the organism, such as hunger and thirst, or emotions, manifests itself as feelings.

The nervous system with its internal receptors is actually a collection of neural pathways and their nuclei in the central nervous system, which work to detect and map homeostatic signals, such as assessing the level of smooth muscle contraction in the gut. This system constantly monitors the internal environment and homeostasis of the body and plans the corresponding automatic autonomic responses. Therefore, it plays a fundamental role in creating the sensory experience in the mind and contributes to the regulation of conscious vital functions[432]. This system has several fundamental characteristics. Firstly, the related nerve fibers are either unmyelinated or have low myelination. From an evolutionary perspective, the presence of myelin, which is primarily a type of fat, separates the nervous system from the circulatory system. It significantly increases conduction velocity from 2 m/sec to up to sixty times, so that the speed of myelinated fibers, such as the A-alpha fibers that transmit deep sensation, can reach 120 m/sec, while the speed of unmyelinated fibers, such as those responsible for transmitting pain, is a maximum of 2 m/sec. Myelin production is very costly for the body and consumes a lot of energy. Metabolically, the process of myelin production and its maintenance is associated with high energy consumption. Myelination allows messages to be transmitted synaptically using chemical agents through the axonal-dendritic segments of neurons and prevents transmission through the electrically ephaptic (side-by-side touching using ion exchange) method, which is a simple form of message transmission, similar to two bare wires stuck together[433]. Why has the nervous system used such a method for transmitting pain and heat? Has the rapid transmission of pain or heat with high speed had any specific value in the evolutionary process for the organism's self-defense and homeostatic control?

The absence of myelin allows nerve fibers to be influenced by chemical changes in the body's internal environment at any point, an event that manifests itself as a feeling of pleasure or discomfort. It is not clear which part of this system is involved and to what extent. However, neurotransmitters (chemical molecules that transmit nerve signals) such as GABA, serotonin, norepinephrine, and acetylcholine can also have an impact on any part of this system[434]. It appears that in the evolutionary process, organisms have accepted myelinated structures in any area where the speed of signal transmission is highly valuable, such as motor or cognitive pathways, while in other cases, they have utilized the nonmyelinated simple method (in the transmission of pain, heat, and internal sensations) to reduce energy consumption. The second prominent characteristic of this system is nonsynaptic signaling. In other words, instead of transmitting electrical messages from one nerve to another through neuronal synapses, the nerves in this system solely respond to depolarization (discharge) of the stimulating nerve[435]. Approximately 95% of the neurons in the dorsal root of the spinal cord are stimulated using this method. Additionally, the neurons present in a ganglion called the nodose ganglion in the vagus nerve can be stimulated by creating chemical and mechanical changes in animals. Endogenous body opioids such as endorphins, enkephalins, and dynorphins can affect different neural receptors (mu, delta, kappa) in these neurons, and depending on the importance of the capacity to perceive chemical factors, they can have an effect on the quality of sensations within a range from negative, unpleasant to positive and rewarding.

The third point in this system is the absence of the blood–brain barrier. Naturally, due to the presence of some sort of cells we call them "glial cells" in the brain, there is no direct connection between the blood circulation and neurons in the nervous system. Nevertheless, from a pragmatic viewpoint, the hypothalamus ought to have the capability to release its secreted hormones directly into the bloodstream without traversing the blood–brain barrier. Consequently, certain brain regions should lack glial cells capable of detecting real-time metabolic changes in the blood. This process takes place in specific points called circumventricular organs (CVOs), which are prominent within the third and fourth ventricles of the brain[436,437]. These centers lack the blood–brain barrier, allowing direct communication between neural tissue and the circulatory system. This facilitates a coherent connection between the metabolic system, endocrine glands, and the body's internal balance. Chemical substances in the bloodstream can directly interact with neurons in the brain tissue, and these chemical changes are quickly responded to in order to maintain homeostasis. Although the exact number of these centers is not well defined, the main ones include the area postrema in the medulla oblongata, regulating the respiratory system, cardiovascular functions, and the vomiting reflex; the organum vasculosum of the lamina terminalis (OVLT) for the release of a hormone called vasopressin and the regulation of body osmotic status; the subfornical organ (SFO) responding to the serum level of angiotensin II and maintaining the balance of vasopressin secretion and ultimately controlling thirst, dehydration, and blood pressure; the neurohypophysis, along with a location called the median eminence, for the secretion of hypothalamic hormones; and finally the pineal gland[438], located in the center of the brain, resembling a pine cone, and being an exception as it is singular, unlike many other brain regions that exist in pairs. René Descartes referred to the pineal gland as the place where the soul resides in the human being and the discourse on the passions of the soul. However, the primary function of the pineal gland is the production of melatonin, which controls the sleep–wake cycle, mood, food intake, reproduction, sexual maturation, and circadian rhythms. In some fish, reptiles, amphibians, and certain fish species, there is a hole through which pineal gland cells have direct contact with light. The emergence of this complex in many vertebrates is responsible for the formation of the neuroendocrine signaling system to control the regulation of water and electrolyte balance, hormones, and sleep–wake cycles. The genetic structure of vertebrates for melatonin production has undergone changes for approximately 500 million years. Overall, these centers create a conscious sense resulting from the combination of multiple sensory factors related to the interoceptive nervous system, deep perception, and external stimuli in different layers of the nervous system. They also establish continuous communication between the object (body) and the subject (nervous system), which is a part of the object (body), leading to the constant flow of the phenomenon of subjectivity.

Monoamines, including dopamine, serotonin, and norepinephrine, are a group of chemical substances that have played an important role in the evolutionary process, development, functioning, and plasticity of the nervous system[439]. They have grown parallel to the nervous system as chemical substances involved in information processing, and they modulate and facilitate the activity of neurons and neural networks. Sensory inputs are transmitted through sensory fibers to the limbic system surrounding the ventricles and the prefrontal cortex at the front of the brain, and they have different outputs from various regions of the brain and spinal cord. These compounds alter the chemical status and the environment surrounding the synapses, consequently changing the activity of neurons and recruiting or dismissing different neural networks. Although this process is quantitative, it has a qualitative and clear effect on the functioning of active neural networks, sometimes leading to unstable brain activities such as cognitive, emotional, mood, thinking, and attention activities. Ordinary and normal experiences indicate the normal activity and flexibility of the nervous system, but this state is severely damaged in abnormal and pathological conditions. Sensory information needs to be reprocessed in the brain, and the brain structure needs to make decisions and reactions. In this process, monoamines determine the time and intensity of neuronal activity in the neural network to specify the internal state of the brain. The cerebral cortex is the site for conscious, prominent sensory experiences and intentional behaviors. Monoamines are secreted even before the formation and evolution of the cerebral cortex in the early stages of brain development. For example, attention as a cognitive process focused on a prominent and distinct side in an individual space is influenced by brainstem monoamines. Norepinephrine intensifies it, serotonin restricts it, and dopamine modulates it. Furthermore, they operate on memory and learning by modulating synaptic strength and increasing dendritic spines, acting as a mental reward as a positive sensation and punishment as a negative sensation, engaging the autonomic visceral system. The positive sensation and the ability to obtain rewards are firmly established in the human mind, often perceiving the future as positive, good, and bright. It constantly plans for the future and often considers what it will achieve in the future. Pessimism and a negative outlook also exist, but usually, people do not dwell on them. A positive outlook on the future has been essential for the survival of intelligent humans based on previous experiences and what we have in our memory. Without it, we would have definitely faltered by now. Biological experiences have gradually taught us that a combination of optimism and a touch of pessimism is necessary for survival. Scientific studies have shown that the stimulation of the anterior cingulate cortex with monoamines plays a prominent role in this business[440].

A significant number of mental illnesses are caused by emotional disturbances, which result in changes in the state of the internal sensory reception system. Diseases such as anxiety disorders, impulse control disorders, depression, autism, and eating disorders have been proven by neuropharmacological findings and, of course, support this theory. At the same time, the improvement of many of these diseases indicates a correct response, emotional involvement, and activation of the internal sensory system. The evolutionary process in this system is very interesting. Having maps of the body's condition in the central nervous system is one of the evolutionary advantages of humans. In fact, signals related to the central body map are transmitted by physiological parameters. These maps indicate the intensity and location of the disorder and hormonal or emotional reactions are performed based on them. A physiological deviation, such as hunger or hypoglycemia, is detected and registered in the central nervous system map, initiating a corrective response. After correcting the blood sugar, the corrective response becomes suspended. It seems that in the evolutionary course, corrective responses have been formed before emotional experience. In fact, many disorders were initially managed through operational programs or simpler physiological mechanisms without having emotional sensations or sensory perception, and with the addition of sensory experience to the bodily map in the evolutionary path, they emerged and have been incorporated into it, contributing greatly to life regulation.

Some types of mermaids are the first organisms to have nonmyelinated nerves. Nonmyelinated fibers in the optic nerve for seeing are found in many mammals, including humans. With the emergence of multicellular organisms, from unicellular organisms, initially, chemical molecules called neurotransmitters were formed. Then synapses were configured, and finally, myelin came into existence. The blood–brain barrier, which is a separating structure between blood circulation and neural tissue, seems to have first appeared in insects like fruit flies, which have a structure similar to vertebrates[441].

Since deviations from homeostatic conditions are perceived as either good or bad, the presence of sensory experience is actually a good proxy for the biological significance of these phenomena and serves as a good natural guide for the adaptive behaviors of living organisms. Emotions, by valuing based on the feelings of pain and pleasure, play a fundamental role in this way. Moreover, they can facilitate the learning process from environmental conditions and utilize this information to predict similar conditions in the future. Predicting future events with the help of emotions and receiving information in the present increases biological flexibility in humans. In summary, sensory experience leads to more corrective actions compared to changes in the central nervous system.

In contrast to the ancient interoceptive nervous system for understanding internal sensations, the nervous system for movement, spatial perception, the five external senses, and cognitive processes utilize myelinated nerve fibers and synaptic transmission of neural messages. At the same time, the continuous and direct connection between the object (body) and the subject (brain) leads to bidirectional modulation and ultimately creates a nondependent two-way relationship. Overall, sensory inputs create a topographic neural map of the body in the cerebral cortex, a map of the object in the subject, which is itself part of the object[426,442].

In this regard, the main function of the nervous system, especially the brain, controls the homeostasis of the whole organism, including its own cells. The nervous system is a part of our main body and has significant interaction with it. In the evolution of the brain, in addition to the existing chemical system, the nervous system has taken charge of homeostatic control. The complex function of the cortical root contributes to the simple functioning of the lower parts of the brain, namely the brainstem. The foremost phenomenon resulting from homeostatic control is the creation of the mind and consciousness by the central nervous system. In other words, a living organism consists of the body, nervous system, and mind. The mind, which is constructed and operated by the brain, has the ability to ascend to places that initially do not seem to be related to homeostasis. This ability ultimately culminates in the creation of culture.

It seems that sensory information, especially emotions, is processed in the right hemisphere, which is predominantly involved in this regard. Patients with damage to the right hemisphere experience difficulties in understanding emotional words and interpreting the tone of speech for expressing emotions. A common sign that neurologists are proficient in is the lack of understanding of left-sided paralysis by patients with right hemisphere damage. Studies have shown that visual–auditory stimuli projected to the left side, which is connected to the right hemisphere, elicit stronger emotional responses. Darwin was probably the first to formally express the difference in the emergence of emotions in two different forms. Spontaneous emotional expressions on the left side of the face have long been recognized[443].

11 Emotion and Self

EMOTIONAL ARCHITECTURE

Democritus, the ancient Greek philosopher (460–370 BC), held ideas remarkably akin to those that resonate with contemporary humans. He posited that feelings and thoughts were rooted in a material foundation consisting of small and delicate atoms. He believed that feelings and thoughts were the result of the movement of atoms in space, which circulated within the soul and body. He regarded the brain as a fortress responsible for protecting the body, the center of intelligence and thinking, and the primary connection to the human soul. He considered thinking to be the result of brain function and the main factor in the exchange of feelings with the external environment. Plato and Hippocrates further developed Democritus' theories and simplified the concept of brain centrality[444].

What fundamental difference exists at first glance between a deer when it sees a Bengal tiger and a bird cleaning mites on a blue horse's body? The answer is clear: the deer is intensely afraid. Therefore, there is a distinct and noticeable difference between what we feel, which is a kind of processed and inferred sensations, and what we merely see, which is a kind of raw, unprocessed sense resulting from the stimulation of nerve receptors by an external stimulus. In the English language, the concept of sensation is expressed with two completely different words. In fact, *feeling* is a type of conscious mental experience that determines the internal state of the organism. It also specifically induces capacity and significance[410,442] to the organism. This capacity and significance are perceived as positive and good to the extent that they contribute to homeostasis and life maintenance, and negative and bad otherwise. Of course, it is the same capacity and significance that places every feeling somewhere on this spectrum. As mentioned, this phenomenon provides a moment-to-moment report of the organism's internal state. It seems that the capacity and significance are phenomena that humans inherently possess for the quality of sensory experiences because this natural process has the ability to assess with the vitality and life of the organism. Contrary to feeling, *sensation* is a physiological type of electro-chemical response. It means that an internal or external stimulus excites the relevant sensory receptors, and thus, electrical discharge occurs through axons and nerve pathways, transmitting to the brain, where the internal sensation or feeling emerges.

The idea of having five primary senses can be traced back to Aristotle (384–322 BC), who was an empiricist (means experimentalist) philosopher. In his famous book "*De Anima*" (*On the soul*), written approximately 350 BC, he addressed each of the five senses separately[445]. However, he mistakenly claimed that humans have no other senses. In reality, humans not only have more than five senses but also often collaborate with each other in many cases. For example, to hear and understand the speech of others in a noisy environment, we use lip-reading with the help of the sense of vision, or, to better comprehend the taste of food, in addition to the sense of taste, we utilize other senses such as smell, vision (to see the presentation of the food), and touch on the tongue (to perceive the texture of the food). The sixth sense, or proprioception (positional sense of muscles), was described by a Scottish surgeon, Charles Bell[446] in 1826. The perception of balance based on the sense of proprioception (for vibration and position) is almost necessary for any creature with motor abilities, such as insects or even worms. Simple nematodes (a type of worm) with a maximum of three hundred neurons in their nervous system also possess proprioceptive sense. However, the type of sensation can vary based on culture. The fact that many people in the Far East enjoy eating various types of insects, raw or even semi-cooked fish, while in the West, they prefer different dietary habits, is simply a learned sensation. Another example of cultural influence on sensations can be observed in the clinical manifestation of anxiety before an exam. Western students may experience "butterflies in the stomach" as a sensation, while Eastern students may experience headaches;

DOI: 10.1201/9781003538400-13

however, students in the Middle East may exhibit dispersed anxiety. According to British neurologist, Oliver Sacks, the central nervous system along with its proprioceptive system tells *who* one is and *what* one is doing, however, the autonomic nervous system which evolved long before the central nervous system tells *how* one is[447].

The human digestive system itself has a type of enteric nervous system, consisting of approximately six hundred million neurons. It is a very ancient classification and the first neural structure to form, often referred to as the "first brain." Neurons without myelin in this system are stimulated side by side, strengthen each other's signals, and transmit them to the central nervous system[448]. This system can significantly influence mood, temperament, and emotions due to its production of approximately 95% of the body's serotonin. It seems that serotonin secretion plays an important role in controlling impulsive behaviors and reducing aggressive states. It should be confessed that the enteric nervous system initially formed to regulate and improve digestion and nutrient absorption in the digestive system and has fundamentally contributed to the control and improvement of homeostasis.

The simple form of feelings regulates all the governing laws of life in terms of good, bad, and in-between is called automatic feelings. The basis of this control and regulation of organismic homeostasis is in the "present moment" state. Some reactions to the external world are real or based on what is stored in memory, known as emotional reactions. Examples of such reactions include heart palpitations when scared or increased breathing when fleeing from danger. The centers for these reactions are located in the hypothalamus, amygdala, surrounding the Sylvian aqueduct, brainstem, and nucleus accumbens[426,449]. These responses occur with the help of nerve fibers and chemical secretions. Although these responses are similar and follow a specific pattern, they are not necessarily inflexible and identical. A diverse collection of these emotional responses can serve as the basis for human socialization. Desires, happiness, care-taking and nurturing of children, pleasure, love, and many other aspects have social dimensions. Socialization and being part of society are a collection of necessary behavioral strategies for humans that ultimately lead to the creation of cultural responses and, of course, play a role in controlling homeostasis. In other words, socialization has long been fully integrated into the cultural mindset of humans. Pleasure and happiness, whether related to sex or displayed in a romantic and passionate form, indicate a desire for care-taking of offspring and the reproduction of future generations. Severing these relationships and emotions can manifest in diseases such as anxiety and depression.

The next question is about the origin of our feelings. While single-celled organisms have the ability to respond to the surrounding environment, with the emergence of nervous structures, protective systems for controlling homeostasis such as the circulatory, immune, and endocrine systems developed. Subsequently, with the evolution of the nervous system, the management of organism–environment communications became the responsibility of the nervous system. In other words, the primary control of homeostasis was placed in the hands of the nervous system, and the practical result of this control is the improvement of life conditions. As a result, lifespan increased, and reproduction gained significant importance. Gradually, genes came into play to assist in natural selection.

Feelings are not solely a neural phenomenon. Although internal senses are transmitted to the brain through non-myelinated or thinly myelinated nerves from various parts of the human body, creating a duality between the stimulus (body) and the sensory receptor (nervous system), the body and the brain are connected and integrated. The brain is a part of the body and is considered a unified entity. Therefore, feelings are a form of mental imagery within a unified system. While emotions occur in the body, they are processed in the brain, which itself is a product of brain function, known as the "mind." The fundamental characteristic of this phenomenon, such as the sensation of pain, is clearly dependent on its role in controlling homeostasis. Dysfunction in this system is always abnormal and referred to as a disease. For example, in the case of fear, symbolic reactions exist. Fear can be triggered by an external stimulus (e.g., a predatory animal) or an internal stimulus, resulting in the activation and initiation of a set of symptoms such as increased heart rate and respiration, bodily numbness, lack of pain perception (analgesia), and increased secretion of steroids (cortisol). This process ultimately manifests in behaviors such as fleeing, fighting, freezing, and behaviors that aim to highlight the effective factor. In fact, this phenomenon consists of several distinct stages. In the first stage, the stimulus must be able to alter homeostasis. The second stage involves the ability to perceive and detect the stimulus. In this stage, the executive system of the nervous system comes into play to regulate and coordinate homeostasis, using pre-determined emotional reactions or action programs to navigate this situation. In the final stage, a neural interface has the ability to correctly perceive the end of the homeostatic response and can issue a command to cease the corresponding reaction[450].

This collection is an experimental part of the somatosensory map in the nervous system, which is referred to as feeling. The topographic map created in the brainstem nuclei, including the parabrachial nuclei (PBG), the nucleus of tractus solitarius (NTS), and the periaqueductal gray matter (PAG) and superior colliculi (SC), can determine the state of sensation and then transmit it to the subcortical regions[451]. In fact, the first simple yet integrated map of internal sensory signals is constructed, which refers from the entire body to the brainstem. The most important stimulating factors for neural pathways are chemical, thermal, pain, visceral, deep sensation, indistinguishable tactile sensation, balance sensation, and vestibular sensation. The vagus nerve and its branches are the most important sensory information transmission system for internal sensations. Alongside it, the spinal pathway (spinothalamic) that carries sensory information from the spinal cord to the thalamus and ultimately terminates in the cerebral cortex is involved in this process. The vagus nerve carries sensory signals, especially related to visceral sensations, such as cardiac, vascular, respiratory, digestive, and urinary-reproductive, to the nucleus of tractus solitaries[452,453] in the brainstem, where the initial topographic map is drawn. These maps have a significant impact on regulating life because they play a fundamental role in the emergence of feeling states. It seems that the collection of visual, auditory, and tactile receptions related to external body senses is transmitted to the deep-central layers in the superior and posterior part of the brainstem, known as the superior colliculi. At this time, three topographic maps from three sensory pathways (vagus, spinothalamic, and external senses) are formed, which overlap with each other. Information from this center is transmitted to the cortical and subcortical layers, where topographic maps are redrawn and reactivated.

The most important part of the brain for understanding internal sensations is a region called the insula. When we explore the lateral sulcus (Sylvian fissure) of the brain, this small brain lobe becomes apparent in its depth. From an evolutionary perspective, the insula is only observed in primates and has significantly developed in humans compared to other primates. The insula has a fundamental connection to emotional feelings and serves as the primary provider of sensory information for human awareness. It is challenging to provide definitive evidence for such a theory. The posterior part of the insula is the site of presentation of primary interoceptive senses, such as pain, temperature, itch, hunger, and thirst. The middle part, through the integration of these senses, works to construct emotional concepts. Finally, the anterior part plays a crucial role in enhancing the relay of information to the anterior cingulate cortex in creating conscious awareness for social interactions[424,454,455]. These functions are accompanied by monitoring the heart performance. This small island does not work alone and acts as an important node for processing sensory information. This information from the interoceptive nervous system (INS) is reconstructed in the association cortex in conjunction with external sensory (exteroceptive) and body position (proprioceptive) information, playing a role in the formation of the "self," body ownership, and the ability to control it. In summary, the primary function of the insula is to regulate homeostasis and learned affective and emotional behaviors[456].

In patients with a viral brain infection called herpes encephalitis, when a specific part of their brain (insula) is damaged, emotional sensations such as pain, pleasure, itching, tickling, happiness, sadness, irritability, gaze holding and care, compassion, hunger, thirst, bladder fullness, and colonic distension can be diminished. At the same time, the insula plays a fundamental role in generating cognitive flow. These types of connections are highly effective for the logical control of motivation and emotions, and their absence leads to simpler behavioral patterns. The sensory-motor cortex, like the insula, also acts as a modulator in the experience of visceral feelings. Those feelings include the system for sensing the muscular condition of the larynx and respiration, which are conveyed to the insula through the vagus and glossopharyngeal nerves. This is necessary for controlling the outflow of air with the help of voluntary striated muscles and involuntary smooth muscles from the lungs during speech. The sensation of breathlessness is manifested by changes in the topographic map created in the insula, especially in its right hemisphere. This network helps individuals adequately and timely demonstrate the correct response to various internal and external factors, such as exercise, dehydration, altitude changes, physical injuries, infections, and social interactions. Meanwhile, the neural generators present in the brainstem correct respiratory control, the status of blood gases in response to airway resistance, and metabolic and biochemical changes. This control can transition from an involuntary state during playing wind instruments to voluntary control and management of breathing. Although damage to the insula does not eliminate the brain map for pain and temperature, it does result in unstable and dramatic emotional responses. Emotional apathy (blindness) refers to the inability to perceive internal sensory (interoceptive) signals related to emotions compared to physiological signals of the body, which are associated with the function of the insula and the anterior cingulate cortex[457]. A classic example is anorexia nervosa, where patients lack a good sense and perception of hunger, misinterpret their internal sensory signals related to hunger, and feel full with minimal food intake. Depression may also be a form of inability to perceive pleasurable physiological signals and involve negative interpretation and evaluation of the "self" and "the others," ultimately leading to suicide. Electrical stimulation of the anterior insula is accompanied by respiratory arrest in macaque monkeys. The ability for speech, memory, reasoning, and visualization in the mind has enriched and refined emotional states in primates, especially in humans.

Finally, how have the unprocessed primitive sensations been transformed into a comprehensible feelings? It seems that the signals related to internal feelings, with their inherent capacity ranging from positive to negative, indicate and evaluate homeostatic changes. In fact, value is a precondition for achieving a behavioral or perceptual response. Signals and neural stimuli continuously contribute to this process by stimulating neurons. As a rule, the orientation of the organism to react to the relevant feelings is based on the capacities of each of the internal emotions. For evaluating external emotions that have evolved later and lack a capacity similar to internal visceral senses, a cognitive system is utilized. The emergence of creating and inventing feelings has been at the same time as the formation of the mind, and this set has caused creatures to be aware of their own existence or life. In other words, emotions are the high-level regulators of cognitive functioning and alertness, which have accumulated in the modern human mind[458]. Dysfunction and apparent disruption of emotions are likely the main factors in the occurrence of modern-day disorders such as depression, anxiety, addiction, and intractable pain. Addiction essentially involves an intense pursuit of mental reward and the release of intense negative feelings (mental punishment) resulting from craving and greed for the substance. To reiterate, in addiction, two sensory phenomena act simultaneously: one for obtaining rewards and the other for discharging mental punishment[459]. These feelings are stored in memory and form a set point that maintains a balance between reward and punishment, which is crucial for effective emotional control of homeostasis. A classic example is the use of cocaine, which initially increases its impact on the prefrontal cortex by inhibiting dopamine reuptake in the synaptic environment, leading to a sense of euphoria. An addicted individual feels capable of doing anything. Increased consumption, in larger quantities, causes anxiety, irritability, and abnormal behaviors, commonly known as the "crack dance," and ultimately leads to psychosis. Cocaine increases synaptic plasticity, leading to compulsive, obsessive, and persistent behaviors in addicted individuals, known as addiction[460]. Structural damage to the brain accompanies various diseases such as recurrent brain strokes, neurosyphilis, and other disseminated brain disorders, resulting in the destruction of the mind. There is a strong association between sensory loss (olfactory, visual, and hearing), especially in midlife and subsequent dementing disorders such as Alzheimer's disease[461]. However, it is possible for the mind to deviate from its proper course while the cellular and molecular structure of the brain remains intact. Hysteria is a type of illness that manifests as the mind deviating from its proper course in a structurally healthy brain. In addition, what we call human soul and spirit is nothing but our internal processed feelings and I think it has no external objectivity. With the disturbance in the understanding of our feelings and emotions, our soul and spirit also suffer from confusion and distress. Perhaps, in this situation, we can consider mental illnesses as just a kind of disorder and/or disturbance in the functioning of our inner senses or feelings.

INTELLIGENT SELF

The capacity to perceive, comprehend, and gauge the passage of time is undoubtedly crucial for executing nearly all of our daily activities. The manner in which humans perceive time and our ability to estimate the duration of tasks is a fascinating and noteworthy phenomenon. It has often happened that in difficult and unpleasant conditions, we feel time passing very slowly, while in good and enjoyable situations, time seems to pass very quickly and always feels like it has flown by in our minds. There are so many daily examples that need not be mentioned. Precise timing not only allows us to have better flexibility in our behaviors but also makes our actions more effective and meaningful. It is a very important question as to how our neural structure has acquired such an ability. It seems that previous human studies have concluded that the neural ability to estimate and perceive time is not related to a specific center in the brain, although certain parts of the brain probably have a more prominent influence on this matter. The cerebellum, basal ganglia, parietal and prefrontal cortices, and premotor areas play an influential role in time perception for humans. Sensory information for the homeostatic system, especially in humans, is the main producer of the sense of perceiving and feeling physiological conditions of the body. This information is processed in the posterior part of the insular cortex and makes the "material me"[454,455]. The limbic system in the brain, for controlling emotions, is divided into the anterior-sensory part for understanding individual emotions (subjective feelings) and the posterior-motor part for demonstrating voluntary agency. The limbic system is considered one of the oldest and most ancient parts of the brain, actively intervening in the dynamic changes of the brain over millions of years.

In the meantime, the anterior part of the insular cortex has also been attributed greater importance. Craig and colleagues[423] have shown in their studies that the insular cortex creates a shared pathway for processing time perception and analyzing emotional and affective information. According to these findings, prominent factors in the body and the environment are encoded moment by moment in the anterior insula, and the insula integrates emotional moments in a temporal unit[424]. Subsequently, consecutive moments of one's inner self, self-perception and body awareness, that serve as indices for the passage of time lead to the emergence of the core perception of time. The subjective bodily feelings are formed in the posterior insula. Sensory fibers, through evolutionary processes, transmit a comprehensive package containing high-resolution sensory representations and images of different physiological conditions of the body and its internal states to the thalamus. Neuroscientists refer to this as "interoception," the sense of the body's internal state[455]. These signals are transmitted via the spinal cord and the vagus nerve to the brainstem and then the brain. These senses include pain, temperature, itch, tingling, touch, muscle tension, breathlessness, peptic ulcer pain, and intestinal distension. This neural pathway, starting from the spinal cord, terminates at the nucleus of tractus solitarius (NTS) in the brainstem. They represent the afferent fibers of the sympathetic and parasympathetic systems, which, in a somatotopic manner, representing the body's shape like a small humanoid figure, project regularly and non-overlapping from the posterior to the anterior thalamus. In humans, these pathways involve distinctly identifiable and/or affective feelings, including cold, heat, itch, pain, stinging, burning, muscle pain and soreness, joint pain, abdominal fullness, bladder and bowel emptying, feelings of illness, nausea, muscle cramps, hunger, thirst, and taste. Collectively, these can be referred to as "homeostatic feelings." These feelings are primarily associated with homeostasis and survival since they are all accompanied by the motivation for survival and a sense of well-being, which are indexed in the anterior insular cortex.

Other studies have shown that in the insula, every kind of primary affective perception from the body is mapped from front to back. According to Craig, these emotions can be considered as a "common currency of awareness"[462]. The sense of taste is likely located in the middle of the insula, and although the sense of chilliness (low temperature) is felt bilaterally in the insula, it shows a stronger response to neuronal stimulation in the anterior right insula[463]. The sense of warmth (high temperature) is also bilaterally represented; however, it seems that the right side exhibits greater excitability. Damage to the insula can be accompanied by a lack of recognition and understanding of illness on the opposite side. This phenomenon also occurs with right-sided insula damage. It should be noted that insular damage on the left side is often accompanied by speech disturbances, thus affecting the ability to clinically evaluate such an experience, posing a challenge.

Ultimately, it appears that the middle of the insula serves as the center for the integration of sensory inputs from all bodily functions. The connection of the insula with other brain regions, including the hypothalamus, amygdala, anterior cingulate cortex, and anterior striatum (means striped body), indicates a central role in sensory processing. This phenomenon mainly occurs with right-sided insular damage, and it clearly demonstrates that the middle insula receives and likely integrates all internal, external, vestibular, premotor, and survival-related senses[464]. The evolutionary progression of receiving and integrating physiological and homeostatic conditions with other frontal lobe functions, such as emotions and cognition, which create motivational, social, and design contexts, has improved energy consumption in the brain, resulting in approximately one-fourth of the body's energy being used by the brain.

Other affective feelings that are not necessarily homeostatic but play a role in the survival of living beings, such as anger, sadness, fear, disgust, happiness, trust, and surprise, can also be understood in the anterior insula[465]. This collection can somewhat justifying Damasio's somatic marker theory for consciousness. He believes that the process of personal feeling of emotions requires the involvement of different parts of the brain in mapping the body or regulating continuous changes in the internal state of the organism in the cerebral cortex and, in a way, directly interfere with homeostasis[426,466]. Understanding feelings guides the brain in making behavioral decisions that impact our survival and quality of life[467]. This guidance is done through constructing a received scene or representation of the surrounding environment and the experienced inner mind that shapes the future. Those feelings, through the neural mechanism that has developed over time in the body, stabilize to control the body's state. Feelings that differentiate the inner world from the external world cause the brain to create a meta-representational model of the relationship between these two worlds for itself. This represented image, reshaped or recreated in the mind, differentiates between the "self" and "the other" (another individual). This hypothesis claims that the degree of awareness and consciousness is related to the successive representation maps of the person (self). Anatomically, Damasio argues that the main role in executing this repetitive representation lies in the anterior insula. However, Craig and colleagues[462] believe that the fundamental basis of integration is the existence of interrelated representations of all salient conditions at any given moment in time, which they call it the "global emotional moment." The homogenized views of all the moments and salient conditions of the totality of an organism are Sherrington's "my material being"[468], and Craig refers to it as the "sentient self"[462]. Together, they constitute the concept of the "self," which is manifested through the activation of moment-to-moment organized topographic representations. These sensory conditions that are effective in controlling homeostasis can be accompanied by energy consumption or acquisition. Studies have shown that the left and right insula are primarily involved in positive and negative emotions, respectively, and while they collaborate with each other, they also oppose each other. In the homeostatic model, the right side serves the sympathetic system with energy-consuming emotions, while the left side serves energy-acquiring emotions. However, recent studies have shown that the right side is involved in initiating actions or switching them, while the left side is involved in monitoring-related actions.

As we have seen in previous chapters, feelings arise from simple sensory signals in the body. Additionally, according to Damasio's model, all feelings are generated based on homeostatic imbalances resulting from changes in the body's state[442]. This can imply that all our behaviors are sensory-emotional experiences and can occur without explicit awareness or alertness, and of course, awareness is not necessarily followed by behavior. Many people believe that their emotions drive their behavior, which JM Lord refers to it as "sound mind." In Craig's cinematographic model, all comprehensive emotional moments form the fundamental and essential basis for the human self's temporal shift. This model has identified the anterior insula, particularly as the receiver of all senses, including the pleasure of music and temporal perception[469]. Music is among the human characteristics and belongs to human knowledge. Specific cases of limited insular damage and the inability to derive pleasure from music have been reported. Music has a special capacity to connect people emotionally to each other and can not only influence our emotions but also alter our perception of time. This system reconstructs itself in every moment and is generally perceived as an unstable state. Therefore, it is compelled to constantly and stably reconstruct itself in such a way that its owner never has a clear understanding of this ongoing reconstruction unless something goes wrong, it doesn't work well, or it breaks down. At this point, the organism becomes sick. When the organism becomes ill, a set of previous background feelings, the current feeling caused by an emotional reaction and our external feelings, at any moment, are the constructor of our "self," which shows itself as an example of the simultaneous activity of different parts of the brain. In this context, our "self" understands and learns the present time (now) a moment later. The present is constantly turning into the past, and when we think about the present, we are actually thinking about the past and we are actually in another present, and sometimes we use the present to plan the future, which is another present, but in reality, they are all stone steps of the past. In practice, there is no "now," and unfortunately, our consciousness is always related to the past. According to Daniel Dennett in 1991, this model is called "Cartesian Theater" in which multiple neural networks produce a mixture of content in parallel that has the greatest impact on the system and enters into the awareness. In this model, comprehensive emotional moments, based on images of the body's homeostatic state, can present all our emotions, including our mental functions, as a sequence of visual frames, internally timed with a frequency of approximately six to eight frames per second (Hertz). In other words, roughly every 125–160 milliseconds, a visual frame (image package) of the entire homeostatic and feeling state of the body is created in the anterior insula[470]. The created images are compared with the images that have already been created or will be created in the future or those we remember and they provide an illusion (virtual image) of agency because the created image that is the "self" is always one tick behind the present moment[471].

In this cortical presentation model, the "sentient self" or "self-awareness"[462] is formed in the anterior insula based on the interference and interaction of all prominent conditions of an organism and its emotional–physical environment at any given moment in time. The significance and prominence of each factor are determined by its effective involvement in the maintenance and advancement of the organism. In other words, the prominence and importance of a factor or situation depend on its ability to control the survival and well-being of an organism. The most important autonomic homeostatic system for controlling the life of the cardiovascular system, followed by other systems, is added to it. In this model, feeling concepts are perceived from the posterior to the anterior insula. The posterior part is the initial and primary site of the overall internal sensation presentation. From an anatomical perspective, from posterior to anterior, the homeostatic system, with the help of the hypothalamus and amygdala, is responsible for motor function, the understanding of environmental conditions with the aid of temporal and entorhinal lobes, pleasure-related concepts and emotions with the help of the nucleus accumbens and orbitofrontal cortex, and finally, cognitive, motivational, and social concepts with participation of the prefrontal cortex, thalamus, and nucleus accumbens, creating a cinematographic image with a frequency of generally 6–8 Hz images in the anterior insula. The famous, large, cylindrical, and spindle-shaped von Economo neurons[425] (after Austrian neurologist Constantin Freiherr von Economo) are responsible for creating electrical discharges in this system, generating a "comprehensive emotional moment" inside people every time[423]. These cells are mostly located in the anterior insula, increase with age, and are more clearly visible under a microscope. The bipolar shape of these cells has been observed in monkeys. In addition to perceiving pain, itching, heat, cold, and other biological homeostatic functions, the inputs to this region also have additional capabilities such as appetite regulation, social relationships, and homeostasis in interpersonal communication. These neurons become active with feelings such as anger, deception, shame, guilt, love, confidence, and empathy. In this model, so-called positive behaviors and good feelings are formed with low energy consumption in the prefrontal and left insula lobes, while negative behaviors (avoidance) are formed with energy consumption in similar lobes on the right side. Psychophysical evidence indicates that negative conditions and feelings work by distancing oneself from the pulses of comprehensive emotional moments and expanding the individual's sense of time, while positive feelings work in the opposite direction, bringing the pulses of comprehensive emotional moments closer and creating a sense of time passing quickly[472,473]. The first shape leads to situations with an awakened sympathetic system and a sense of time lengthening and significant delay. In the second form, awareness is placed in a positive and calm state, with the left prefrontal lobe and the dominance of the parasympathetic system. The pulses of comprehensive emotional moments decrease, the sense of the speed of time increases, and the sense of time passing is rapid. This set corresponds to Damasio's description of consciousness as it corresponds to the meaning of "the feeling of knowing that we have a feeling."

While the presentation of the autonomic system in the insula is not prominently asymmetrical in the two hemispheres, in general, this relatively asymmetric model can describe the opposing effects of internal sensory conditions on our perception of time. For example, if pain causes a sense of time lengthening, slow breathing and a feeling of calmness (increased vagal tone) create a sense of time shortening. Similarly, if fever and hyperthermia give a sense of time lengthening, reducing fever (hypothermia) has a completely opposite effect on the perception of the passage of time.

12 Consciousness, Mind, and Molecules

MIND AND LIFE

Some words may seem simple, clear, and straightforward until we endeavor to provide a precise definition for them, at which point we frequently encounter a fundamental challenge. One such concept is the mind. What is the mind, and since when has it taken its place in the life of human being? The first nucleated cells (eukaryotes) likely appeared about two billion years ago[474]. Multicellular organisms probably emerged around 600–700 million years ago, and the simplest nervous system arose around 500 million years ago (during the pre-Cambrian geological period). The primary significance of the formation of a simple nervous system during this period was the control of the organism's homeostatic system.

The nervous system that emerged from the beginning of its formation until now has had and continues to have two prominent and essential properties. The first property is excitability, meaning that a nerve cell (neuron) can become active and transmit electric discharge through its membrane to the axons by changing the ionic state of the surrounding environment. At the end of an axon, with the release of chemical substances, the electrical message is transmitted to the next neuron. This system forms the basis of bioelectrical signals. The second prominent property of the nervous system is the presence of sometimes elongated extensions that stretch to the far reaches of the body. Afferent sensory fibers transmit molecular, chemical, and sometimes mechanical stimulations through electrical messages to their own conducting efferent fibers, and these messages ultimately reach the brain. Initially, this system was formed to control messages within the intestines and was called visceral perception, but later, considering that it encompassed a wide range of internal sensations related to tissues and organs and formed the sensory capacity of the body's homeostasis, it adopted the term interoception[454]. Homeostasis is a set of intertwined functions of the autonomic system, endocrine glands, and behavioral mechanisms. One prominent example is the control of body temperature for optimal energy consumption. Although it may appear as a kind of external sensation, it is a type of internal sensation based on the creation of pleasure according to the individual's needs. In other words, when you feel a lot of warmth, a glass of cold water is very enjoyable, but when cold takes hold of you and makes your body shiver, it does not bring a pleasant feeling. This neural network forms the basis and foundation of the intense and complete supervision of the nervous system. These two phenomena together contribute to the improvement of homeostasis. In other words, the basis of the nervous system's function is homeostatic control for survival. One of the results of constant access to environmental information is the ability to predict future events. Initially, a simple nervous system was formed for the internal organs of the body, so it was called the visceral nervous system, which can be considered as the "first brain"[448] of the living organisms. This system simultaneously creates a kind of map of the body's interior. This type of processing requires not only the presence of a brain but also the body itself, and of course, it will only be accessible and comprehensible by the organism itself.

The main question is, where does the origin of the mind lie? How does a single-cell organism eventually acquire the abilities of a complex system? Perhaps factors other than natural selection and genetics may play a role in its development. It seems that homeostatic pressure may be an influential factor in this process. Although initially, the understanding of single cells about the external world was through sensing an external agent and responding to it, in the multicellular world, the presence of an endocrine system with the secretion of specific molecules (hormones) is considered valuable.

 DOI: 10.1201/9781003538400-14

This framework is of great importance for controlling homeostasis. The advancement of this system is accompanied by the expansion and development of the central nervous system with more precise control of homeostasis. This structure for homeostasis control has initially gave rise to the formation of a world that is mainly responsible for controlling the metabolism and functioning of the internal organs (heart, lungs, intestines, skin, smooth muscles, kidneys, and sexual organs). The created images of this old internal world represent a spectrum from bad to good, negative to positive, such as pain, pleasure, fatigue, feeling of well-being, tiredness, heart palpitations, stomach burning, colicky spasms of the intestine, and sometimes shortness of breath and spasms of the throat. These are all the main components of our feelings, and the assessment of their intensity as good, bad, or in-between is due to the existence of a phenomenon called "feeling capacity," which has formed in the brain[410,442].

The emergence of skeletal structure, muscles, skin, and sensory portals (inputs) such as the cornea of the eye, ear, tympanic membrane (eardrum), cochlea, nose and olfactory receptors, tongue and taste, and tactile receptors had the ability to capture the sensations related to our old internal world. By creating a perception of space in our minds, they took a fundamental step in creating a general image of the organism in the nervous system. In fact, we can call it the creation of a general image of the organism of "self," which is the main factor in the formation of subjective mental life[426] and is effective in forming the mind. Ultimately, the five senses establish human connection with the external world, and the brain can create different sensory patterns for itself. The combination of these three worlds that is internal, external, and proprioceptive senses has the ability to form a map of images in our minds, which can help us understand the concept of human subjective life. These images are not necessarily visual; they encompass all types of sensations and can be a collection of visual, auditory, and tactile sensory information that, as a whole, give us a general picture of a living being or its surrounding environment. Presumably, the basic unit of the mind consists of images that encompass various sensory characteristics.

The image that one forms of their surrounding environment depends on the information received from the environment and their ability to analyze it. Solving this problem forms the basis of the dynamics of survival. Almost all living beings have the ability to problem-solve for their own survival. However, the first step in problem-solving is recognizing and adapting to the environment. Neurons in the mammalian nervous system function as analog units, but the information transmission system is digital (electrical transmission). In other words, information is inputted and then selected. After entering the stages of analysis, integration, and ultimately storage, it is executed. The cerebral cortex acts as the neural base of the brain by creating the ability to predict in an integrated manner. The difference in the cortical structure is the fundamental reason for the diverse recognition of living organisms. The structure of the cortex shows a disproportionate growth in mammals with a large brain. Images of the internal world are transmitted through the brainstem to the insular and cingulate cortex, and in order for the images to be temporally aligned and have the correct chronological sequence, images of the external world are transmitted through the superior colliculi to the cerebral cortex. These images are transformed from a concrete to an abstract state and are used as metaphors and symbolic representations of objects. These all form the foundations of our mental life. The structure of memory, which is based on chemical and molecular changes, creates a new ability for the mind called "recall"[475,476]. This ability enables the mind to react and respond to similar phenomena by reconstructing the memory data. The interaction and juxtaposition between current images and recalled images have brought about another ability called "reasoning" for our minds. In other words, recalling past images in the mind and the ability to compare and symbolize them constantly and fundamentally assist our conscious mind in the process of reasoning, thinking, judging, and decision-making.

ALERT MIND

Several fundamental questions persist without clear answers. What is the "mind"? How did the brain construct this entity? How does the mind achieve consciousness, and what precisely is consciousness itself? Does consciousness, as Descartes suggested, refer to the flow of thoughts? What is the importance of this process in evolutionary progress? Is it intended to improve homeostasis? Is it meant to regulate the laws of life? Where did this process begin, and where will it lead us? Is reality truly separate from the mind? From a historical perspective, Darwin practically had no opinion on the concept of the evolution of the mind. According to him, the mind is the practical and functional representative of the brain, and it does not differ from mental forces in other animals. He believed that the capability of mental abilities is directly proportional to the size of the brain in primates. Furthermore, he evaluated intelligence against instinctive reactions[477]. After the materialist Darwin, the analysis of the mind and mental powers was fallen into the hands of behaviorists, and at the head of them was the Austrian psychiatrist, Sigmund Freud.

While adapting to the environment is the primary process of a living organism in biological evolution, the development of intelligence in an organism can gradually make it independent of the external environment by modifying the environment according to its own needs. Recent studies suggest that in the intelligent human, the development of wisdom and thinking occurs in 60% of cases due to ecological challenges, followed by general social challenges in 30% of cases, and finally in 10% due to intergroup social challenges, most likely due to developed culture. In other words, the foundation of cultural innovation is specifically based on dealing with ecological challenges to improve realistic models in humans. Cultural evolution is, in fact, an emergent process of mind evolution constructed based on biological facts. Additionally, it should be noted that cultural evolution has transcended the path of physical evolution and demonstrates its autonomous nature[478]. Biological evolution alone is insufficient without the mind and wisdom. If we accept that culture contains ideas and opinions that can be transmitted from one generation to another, we will realize that cultural inheritance occurs much faster than genetic inheritance because information easily exchanges through social communication, namely, language, writing, media, and art.

The capacity of human mind and thinking is not assessable without considering the ability to speak. Previously, the prevailing view was that the ability to speak began in the human species about 70,000–100,000 years ago. However, it seems that a "co-evolution" in language and culture began about four to six million years ago in the *Pan* species and its simple neural model can be seen in monkeys. Of course, it has been greatly enhanced in humans and has resulted from the evolutionary period of hominins. Overall, biological intelligence can be defined as the ability and capacity to solve problems. The enhancement of sensory abilities and the increase in the ability to process sensory information require the existence of a nervous system that can select, integrate, and ultimately store information to construct complex models of external reality. This cognitive system can enable an organism to perceive a mixed and disordered world with accurate sensory interpretation.

The process of what we call mind creation is a highly specialized function of the brain[479]. It seems that the main raw and unprocessed material (substrate) for creating the mind is visual images that are mapped by the brain. The mapping of phenomena and frameworks for internal and external bodily sensations is presented to the brainstem. The body is also visualized by the brain, which itself is a part of the human body[480]. In fact, the voluntary attitude and intentional perspective of the brain to the body is bringing the body into the mind. In other words, the main and fundamental property of the mind is the ability to represent the surrounding world in a way that can simulate its potential effects on behavior. This capability consists of the capacity to convert sensory stimuli into neural patterns, reconstructing them into new, different, and understandable symbols by the brain, and finally transforming the mental representations into motor activities, creating movement, and constructing mental models of reality, what Harry Jerison called "biological intelligence"[265]. It is the intelligence required for self-recognition and understanding of the surrounding world. However, it should be noted that both the concepts of "self" and "world" are products of our minds, an unreal image of reality. Creative intelligence is the creation of new forms of actions and expressions, which has been developed with the enlargement of the prefrontal cortex (PFC). The function of intelligence, like many other functions of the frontal lobe, is forward-looking, and with age and maturation of the neural structure in the PFC, imagination is added to it. Creative act is actually a kind of re-creation of the past with the help of imagination. But this unreal mental image is crucial, necessary, and beneficial for the continuation of life, homeostasis control, reproduction, and survival. Although it may not always work properly, such as imagining entering or being present in someone else body, which is called "out-of-body experiences." It appears that the comprehension of the concept of space and subsequently the scene, with the self at its center and an external or other object in front of it, emerged with the improvement of visual structure in vertebrates during the Cambrian explosion. This phenomenon not only led to self-awareness but also initiated the ability to move. Additionally, the self represents emotional, cognitive, and physical experiences of the past and present, and has a special connection between the interoceptive nervous system (INS) and dysfunctional schemas, which can be involved in the occurrence of certain mental illnesses such as anxiety and depression.

The body is in confrontation and interaction with the surrounding environment provides the relevant sensory signals to the brain. As mentioned in previous chapters, this interaction ultimately manifests itself as a topographic map in the brain, and the brain interprets the external world through the body, which is part of it. Not only information is sent from the body to the brain, but the brain also sends its commands to the body, but this phenomenon is not completely consistent and unidirectional. The brain perceives information from the body, which has happened in the past, and then instructs the body on how to behave in order to maintain homeostasis in the future. In other words, the body tells the brain how it is constructed, how it has been, and how it should be perceived, and the brain responds to the body, instructing it on how to be in order to establish homeostasis. The brain and the body are in constant interaction and dance, some sort of duet, exchanging and receiving information from each other. Any small change in the body can manifest as a fundamental and significant change in the brain, and likewise, any small change in the brain (such as hormone secretion) can be accompanied by substantial changes in the body[481]. The brain is both a receiver and an executor, giving commands, and since the brain can simulate sensory inputs, it ultimately has the ability to understand sensory phenomena in the minds of others. This capacity contributes to the construction of a mental phenomenon called "empathy," which means the ability to share unpleasant emotions with others.

It seems that the basic unit of the mind's structure is an image[482]. The information presented to the brain is reconstructed as a map, a map that can encompass all sensory properties. Maps are not always the same and can change depending on the situation, time, or movement. Mind is probably a product of brain, body, and information received from the external world. The first step in constructing these maps is the responsibility of the brainstem. Geniculate bodies shape the visual and auditory processes and, along with the superior colliculi, design visual maps in a different form. They also serve as the main interface between visual maps and auditory maps and other body parts. The inferior (lower) colliculi contribute to the creation of auditory maps, and the nucleus of tractus solitaries (NTS) and parabrachial nucleus (PBG)[483] play a powerful role in creating the sensory-visceral map of the whole body. PBG acts as a hub for senses of pain, disgust, bad taste, regulation of swallowing behavior, and other sensory-modal behaviors, which is considered a type of defensive behavior[451]. Bitter foods are generally perceived as dangerous and toxic, leading to a type of avoidance behavior in animals. Damage to this nucleus is associated with an inability to perceive taste, control swallowing behavior, and avoid unpleasant tastes. While pain perception is important for controlling body and physical integrity, and taste and flavor perception are important for controlling the entry of harmful substances into the digestive system, temperature control within a specific range using heat receptors on the skin is also crucial. Simultaneous reduction in body temperature along with reduced food intake can be deadly, so some animals enter the hibernation cycle to escape from this situation during the process of evolution. With this method, by significantly reducing metabolism, they can keep their central body temperature stable to overcome the harsh winter. It seems that the accumulation of sensory information related to the cardiovascular, respiratory, metabolic, and pain systems in a simultaneous and coordinated manner to the PBG has been useful and effective in constructing a behavioral strategy such as self-regulation of body temperature or hibernation[484]. The next nucleus, NTS, is also located in the brainstem and receives numerous sensory fibers related to the internal organs of the abdomen and the cardiovascular and respiratory systems through the vagus nerve.

At the same time, two fundamental properties can be seen in maps. Firstly, maps are directly transmitted to the brain and are only comprehensible by the owner of the mind. These fleeting and transient maps can be created from anything inside or outside the body, tangible, and concrete or abstract, real or recorded in memory. The continuous flow of fleeting moment-to-moment maps or sent images constitute the process of the mind. These images are logically related in the mind, but in dreams, this logic and connection are disjointed. Mental images can be forward or backward in time, their pace can be fast or slow, and they can be continuous or have gaps. These images can be in the form of a sequence or multiple sequences, simultaneous and parallel sequences, or overlapping sequences that interrupt each other. This phenomenon is a biological process that is evaluated based on intrinsic value for each individual, not based on a sequential order. The mind can be present with or without consciousness, although many images never enter our conscious mind. Both forms of sending images interfere with our decision-making process. As noted earlier, fundamental internal sensations travel via the interoceptive sensory system, while sensory experiences tied to the external environment are relayed through the exteroceptive sensory pathways. Additionally, the sense of equilibrium is communicated via the equilibrium sensory pathway[426,449].

Processed and inferred sensations (feelings) and perceptions through these pathways turn into emotional reactions. Emotions, the most important and intelligent offspring of biological value, are the result of the fundamental valuation. Emotions are pre-determined executive programs that have been evolved and completed by cognitive mental programs[485]. Their symbol in our body ranges from facial expressions, various changes in appearance, to changes in the external and internal environment of the body. Emotions are both felt and perceived in the body and mind. However, emotions have a fundamental difference from unprocessed sensations. External or internal stimuli are received, transmitted, and perceived by the sensory nervous system, while emotions are both felt and perceived in the body and mind and have an impact. Sensations form an interaction due to the existence of an external or internal substrate, but emotions themselves are the actions that manifest in the mind. Therefore, emotions appear differently based on beliefs and specific thinking methods, but sensations are not accompanied by such phenomena and are manifested uniformly everywhere[486]. After processing images in the mind, emotions give orders to the emotional response centers, which can be completely involuntary or sometimes undergo changes under individual will. Emotional reactions can change our mindset toward the surrounding events, reduce the speed of our thinking in response to discomfort, and increase it in response to happiness, ultimately altering our perception of the passage of time. At the same time, when one emotional pathway becomes active, it can completely influence another pathway. Disturbing phenomena also hinder the enjoyment of ordinary pleasures. The most important forms of emotions are fear, anger, discomfort, happiness, disgust, and surprise. These general emotions are almost universal, inherently self-existent, and do not require learning[487]. They are essential for maintaining life, homeostasis, and intellectual maturity in individuals. A classic example is being in a dangerous environment. In this case, through external senses such as vision and hearing, the phenomenon or dangerous situation is sensed, then weighted in the brain and perceived as an internal sensation, followed by an emotional response such as increased heart rate, cortisol secretion, and muscular readiness for escape.

Rene Descartes (1596–1650) believes that the essential characteristic of matter is occupying space, and the essential characteristic of the mind is consciousness, or in other words, thinking. Descartes believed that emotion and reasoning are two independent and antagonistic substances. Furthermore, he assumed that mind is ascertained before and independently of the body[488]. He thus adheres to dualism. The brain, res extensa (the extended substance), is a manifestation of matter that produces mental substance, res cogitans (the thing which thinks) and consciousness, which itself is a manifestation of the mind. In fact, abstract "conscious substance" arises from concrete "brain matter" and comes to life[489]. However, the story is much more mysterious and complex than this.

According to Colin McGinn, a British philosopher, we still do not know exactly how the juice of the physical brain has turned into the wine of consciousness[490]. In addition to Descartes' religious–spiritual view based on duality, there are other views on the concept of consciousness. It seems that the discussions on the relationship between mind (soul) and body (matter) were first in the period of Plato, Democritus, and Aristotle. They were believed in emotions called passive states and located in the body. They actually thought that emotions are psychic states inserted into the bodily feelings[491].

David Chalmers, a professor of philosophy at the University of California, proposed the term "the hard problem"[492]. In his opinion, consciousness is not about a special content, but rather a kind of personal experience. Fundamentally, these experiences are considered a private experience because there is no foreign object or substance in it, so they are outside the scope of objective science. Another theory is based on the theory of "self-model" by Princeton University psychology professor Michael Graziano[493]. Unlike Chalmers' theory, information processing in this model requires the existence of external objectivity what we call it "matter." Here, the understanding of feelings and emotions can also arise only with the pre-built information of the brain. In this model, "attention" is used as a mechanism used by a specific agent to boost and strongly stimulate the mental process of the relevant agent in a deeper and more detailed way, and finally affected on its output system. Attention is due to the use of limited brain resources to interpret a very important phenomenon. Unlike attention which is a physical process (such as an insect entering the visual field) and measurable by the brain, "subjective awareness" which is almost always associated with attention is a property that only the individual has and know about it. He knows about it personally. Without subjective awareness, attention can no longer be controlled and is accompanied by behavioral collapse. It should be noted that in the evolutionary course of individual attention and awareness, it was not a chance event, but an adaptive process that was gradually developed millions years ago with the simultaneous ability of animals to control their attention[494]. The evolution of brain networks has assumed its basic responsibility and the damage of these networks in various diseases can be associated with spatial neglect.

Consciousness, linguistically speaking, is derived from the Latin word "conscientia" meaning sharing awareness with others. Consciousness appears to be a dynamic process resulting from a kind of re-entrant functioning of the thalamo-cortical system, which interacts with the surrounding environment[495]. It is a step beyond constructed emotions, may be born from the mind and likely a part of it, propelling it forward. Thinking is a conscious process that operates with unconscious mechanisms related to the cerebral cortex and subcortical regions. It is expressed through language and gives rise to the concept of the "self" which is essential for survival. Consciousness is not a constant phenomenon, but a dynamic and evolving process caused by the electrical activity of neurons in the brainstem and cerebral cortex[496]. It is constantly changing, begins from the embryonic period, and evolves until about the age of five. Consciousness can be defined as a state of the mind that gives us the awareness and knowledge to understand our existence (our "self") and distinguish ourselves from "the others." This collection includes consciousness, thinking, memory and knowledge, intelligence, and genius, which form the structural nature of the mind, have been emerged simultaneously with the evolution of a *sapiens*. Consciousness shows itself when it is accompanied by a form of knowledge that binds it to its specific organism. Thinking is the ability to solve problems and intelligence is the capacity or valence that the process of thinking has. Consciousness conveys feelings to the mind in a regular and continuous manner and connects the mind to the body. Of course, homeostatic feelings, which are related to internal homeostasis, are not conscious for humans, but they should be presented in order for the images from the outside world created by our exteroceptive senses to be understood in the conscious mind and added to our mental belongings. Transmission of homeostatic sensations through the internal senses that is interoceptive nervous system, first to the stem and then to the brain itself, is necessary and inevitable. The images created in the world of our homeostatic feelings are created as maps in the brain and with the body of the organism, which was the origin of them appear as a hybrid, but in the world of the external senses, the mental map and the object are separate and do not blend together.

As a result, the imagery feelings of these two systems are completely different. Regarding the formation of homeostatic feelings and the concept of "self" in the brainstem, damage to the NTS, PBG, and PAG nuclei in this part of the brain disrupts the integrated formation of an image of the body. This disorder severely affects our understanding of the external environment formed by external senses. Although the cerebral cortex creates the visual content of our mind, the subcortical areas help us feel that this content is connected to a specific body. As a result, contrary to imagination, small lesions in the structure of the brainstem will have a significant impact on human consciousness, and with the damage of that, the world made by the cerebral cortex cannot ride on our body where it belongs[497,498]. However, consciousness with an uncertain nature and gender is not a fixed phenomenon. In animals that lack the ability to speak, consciousness takes the form of constructing a mental image of oneself. However, in humans, it is a unique and coherent but fluid, interconnected, and fully personal process with limited capacity. The most important phenomenological characteristics of consciousness are its unity and ownership. Although it is unitary, it is not fixed, and it is fluid and constantly changing, a point that William James in 1890 interprets it as "the stream of consciousness."

This definition indicates that consciousness is an integrated process rather than a concrete structure, which can connect mind and body to each other. Several neurons, called the dynamic core, are continuously involved in its creation, with rapid and simultaneous functioning. These neurons transmit signals that is information from the thalamus to the cerebral cortex using thalamo-cortical pathways and then provide feedback through the basal ganglia, returning to the thalamus[496]. The collective functioning of these neurons forms an indivisible unit, and it cannot be decomposed into separate components. Neurons present in the dynamic core constitute a functional group or cluster specific to an individual's brain, belonging to a particular perspective, which leads to the "personalization of consciousness," which can be considered as inherent or intrinsic subjectivity[499]. Coherence refers to the inability to focus on two incoherent phenomena simultaneously. In other words, following an event, the occurrence of an incident related to the initial event is comprehensible. For example, if numerous images are simultaneously displayed on a screen monitor, although we see them all, we can only choose to focus on a few of them at will. Studies show that consciousness has a limited capacity, and approximately can process between 1 and 16 bit/sec of information. This capacity continues along a continuous and connected trajectory but can enter a new path and embark on a new course at any moment.

Every brain, through its numerous and diverse synapses, possesses a unique virtual personality. This diversity and differentiation are key elements in human consciousness. Additionally, we can contemplate our own and others' consciousness, discuss it, feel our own existence, differentiate ourselves from the others, be aware of our own wakefulness, connect ourselves to the past and future during wakefulness, or construct it in our minds. To construct consciousness, we rely on sensory classification and categorization, concept formation, and the use of short-term memory based on the valuation of sensory categories. The ability of animals to construct a conscious scene is essentially the creation of a remembered present, which occurs in a fraction of a second. That's why often, external stimuli such as unidentified sounds, wind rustling, the shaking of leaves and bushes, and the sight of shadows all happen in a fraction of past time, but we remember them to construct the present[500,501].

Consciousness in humans is primarily related to personal experience, the ability to sense and perceive it. In philosophy of mind, this is referred to as "Qualia," which essentially is a unit for the perception of the qualitative aspect of a sensation. For example, why is red perceived as red and evokes a particular feeling in us? Why does the color blue mentally feel different to us? How intense is each color? How red is red, and how blue is blue? How hot is the heat, and how do each of us perceive and react to similar temperatures? We can consider "Quale" as the unit of the quality of our sensations[502]. The fundamental challenge in understanding Qualia lies in comprehending the difference between our own perception of the environment, others' perception of the environment, and the difference between the two. Temperature differences or the qualitative aspects of physical phenomena are present, but consciousness leads to the activation of different neurons and the discharge of electrical signals, which varies among individuals. Different neurons with distinct functions become active, and with the help of a dynamic core, this difference is reconstructed in the conscious mind. The degree of perception, even within an individual at different times, will vary depending on levels of wakefulness. In other words, it is not a physical structure that creates Qualia for us, but rather the difference in sensory perception between two objects plays a fundamental role in its emergence. This difference can manifest in infinitely diverse forms in the minds of different individuals, resulting in multiple Qualias and consequently differing opinions and perceptions. This innate ability to understand asymmetry and similarity, both in our own perception of the environment and in others' perception, is called "epistemic asymmetry" which in philosophy means related to the knowledge. Wakefulness is inherently related to it and always pertains to something or is connected to something that forms its content. That something is an obligatory aspect of our perception, an aspect that indicates its similarity to us or its dissimilarity. In the words of Franz Brentano (1838–1917), a German philosopher and psychologist, wakefulness always includes content, with its main focus on the surrounding world of the individual's body and our mind. It is essentially awareness of something in which the primary property is hidden through intention and will, and a perspective that is in either the first person or the third person. According to him, which is stated in his famous book "*Psychology from an empirical standpoint*"[503], originally published in 1874, the mental phenomenon is completely separate from the physical phenomenon due to its content. The mind is a dynamic system, action-oriented and context-specific, which by internally re-reading and rendition of sensory inputs and perceptions and conceiving their representation in the form of abstract and disembodied phenomena. The intentional model of the mind is based on the idea that the mind is a dynamic system that is embodied, active in the surrounding world, and permanently pointed at an object. Actually, there is not any external world object, but the object is a mental tool to perform a given action[504,505]. This system may be acted and work in concert with mirror system neurons and prefrontal regions to provide information about other people "why" perform some certain actions[506].

It is likely that simpler forms of consciousness exist in animals, especially vertebrates, apart from humans, but proving this requires communication and linguistic establishment[507,508]. However, it is only when wakefulness is in higher-order states of manifestation that "language" comes into existence and, through the formation of social and human relationships, the "self" takes shape. Gradually, it creates emotions within the realm of "thought" and "beliefs," and then, with the liberation of our minds, "awareness" accompanies the formation of images. At that point, the enlightened human is compelled to learn "writing" to record the images. The storytelling, myth-making, glimpses into the future, and exploration of the past are the consequences of this adventure[509].

In order to be properly alert, in addition to being awake, we must have an active mind, in which there is a spontaneous, involuntary, uninferred, and not comparable feeling of "self" as the main actor of an experience. Wakefulness, having an active mind, and a "self" that operates as a dynamic process each have separate structures in the brain that are intertwined and necessary for creating alertness. Furthermore, a disturbance in alertness can affect emotional reactions. Emotional reactions occur in the conscious mind through purposeful behavior. Before these reactions, which are systematically associated with emotions, take place, the majority of sensations are initially processed in the brainstem[510], in centers such as PAG, NTS, and PBG.

In addition, the various social and cultural dimensions of modern humans set them apart from other primates. The core of human sociality is related to the mother–child pair, which forms the core of the family. The expansion of this small structure leads to larger families with ten to fifteen members, then to groups with thirty to forty members, and eventually to a clan, tribe, or village with a maximum of 150 individuals. Robin Dunbar, a British anthropologist, has shown that the size of human groups is a universal and global characteristic that reflects any type of social structure of gatherer-hunter societies worldwide[511,512]. There is some sort of cognitive threshold for an individual to the number of people whom can make stable relationship. This number which approximately is around 150 people shows human cognitive ceiling. Beyond this number, the neocortex capacity to make relationship with other individuals has prominent limitation. He has demonstrated that the growth of the brain cortex has a direct correlation with the base population of humans, equivalent to a tribe or village, during past periods. One of the most distinguishing characteristics of human groups from other close relatives is their strong interest in sharing information within a social group[513–515]. This interest is more pronounced in children who have recently learned something or in elderly individuals who have abundant knowledge in their memory. According to Nicholas Humphrey, the need for socialization is the most important environmental pressure for the evolution of complex social behavior and the increase in brain size[516]. This concept, which Andrew Whiten insists has political and strategic implications for behavioral evolution, is referred to as "Machiavellian intelligence"[517]. The ability to speak greatly intricates and enlarges the human social network. Machiavellian intelligence was first described by Frans de Waal in 1982 in monkeys. This theory is based on the fact that interpersonal relationships in monkeys are based on individual needs, and friendly relationships can turn into hostile relationships if the needs are not met, and of course, it requires a degree of social intelligence to communicate. At the same time, social complexities especially require a degree of general intelligence and social cognition. Machiavelli's intelligence is based on the premise that individual success is possible if there is cooperation and competition between people with having a degree of subtle and clever social skills is associated and is not necessarily dependent on individual strength or power. Prior to the 1980s, what we thought of as a smart person was defined as having the ability to face physical challenges in the environment, such as the ability to find and access food, remembering safe places to hide and early understanding of danger signs[518,519].

One of the human capabilities that has played a role in the formation or enhancement of human consciousness is the inclination toward cultural learning. If we consider the most important function of the human brain as the protection of the human race by using cognitive capabilities and influential memory to find sources of energy (food) in difficult conditions, extended families contribute to this process simultaneously with the increase in brain size[519]. John Allman believes that the family economy is an information that is continuously transmitted from one generation to another, and thus, longevity is necessary for this matter[520]. According to him, extended families are necessary for nurturing a large brain and simultaneously constructing and maintaining information in it. Genes that prevent the onset of Alzheimer's disease are evaluated in this regard. Three forms of the *ApoE* gene, which are involved in cellular lipid metabolism, are observed in humans. The *ApoE4* allele has a very close association with the onset of Alzheimer's disease, but the frequency of the *ApoE3* allele, which is considered a protective type against this disease, is higher, up to half of the cases, in the human species than the other two forms. Therefore, elderly individuals can pass on their information to the next generation[521]. Stephanie Fullerton's studies show that the emergence of *ApoE2* in *sapiens* dates back to about 200,000 years ago[522,523]. The development of tools, which signifies the construction of mental images and ultimately mental travel, along with the ability to use fire for cooking food and optimize energy consumption, the use of episodic and semantic memory, and cooperation to connect beliefs and thoughts, the emergence of sign language, and later spoken language, and the transformation of speech into script, all of these have been important factors in the development of the state of consciousness and cognition in modern humans[524].

Consciousness can also manifest itself in various domains and with different intensities. For example, simply eating an ice cream in a friendly gathering can be strongly focused on the taste and aroma of it or the friendly conversation in that gathering. The stronger the impact and emotional reaction, the more likely it is to be stored in memory[449]. It seems that due to the abundance of images during the lifespan and the lack of space for information storage, there is a tendency to organize images in a narrative form, and the organism automatically selects the necessary images for its survival. Those images have higher emotional value and elicit a greater emotional response, and they can serve as a bodily marker. These markers not only refer to higher cognitive levels but also have an effect on lower stages related to evolution. The creation of mental images is a result of the operation of circuits with repetitive loops in the cerebral cortex and thalamus, which creates a "state of remembering or recollection." The present moment represents an understanding of a sensory phenomenon that has occurred immediately in the past[525]. Initially, it seemed that the thalamus was merely a sensory relay center, but subsequent studies have shown that it also has somatotopic regions and plays a major role in sensory information processing[526].

Animals that possess basic and primary levels of consciousness can construct a mental image of their environment in real time using sensory inputs and neuronal activity in the dynamic core and varying degrees of unconscious activity under the cortical level. Although they have biological individuality, a true self has not been constructed for them, and therefore, the concept of future and past has not formed[527]. These concepts have emerged in the course of evolution after acquiring linguistic–semantic abilities to express feelings and using symbols to refer to objects, events, and the surrounding environment. With the formation of grammatical structure in the human mind and the expansion of vocabulary through social interactions, consciousness has evolved to higher levels. Now, the conscious human had gained self-awareness and insight into their consciousness. Initially, the advancement in self-awareness and then awareness of others occurred primarily between close companions, and it is not known for certain whether they were mother and offspring or sexual partners. Awareness and knowledge of the future and past could not be fully developed without transcribing them into written form. Thus, the conscious human liberated themselves from the constraints of the present moment. The conscious human became the greatest intellectual creator of information in the form of stories, myths, and other types of texts for the present, future, and past[528,529]. In fact, information is a biological concept and a product of the conscious process in the human mind, a process that has emerged through natural selection in the course of evolution.

Ultimately, the image processing system begins its operation based on effective prediction of situations, examination of probable outcomes, information navigation for possible futures, and the invention of an acceptable method. This system plays a significant role in the advancement of culture and civilization. In this state, the organism not only has sensations that are solely sensed or felt but also possesses sensations that are recognized by their specific and distinct content. Thus, the preconscious life system, which was merely self-induced reactions to events with good or bad value, transformed into intentional reactions directed toward the "self." But the consciousness mounted on the mind not only has caused that spontaneous reactions to remain, but also to be deliberately focused on "self." Although the mind is a combination of tendencies, including encouragement, punishment, improving motivations, movements, emotions, and creating maps and images, the integration of consciousness with the mind has propelled the entire capacity of the mind a big step forward. Although consciousness provides a direct experience of the mind, the mediator of this experience is the "self" of the individual and not an external observer. This "self," however, is an incomplete internal messenger. Moreover, while consciousness is created by the brain, the brain, itself, must also construct the mediator. Essentially, the "self" is fashioned itself by information mainly derived from internal senses, while "the other" is only formed by external senses[530]. This process of self-construction in the awake mind has several stages. The first stage is the construction of the "proto-self" followed by the formation of the "core-self" and, finally, the emergence of the "autobiographic self." Before that, two questions are of importance. The first question is about ownership and how do I know that this body belongs to me? And the second question is about the agency and how do I control the body? Philosophers like Friedrich Nietzsche believe that there is a very close and intimate relationship between the self and the body.

In 1785, a French anatomist and physician named Jean-Joseph Sue claimed that an Englishman asked him to amputate his healthy leg in exchange for one hundred Guineas. The doctor initially refused, stating that he lacked the necessary tools, but when threatened at gunpoint by the Englishman, he proceeded to amputate his leg[531]. Later, the Englishman sent a very grateful letter acknowledging that with the amputation, he had become the happiest man in the world, and he also sent two hundred and fifty Guineas as payment to Jean-Joseph Sue. The concept of body schema was first introduced by an English neurologist named Henry Head, a century ago[143,532], and the term body image was first coined by a German psychiatrist named Paul Ferdinand Schilder[533,534].

Experience had shown Head that damage to the right parietal lobe is associated with lack of recognition of the body. He concluded that this part of the brain has the ability to display the dynamic representation of the body or a sketch of the "self." This image can be either live or based on previous information. The body image is, in fact, a social phenomenon, and from the very beginning, it establishes a strong connection between our own body image and that of others. The map of sensory perception in the brain cortex, depicted as an idealized inverted human shape called homunculus (little man), was first illustrated by Canadian neurosurgeon Wilder Penfield[535] in his famous book "*The cerebral cortex of man*" in 1947. As most of Penfield's patients were male, his sensory homunculus took on a more masculine form[536,537]. The represented body was not normally proportioned and was dependent on the amount of sensory innervation. Naturally, with the damage to one of the limbs, the representation of the body in the cortex is diminished. Such representations of the body maps in the cortex not only serve to receive sensory input but also assist in performing actions, although it is not clear how many topographic maps of the body exist simultaneously or how they interact. Viktor Hamburger, an eminent professor of embryology and neuroscience, together with the Italian Rita Levi-Montalcini, showed in their experimental studies in the 1940s that by amputating the limb bud of chick embryos, the sensory fibers degenerated and finally disappeared. Subsequent research by Levi-Montalcini and Stanley Cohen, an American biochemist, led to the discovery of the nerve growth factor, which is secreted from the embryo's limb buds, binds to the neuronal membrane and stimulates the genetic structure inside the cell nucleus[538]. The stimulation signal acts for the growth of peripheral nerves. This collaboration brought them the Nobel Prize in 1986. The importance of sensory representation in the brain cortex, in addition to the initial perception of various senses, is secondarily effective for understanding the body size and shape and finally the presentation of the body posture. The human brain has created a body model for itself based on sensory input and genetic structure. But the stabilization of the body map on the cerebral cortex depends on the number of neural endings or sensory receptors, but their size is not very effective. Therefore, the area of cerebral cortex related to the function of an organ reduces with rest and expands with activity, which is called remapping. Cutting the sensory nerve fibers causes its sensory perception areas in the brain to fade away or disappear, and the surrounding sensory areas in the cerebral cortex take its place. The same thing happens for motor nerves. Limb amputation and phantom limb phenomenon are usually associated with pain, which is probably caused by cortical reorganization of the brain[419].

The ability to control or "agency" is completely separate from the concept of "ownership." It refers to the brain's ability to establish a causal link between an action and its resulting effect, and what matters is the manner in which this connection is made. For example, in alien hand syndrome, the patient has no doubt about the ownership of the limb, but the problem is with the agency, and they believe that the limb does not obey them and they consider the limb's movements to be beyond their control. However, in the case of Joseph Sue, now referred to as body integrity identity disorder (BIID), the fundamental issue lies in the lack of ownership perception. The patient suffers from having the limb and finds joy and happiness in its amputation[539]. They are willing to pay for achieving their goal and even force physicians to amputate the limb by causing harm to themselves. Furthermore, this multifaceted disease can be associated with disturbances in emotions, mood, sexual and/or erotic attraction, self-harm, and amputation. In fact, this mental disorder is caused by dissatisfaction with a part of the own body and its function. At the same time, there is a kind of mismatch between the internal presentation of the self (bodily self) and the physical state of the body. Due to the strong feeling that a body part does not belong to them, patients have a strong desire to cut it off[540,541].

According to Joseph Gerstmann, Austrian neurologist, the body and the self are something beyond the mind[542,543]. Comparing what we intended to do and what actually happened creates a sense of control over our actions and can be seen as a kind of prediction of future sensations. One of the most important tests related to agency is the comparison between our own actions and those of others. Many schizophrenic patients firmly believe that their thoughts, actions, and speech are controlled by an external force. They consider their visual and auditory hallucinations to originate from the external world. The more prominent the delusions and hallucinations, the more distorted the sensory feedback becomes, and the ability to predict the future decreases. The heartbeat, during the systolic phase, also generates rhythmic movements with a very small amplitude (about fifty microns) in brain tissue, which affects neuronal activity. This suggests that understanding the concept of control over movements is essential.

"Proto-self" or primitive self is the foundation for building the "core-self" and encompasses the entire "self"[544,545]. The proto-self is created by the neural cores located in the brainstem that are associated with feelings. The brainstem not only functions as a conduit for transferring information and signals from the spinal cord to the brain but also processes the perceived sensations from within and outside. The most significant product of this processing is the emergence of spontaneous feelings. These so-called primal feelings have a specific quality called "valence," which ranges between the negative end of the spectrum (pain) and the positive end (pleasure). This most primitive form of feelings is the primary basis for creating emotions, which serves as the basic foundation for an organism's interaction with the outside world, other living beings and objects. The feeling of just what is happening is not the whole story, the feeling that my own body exists and is completely independent of objects and the outside reality, but at the same time has the ability to interact with the external world, confirming the notion of "I am alive." This set is generated and understood through the proto-self. Perhaps the most important characteristic of the mental images constructed from perceived sensations and feelings are their ownership, which they belong to the same living entity[544,546].

The proto-self is the neural description of a stable blueprint of a living organism. This stable blueprint is constructed from the organism's physical structure, creating a map of its body that not only constructs it but also feels it[547]. These feelings spontaneously emerge in the awake, natural, and normal brain based on the estimation of internal sensory inputs and external sensory portals. Internal sensory signals are necessary for physiological correction and homeostatic control, such as hunger and thirst, while also contributing to the construction of pleasure states, such as engaging in eating and sexual behavior. The brainstem is involved not only in transmitting electrical signals but also in modulated and pre-determined methods of constructing primal feelings that arise before other feelings[548]. Furthermore, they serve as a symbol of the feelings of the living body, while also being an observer of external feelings. This set creates the concept of a body, a mind, and a "self." At the same time, the proto-self is a collection of maps that are in connection with its origin, namely, the sensory information of the organism[426,497,549].

The "core-self" or central self constitutes the second stage of constructing the "self," and it occurs through the interaction and collision of images of proto-self with images within the mind or brain[550]. These modified images from the organism are perceived as a continuous pattern and then referred to the cognitive structure of the brain as repeated visual pulses. This collection essentially represents the feelings of what we know, the feelings related to the main personality traits along with their accentuation, which manifest themselves as the core-self[551,552].

The final stage in the formation of "self" is the formation of a story of own life, that is, "autobiographical self"[426]. Autobiography includes personal memories, summaries of personal experiences and individual patterns, plans for the future, and when they enter the consciousness system, they create autobiographical self, which are actually our conscious memory and cause us to live a double life; the first is related to the conscious, open and obvious mind, and the second is hidden in the memory and of course has the ability to be activated again[553,554]. The primary strategy of the brain in constructing the autobiographical self is the classification and grouping of autobiographical memories to present to the proto-self. The result of this process is the modification of autobiographical memory and its representation as a pulse from the core-self to the thalamus and cerebral cortex. This system undoubtedly requires a robust neural apparatus capable of receiving and transiently holding multiple core-self pulses[551,552]. The most important function of autobiographical memory is to simulate the future. In fact, the evolution of the memory system requires a kind of reconstructive process, so it glues together real pieces of the past to simulate future events. This makes modern human has the ability to predict the future with some degree of error. In other words, we can remember the past, imagine the future, and make decisions about it. Memory disorder is a disorder of future imagination[555].

The coordinating regions in the brain for creating the autobiographical self include the thalamus, the claustrum, and, most importantly, the posterior-medial cortex (PMC)[556]. The PMC exhibits activity during wakefulness and significantly decreases its activity during non-REM sleep, but it remains somewhat active during REM sleep when dreaming occurs. This area of the cerebral cortex is the basic place of creating subjectivity or mental experience. Subjectivity arises from the interplay between any phenomenon that takes shape in the mind following our receptions from the internal or external environment and what exists in external reality[557]. This formation may not perfectly match external reality. Anatomically, the blood flow to the PMC is supplied by vessels from various parts of the brain and usually remains functional without significant impairments following vascular lesions and strokes.

Different senses enable us to recognize the boundaries of our own body. Sometimes, having a mental image of ourselves and introducing a new sensory stimulus, often touch, can mistakenly create an illusion that we possess a new limb. The rubber hand illusion, first published in *Nature* in 1998, represents a process that interprets body awareness[558]. In this test, simultaneous stimulation of the real hand and the rubber hand, with a focus on looking at the rubber hand, eventually leads the individual to perceive the rubber hand as their own, and the stimulation of the rubber hand alone becomes perceptible to the subject. Basically, this test is used to examine body ownership. Similar tests have been designed for the lower limbs (rubber foot illusion)[559] and even the whole body (full-body illusion)[560]. The insula along with the frontoparietal cortex by mixing information from the external senses (exteroceptive) and the senses inside the body (interoceptive) supports our experience from our body, in a way that makes us have a correct understanding of "self" and how we feel about the image or representation of our body[561]. Subcortical areas are older than cortical areas of the brain and their damage is shown by a clear unawareness of our body[562]. Body awareness plays a significant role in memory function. Awareness of the body forms a personal space around us, which, without official and distinct boundaries, threatens our personal space when crossed. The sense of having a limited space gives humans a sense of "self." This boundary is usually about the size of a circle with the radius of the upper limb. Individuals respond to each other based on their distance. This distance is encoded in the right parietal lobe. The brain integrates visual and tactile sensory information and sometimes, with the help of auditory perception, determines the peri-personal space. The primary function of visual and auditory senses is to enhance tactile sensation[561].

In Alzheimer's disease, initially, the hippocampus, which is mainly responsible for virtual memory control, separates clearly from its neighboring entorhinal cortex and shows difficulties in learning new concepts, impaired judgment, and spatial and navigational perception[563,564]. When the fundamental basis and foundation of autobiographical memory, which is the learning of new things, are affected, the connection between current and past personal events is lost. The individual's connection with the outside world gradually diminishes, and they respond less to environmental stimuli. Emotions become muted, and the individual's behavior becomes calm, quiet, empty, absent, and sometimes restless and unfocused. The vascular damage to the right hemisphere's posterior cingulate cortex, although accompanied by weakness on the left side of the body, the patient is unaware of their condition and gives vague reasons for their inability or weakness in movement. This impairment almost never manifests itself with damage to the PMC in the left hemisphere of the brain, suggesting that the autographic controlled neural pathways may differ between the two hemispheres[565].

The anatomical structure of the human brainstem and other similar animals is alike, but the main difference lies in their size, which is directly related to the body size of each animal. However, the cerebral cortex has significantly evolved in humans. The ability to create images, increase memory capacity, dream processing, reasoning, and ultimately the emergence of language are characteristics of the functioning of the cerebral cortex in humans. Additionally, the function of the brainstem has not doubled compared to the cerebral cortex, and they each have separate functions. The fundamental importance of the brainstem lies in regulating vital functions, forming the foundations of consciousness, influencing the functioning of the cerebral cortex, and, most importantly, constructing the proto-self and core-self. The thalamus plays a regulatory and signal analysis role in the middle of the brainstem on the way to the cortex for the construction of the core-self[566]. The absence of the brainstem is associated with death and requires constant collaboration with the cerebral cortex.

The conquest of consciousness over the mind and the advancement of the game of intellectual evolution began when the "self" was placed at the center of the mind process. The influence of consciousness began when the "self" encountered external events, living and non-living organisms, stationary and moving, considering the fundamental principle of survival[567]. The structure of consciousness becomes clearer and more pronounced when combined with memory, reasoning, and language. The benefits of mindfulness, which are mainly related to the possibility of examining and evaluating possible futures, can inhibit and control spontaneous responses. The dominance of consciousness in the course of evolution led humans to accept the phenomenon of "deferred satisfaction" or "delayed gratification." In other words, a calculated trade-off for giving up something good in the present for gaining something better in the future or avoiding a negative consequence in the future. This phenomenon has actually laid the foundations for the socio-cultural homeostatic process on top of the physiological homeostasis. This set of processes has made conscious learning relatively slow but stable, and it transfers to the non-conscious mind after learning. Playing ball, writing, reading, and many other activities are part of this context. Conscious intention, in the sense of the ability to make accurate decisions, learns over a long period of time, under the guidance of an authoritative "self," and structured autobiography and clear identity are the basic consequences of the conscious mind.

If we assume that intelligence is the capacity for problem-solving, we should note that comparing the intelligence of two beings that are genetically and environmentally distinct is not practically valid[568]. In other words, each ecological niche requires its own biological intelligence. It should be considered that the cognitive and emotional behavior of each organism arises from the interaction of genetic traits and environmental factors to create a functional, natural, normal, and biological performance. According to John Hughlings Jackson, the highest level of brain efficiency is the most complex, intertwined, and extensive network of sensory-motor coordination[569,570]. His neurophilosophy was based on the relation between mind and brain. According to him, reflexive function, unlike mental function, is not limited to automatic or stereotypical processes but continuously reconstructs itself throughout its life to become an organ called the mind, which is a dynamic and evolving pattern of sensations and responses. An organism in motion, for example, can explore its surrounding world and modify its structures by generating and giving birth to new forces. This evolutionary process occurs consecutively and after obtaining results in previous performances. The new learned behaviors themselves can also be effective forces for change, in which the complex forms and structures of stereotyped behaviors disappear and are replaced by responses based on experience. It should be noted that part of the brain enlargement in evolutionary terms is due to the physical need of the brain to integrate information, and this integration of information is referred to as intelligence[571].

In addition to the mentioned neurological concepts, Leslie Brothers[572] believes that the social networking referred to as "culture" in our minds emerges from the shared activities of different human brains. No one can accomplish this alone or in isolation. In other words, this network is the constituting element of living content in the mind or wisdom, so the mind, by its nature, is a communal phenomenon. Each of us has specific sensory experiences that create a personal image or "self" for us. Over time and through various experiences, our "self" changes. Therefore, our perspective on the world becomes constructed in a particular and unique way. We assimilate all of these to construct our own reality. These observations have fundamental importance in linking the individual and society.

Finally, when we construct the entire network, its components only gain meaning within that network. The internal connections between individuals create an image of the entire society, which can be called a social schema that creates external reality for us in practice. This set of phenomena, on which there is agreement among members of a society, such as social laws or language rules, is established. However, these laws also undergo changes and transformations over time. Ultimately, this set creates a new concept called mimicry. Through this process, units of believes, symbols, or experiences are transmitted from one mind to another with the help of social schemas. Social schemas are preserved in society as a collective entity, transmitted from one generation to another, and expand.

13 Speech and Language

SINGING CREATURE

The emergence of language can be viewed as one of the most significant events in the evolutionary trajectory of humanity. As a cognitive tool, language has played a pivotal role in establishing and fostering both unity and cultural diversity among humans. Its complex and evolved form enables the encoding of thoughts and experiences for communication with fellow beings, and it is practically the point where humans separate from other animals. In the other word, language is not only a tool for communication but also a means for expressing thoughts. The structural foundation of language lies in its signs and symbols, which are called sentences. It seems that humans have an instinctual ability to speak, with its basis and foundation being syntactic structure. Actually, syntax is the ability to combine words according to completely different systems to form a huge number of sentences with distinct messages, a system just works in the human beings. However, scientists have attributed other characteristics as the most prominent human traits. René Descartes considered free will, Thomas Willis considered the ability to laugh, Martin Luther considered the ability to possess relative ownership, and Benjamin Franklin considered the ability to create tools as the most important and prominent human characteristics. Aristotle, on the other hand, regarded humans as political animals. This human characteristic has led to increased socio-cultural growth, the ability to record past events in written form, and the ability for individuals to communicate with one another. However, the formation of such a process in the mind requires multiple infrastructures.

Based on Ethnologue profile which is an encyclopedic reference catalogue for languages, it is estimated that there are 7099 living languages, which can have around 39,000 different names or dialects[573,574]. Most languages are spoken by a small number of people. Approximately 300 languages are used by 90% of the world's population, indicating that the majority of linguistic diversity among contemporary humans is used by around 10% of the population. Languages are constantly changing, making it difficult to define what we mean by "language." A dead language is one that has no speakers and is not spoken by anyone, even if it is recorded or preserved in some way. If we consider languages with over 20,000 speakers as living languages with a low risk of extinction and death, then approximately 600 languages in the world are safe. Based on available evidence, we are only aware of 75 dead languages.

Studying these infrastructures is clearly difficult and complex. In 1860, the Linguistic Society of Paris (Societe de Linguistique de Paris) banned all discussions regarding the origins of language and its evolutionary development[575]. They claimed that there was no reason for animals other than humans to have speech, while considering the ability to speak as exclusive to humans. However, they might have wanted to exclude biologists, doctors, and other researchers apart from linguists from this realm, an endeavor in which they were not successful. Additionally, the church strongly resisted accepting Darwin's theory of evolution at that time[576]. They believed that language was a divine gift. These events coincided with the publication of Darwin's book "*On the origin of Species*" in 1859 and Paul Broca's theory regarding speech disorders caused by brain injuries in 1861. Darwin extensively addressed this issue in his subsequent book, "*The descent of Man*" in 1872. Although Darwin believed in the theory of continuity based on natural selection, he thought that humans were the only beings with a sophisticated and articulate language, while the vocal communication of other animals consisted of inarticulate cries. He also believed that language was an instinctive tendency for humans to communicate. However, studying language is also very difficult and complex due to the lack of fossils that can provide us with evidence. Perhaps the only remaining fossil related to the hyoid bone, located in front of the larynx, indicates that it was larger in Neanderthals

DOI: 10.1201/9781003538400-15

compared to their intelligent counterparts, *sapiens*. In this chapter of the book, we are obliged to discuss and explore these infrastructures separately.

Before that, it is necessary to clarify the difference between two fundamental concepts: communication and language. Dr. John Dolittle's parrot Polynesia reminds him that his lovable and dignified dog, Jip, always asks the doctor questions with his nose movements, and talks by wagging his ears or his tail[577]. Signals are not just audio and can be tactile, visual, chemical, and even in other forms. This method is used not only by human beings but also by other living beings. Many organisms communicate with each other in various ways. For example, plants communicate by releasing chemical substances called pheromones. Some animals use different methods. Dogs, in addition to the smell of urine, use various vocalizations such as barking, growling, and sometimes symbolic visual cues. Velvet monkeys use different alarm calls to alert other monkeys about the presence of predators[578]. These calls are usually monosyllabic and vary depending on the living environment. These calls do not require learning, and if they are placed in sound isolation, they will still be able to produce them. Birds are more related to each other through complex vocal structures. They mostly appear as duets between a pair and sometimes in the form of songs. Unlike cries, songs are multisyllabic and require learning. The main function of songs is to choose a mate, determine the range and territory of life, and establish social connections, as seen in parrots, magpies, and thrushes. Elephants utilize subsonic sounds for communication, which is inaudible to humans. Electric fish use the production of electric current, while compound fish use color changes. Apart from speech, humans also use facial expressions such as smiling, laughing, crying, and body movements like hand gestures for communication. Actually, communication is what we do and language is just a tool to do it. We can convey different messages while using the same language and using similar words or sentences. This is quite popular among politicians.

Although linguists had been working on the structure of language for many years, in 1967, Eric Lenneberg[579] introduced language cognition to the world of biological sciences by writing "*Biological foundations of language*," which featured an introduction by Noam Chomsky. In the 1990s, many scientists such as Steven Pinker, Michael Corballis, Derek Bickerton, Terrence Deacon, Michael Tomasello, Philip Lieberman, Jean Aitcheson, Sue Savage-Rumbaugh, Peter MacNeilage, and many others entered into this field. Language is a fundamental part of human nature, and increased awareness of its impact on culture, politics, personal relationships, and the relationship between dialects and socio-economic levels has significantly augmented their interest in researching it[580].

According to Charles Hockett, American linguist, language is a designed set of features suitable for performing various tasks, some of which are similar to animal communication, while others are specific to human language[581,582]. Hockett's proposed set consists of thirteen different properties, some of which are highly significant. One prominent feature is the ability to combine an unlimited number of words and create different meanings. Thanks to the flexible and abundant nature of language, it allows us to express many things that have not been said or heard before and is suitable for expressing all the thoughts we have in our minds. This descriptive power is facilitated by the openness of language's descriptive capacity and exists in sign language as well, although it lacks music and sound. Sound enables us to easily express many human thoughts[583] and abstract concepts.

Finally, it should be noted that language is a form and tool of communication. Understanding the difference between these two aspects is important in examining the evolutionary process of language. One of the important characteristics of the language is broadcast transmission, directional reception, and rapid fading. The sound fades quickly even where there is a good echo in the caves. Basically, male birds sing to win the hearts of female birds and find a mate; however, both female birds and other males are able to hear it. To establish a non-verbal communication, the female-worker bee dances after finding a flower, but the queen and the male-drones can only understand what happened but cannot dance. Words are created by the mind and only get meaning by convention; it means they are arbitrary words. Even the expression of animal sounds can be dissimilar in different languages. For example, turkey says gobble–gobble in English and glu–glu in Turkish. Finally, language is something learned and it was not engraved in our brain since birth.

However, in practical terms, this concept received significant attention with the publication of the book "*Syntactic structures*" originally published in 1957 by Noam Chomsky[584]. He believed that the sequential predictability of speech follows directly the innate and universal grammatical rules and is not dependent on probabilistic guidelines. It means that everybody has the wired brain system genetically since birth. While he claimed that the study of language naturally relates to the biological structure of humans, he never explicitly stated that language is a product of natural selection. According to Chris Knight, who is a senior research fellow in the Anthropology Department at University College of London, Chomsky remains an anatomist who only focuses on the function of an organ and does not consider how it came into being. Actually, Knight think that his work is a kind of "contribution of baffling incomprehensibility which, on close examination, turns out to make no sense at all"[585]. As indicated by Ray Jackendoff, Chomsky's view of language is based on two concepts: mental grammar and its innate basis. In other words, children deduce the rules of language unconsciously through their inherent ability, which is likely genetically encoded within them[586,587].

The fundamental principle in the evolutionary process of reproductive success is limiting energy consumption. Without it, evolution has no meaning. The sexual behavior of male organisms in caring for offspring and the same behavior of female individuals in close and long-term contact with offspring have led to three essential occurrences that have been influential in the evolution of human language: firstly, cooperation and coordination among adults; secondly, the involvement of mothers and children in learning and entering into the larger groups; and thirdly, the improvement of social relationships for enhancing social intelligence and pragmatic inference. For survival, it was necessary to share what is in one's mind with others. Therefore, intelligent humans utilized language to express thoughts and used speech as a system of transmitting auditory signals. Although these two concepts, language and thoughts, are often used interchangeably, speech is actually a kind of complex vocalization that is created through multifaceted and proper coordination between the throat and the upper vocal tract, which is technically referred to as articulation[588].

Many animals try to solve the problem of survival continuation by increasing the number of offspring, in other words, by increasing the number of cubs. However, this method was not feasible in primates due to the period of fertility cycle. In fact, the reproductive pattern in *hominids* or early humans was the main factor for the emergence of language, although the exact timing of its onset is not clear, but it probably began with the sharing of food, game meat, and animals. This phenomenon may have started around two million years ago, and there are certainly indications that it is more prevalent in animals with sexual dimorphism, meaning animals where males and females have more pronounced physical differences. This type of variation is observed not only in the reproductive system but also in the size of canine teeth and body size.

Another form of language is called pidgins. These languages usually borrow their vocabulary from different languages and do not have complex grammatical structures. They were previously used by traders for business communication. The start of this speech model dates back to the 16th century when people from the new world wanted to communicate with people from the old world. Slavery had started during this period, and there was a need for a common language for communication for slave trading between the two groups. Really, when several groups are forced to live together in different languages, they have to speak in pidgin[589]. The process of constructing such languages is called creolization, and the speakers of this language are called creoles. This linguistic process causes language and culture to emerge in a novel way[590].

THE MIRACLE

Edgar Rice Burroughs, an American writer, depicted Tarzan, a child raised in the jungle, attempting to learn language skills in the first volume of his Tarzan book series, "*Tarzan of the apes*," which he penned in 1912[591]. Although his girlfriend, Jane, at the outset couldn't communicate verbally with him, it seemed that Tarzan was a very clever and intelligent man who quickly learned the language.

The speech impediment in the Tarzan films of the 1930s was easily overcome, and he easily learned to speak different languages. However, in fiction books talking like a human being and communicating with them for animals is not only difficult, but also it seems obvious and inevitable. "*Animal farm*" by George Orwell with its extremely political theme is one of those books.

Monkeys also have a communication system specific to their species. Its basis is a facial gesture system primarily used to convey emotions. Viki, a chimpanzee, had been living with Catherine and Keith Hayes since her birth in 1947, but despite their efforts, she could only learn three words. She was one of the first experiences of teaching language to apes[592]. Among animals, chimpanzees have the closest familial and genetic relationship to humans, but they cannot speak. However, human children learn many words and easily convey many concepts until the age of three. There are many stories about children who have been raised by wolves, bears, or monkeys and initially lack the ability to speak but quickly learn to do so. Examples include Mowgli fictional character of "*The jungle book*" stories written by Rudyard Kipling and *Peter Pan*, a young boy who can fly but not grow up (Half-Bird, Half-Human), written by Scottish novelist J.M. Barrie. However, the reality is different from the world of stories.

Language learning has a critical period. Human children, if not exposed to an environment where other people, especially their parents, speak to them from an early age, will eventually lose the ability to learn language. Real cases have been observed. The most famous ones in 18th and 19th century were the wild girl of Champagne, Marie Le Blanc; Victor the Savage of Aveyron; Peter the wild boy of Hanover; Kaspar Hauser, probably a member of the grand ducal House of Baden;[593] and recently Genie, American feral child[594]. Genie, known as the "wild child," was born in 1957 in Arcadia, California, but due to the violence of her parents, especially her father, she was effectively isolated and imprisoned. When the Los Angeles police found her at the age of thirteen, it became clear that, in addition to physical abuse and isolation, she also lacked the ability to speak. Although Genie is still alive, unfortunately, she has not been able to learn natural speech despite extensive training. Another example is Kaspar Hauser, a German child born in 1812. According to various accounts, he was severely isolated and was first seen in Nuremberg's Hailer Gate, Germany, during the Whitsun Holidays, in the spring of 1828, the same year that Victor, the savage boy of Aveyron was found and captured in the southern forests of France[595]. At first, he was living with the family of the superintendent of prison called Hiltel. During the short time they lived together, 12-year-old, Julius Hiltel, tried to teach him language. Although he had better progress in language learning compared to Genie, his death in 1833 due to a stab wound to the chest ended the process of his education and learning.

Generally, if a child is not exposed to an environment where they can learn to speak until the age of 12 or 13, the likelihood of language acquisition will be very low. Personal experiences of individuals have also shown that learning a second language and be fluent in verbal communication in a language other than the native language is highly dependent on the age of learning. These factors indicate that for individuals to learn a language, they not only need to be in an appropriate environment where other members speak the language, but there is also a critical period for language learning, and missing this period will result in significant difficulties in language acquisition[596,597]. The difficulty of language learning in later years, especially spoken language, supports this point. Furthermore, inability to learn a language does not imply lack of intelligence. For example, John Ssebunya, a Ugandan, took refuge in the jungle after his mother was killed and lived with chimpanzees for about three years. However, when he was found again at the age of six, barely alive, he could only imitate bird songs and monkey warning calls, but couldn't do anything else. Due to his young age when reintegrated into human society compared to Genie and Hauser, he was able to learn French and the local language, Hutu. These findings indicate that the early stages of language learning are crucial. In fact, the critical period of learning is a time window when the growing brain needs a wealth of sensory information to expand its neural connections. This phenomenon is also important for vision. Kittens are generally born blind. If they grow up in darkness, they will definitely be blind in adulthood. In fact, a protein called *Otx2* is transferred from the retina through the optic nerve to the visual cortex and activates basket cells; the reason that we can see[598]. In other words, sensory experience changes the cellular structure in the visual cortex, which is also necessary for human speech ability. In addition to the critical period for vision, establishing a connection between parents and children is necessary from the very beginning. The presence of parents–child bonding is essential for the child to become familiar especially with the voice of the mother and other family members, and it greatly facilitates language learning. In many cases, the period of sexual maturation is also considered a critical period because the organism seeks to enter new social groups.

In general, it seems that animals other than humans do not have the ability to vocalize. Large primates can learn two to three words at most, and even that does not have a specific syntactic structure. Humans can change the meaning of compositional syntactic units, but bird songs can only be changed phonologically, which probably does not create a change in meaning. Although some of them have a special form of call such as "aerial predator" or "flight" calls, they do not have a good anatomical structure in their larynx, mouth, tongue, and lips, and there is no good coordination in the movements of those organs[599]. In addition, the innervation of the larynx in humans is from the motor cortex, but in monkeys, it is from subcortical areas that mostly generate emotional cries rather than clear speech. At the same time, the innervation from the cortex to the larynx in humans is direct, while in monkeys, it is indirect, and of course, it has better and improved function in humans.

Long before, the Royal Prussian Academy of Science was established in 1769, it awarded Johann Gottfried Herder a prize for writing an article on the origin of language[600]. Since then, the main dispute among scientists has been whether language is a human invention that has evolved and emerged over time, or if it is a biological function of the human brain, existing since the beginning of creation. Supporters of the first theory include French philosopher Jean-Jacques Rousseau and Scottish thinker Adam Smith. They believed that language has evolved for expressing feelings and moral values[601]. However, advocates of the second theory, such as Johann Herder and Adam Ferguson, believed that humans are not truly human without the ability to articulate and use language, which is a manifestation of human existence. The fact that we can construct almost infinite sentences with a limited set of letters is one of the remarkable features of language, as Noam Chomsky put it, language is a discrete infinity in which words are discrete with a finite number, and can made unlimited number of sentences. But what is language fundamentally?

According to Michael Arbib, language is most likely a statistical collection of sentence structures and word usage, representing our social framework, defined by shared meanings and a grammatical structure. In other words, in every society, language consists of a dominant statistical grammar, individual meanings, and vocabulary, while recognizing that language never exists in isolation and can undergo changes in each generation[602,603]. Multiple theories have been proposed regarding the process of language acquisition. Probably the earliest known account is found in ancient mythology, Old Testament, the story goes that after the creation of man, Hebrew's God came down from the Tower of Babel (gate of God) and taught different languages to men, fearing that men would be overcome by his almighty power. Whether this creation was merely a form of punishment for humans or a process that brought humans closer to a complete language is still a subject of debate and discussion among religious philosophers. Nonetheless, the teaching of multiple languages to humans has led to a lack of understanding and separation between them, allowing divine power to prevail. The prevailing theory among philosophers such as John Locke and René Descartes in recent centuries is based on the notion that the capacity for speech and language is endowed by the essence of existence, and it does not differ much from the writings of ancient mythology. Linguistic diversity makes the cultural niche of humans more capable and resilient. Essentially, the survival of living organisms depends on their biological diversity. At the same time, each indigenous language signifies the identity of a nation. A Welsh proverb states that a nation without a language is a nation without a heart. Moreover, different languages serve as a historical repository and genealogy of a nation, enriching human knowledge. Without language, nations would lose their history and their past[604].

Alfred Russel Wallace, a rival of Darwin who had conducted extensive research supporting the theory of natural selection, never accepted that language was part of the evolutionary process. He believed that natural selection fundamentally allowed for the creation of such a phenomenon. Along with many other intellectuals of his time, he considered Darwin's notion of language evolution as flawed[605]. Noam Chomsky, in the 1960s, gave a seemingly scientific explanation for the miracle of speech[584] in his book "*Syntactic structures*." He believed that the capacity for language in human thought and cognition was inherently linguistic from the beginning, either in the form of internal language or through gestures. Later, he proposed that an external language structure had emerged or been constructed. In the first version of Chomsky's book, the theory of grammatical structure was constructed in a manner similar to mathematical foundations, in which people could use them to communicate. This model is widely used in the digital world to generate well-formed sentences. Chomsky believed that this phenomenon was entrenched in individuals' biological structures and had deep roots, although later evidence proved its inaccuracy. Nevertheless, new methods were needed to solve the problem. According to Chomsky, language is like the taste of food. Although all races and individuals have the ability to perceive the four primary tastes, their food has distinct flavors. At the same time, he believed that language was more powerful than being merely a means of verbal communication. He considered it an environment for generative and creative thinking. The question arises as to whether this ideal environment for thought has dropped to a lower level of communicative discourse.

Chomsky's model, based on universal grammar, sparked debate and controversy regarding European languages in the 1960s. These languages include a noun phrase, such as "The nice girls," and a verb phrase, such as "like beautiful clothes." However, it was gradually realized that some languages do not follow this pattern, like the Warlpiri language in Australia[606], where grammatical elements are dispersed throughout the entire sentence, and some languages lack a verb phrase. In certain ergative languages like Indo-Aryan languages, which are a branch of the Indo-Iranian languages like Kurdish, Urdu, or Sri Lankan, the subject of the intransitive verb can also be the object of a transitive verb, presenting similar challenges.

The fundamental basis of universal grammar is the pattern of discrete infinity, allowing language elements to blend together, with smaller elements gradually forming larger ones. This phenomenon continues in a recursive manner. In blending, letters combine to form phonemes, phonemes combine to form morphemes, and then, they become words, phrases, and ultimately sentences. It seems that Chomsky's theory is not much different from the opinion of believers in this field that God taught or gifted the inner language to Adam, and later, when God came down from the Tower of Babel, he spread external language among people. Chomsky uses the term "externalization" to refer to this phenomenon. Chomsky believed that language could not evolve through natural selection because the internal symbols of language have no reference to the external world and therefore cannot be selected for evolutionary progression. He also believed that thought and language emerged almost simultaneously and suddenly, possibly through a genetic emergence in the human mind about a 100,000 years ago. Perhaps, through a genetic mutation, in colloquial terms, a kind of rewiring occurred in the brain that has the ability to blend without limits. This phenomenon has given individuals the ability to construct and express infinitely many sentences. In the process, grammar has gradually evolved at a slow pace. Grammar has changed the role of words and made the expression of concepts more prominent and efficient.

The fundamental difference with Chomsky seems to lie not in the conclusion but in how language began and came into existence. We have a mental construct called grammar (language rules) and vocabulary that have been encoded in the brain to understand words and sentences. Words are placed together using grammatical rules to form meaningful sentences in the mind, and then, their meanings are interpreted and analyzed. Grammar allows us to construct messages and sentences from various elements rather than from pre-determined lists. This enables us to express infinitely many new sentences. As the reporters say, "when a dog bites a person" it is not considered news, but when "a person bites a dog" that becomes news[607]. The rearrangement of words in the grammatical structure changes the meaning of the sentence to such an extent that a simple sentence becomes a global news. Grammar allows us to convey news to one another. In this way, language can make numerous structural combinations economical for us. As Stephen Pinker says, the number of sentences a person can construct or comprehend depends directly on the number of words that can be positioned to the power of sentence length. This number is usually around 10^{20} separate sentences for an English language learner[608–610].

Hockett also presents a similar theory to Chomsky's theory. He believes that the design of language structure is based on two principles. Firstly, the decomposition of words in a way that each has a standalone meaning, and secondly, their combinatory pattern that determines their specific meaning. Although longer sentences are analyzable, their analysis is more difficult. Furthermore, in practice and in everyday speech, many sentences do not follow syntactic structure and lack grammatical correctness. In these cases, many sentences are incomplete and sometimes incomprehensible, but they are understood or easily comprehended by the listener. In other words, in such cases, shared understanding of expressing or understanding sentences is helpful and can ultimately shape the conversation and lead to common goals. Hockett also emphasizes the duality of language, meaning that meaningful units of language are formed from meaningless units. For example, there is a limited number of basic sound types (phonemes), but the number of words formed from those sounds is much greater. Take the letter "a" in the words bat, father, eat, and state, which create different words in terms of pronunciation. Additionally, almost all commonly spoken languages have conventionalized meaningful signals that are not derived from the phonemes of that language, such as "tsk-tsk" (noch-noch means no no) used as a sign of disapproval. Although these sounds cannot be accurately transcribed, they carry meaning in a traditional sense[581,582]. Moreover, sign language also exhibits duality in its pattern. There are many meaningless hand movements or shapes that, when combined, create various signs and can vary in intensity among individuals who understand sign language. In fact, human vocabulary is virtually unlimited, and in sign language, this issue is addressed through a combination of arm, hand, and facial movements.

However, recent studies have shown that certain individuals or groups, such as the Piraha and Iatmul tribes, lack the ability to learn other languages. These languages do not possess the property of internal recursion and blending[611]. In many languages, it is possible that there are no prepositions, adjectives, or adverbs, and in others, there may be no clear distinction between verbs and nouns. Overall, these languages do not adhere to the boundary-less blending property proposed by Chomsky. Supporters of Chomsky have introduced another concept called "point balance," which means that after a sudden period of stability, a new species emerges. Usually, this phenomenon occurs as a result of environmental and ecological events and applies to the emergence of new species, not phenomena like language. However, is there a fundamental difference between inherent grammar, which Chomsky discussed, and the inherent ability to acquire grammar? Although Chomsky considered the concept of inherent grammar, it practically did not assist linguists but proved helpful for computer specialists in constructing computer languages. He emphasized the idea of transformation in his book, which was also very helpful in this regard. Meanwhile, Michael Arbib believes that the global grammar structure is merely an umbrella for a diverse set of human languages, rather than a biological or genetic reality[604].

In total, four main hypotheses have engaged linguists, excluding Chomsky's theory. Ralph Holloway[612,613] (1974) considers toolmaking and dominant manual dexterity as prerequisites for the emergence of language. The proximity of speech centers to hand motor centers in the cerebral cortex leads to their shared function. William Calvin[614] (1983) also confirms the effectiveness of the ability to grasp and accurately throw objects, especially during hunting. However, Robin Dunbar[512,615] (1993) has focused on the impact of social relationships. All primates, from prosimians to large apes, live in social groups. Group size is directly related to the size of the cerebral cortex, and they dedicate most of their time in these groups to grooming each other. Based on the size of the cortex, it accounts for about 40% of human waking time, although in practice, it is not more than 20%. Humans attempt to replace whispering, hiding, touching, and two or multiple-person conversations with such time-consuming activities. Conversation contributes to social cohesiveness. Although this theory has emphasized the relationship between cortical size, size of social groups, and grooming time, it does not explain the root of language emergence. Terrence Deacon[616,617] (1997), on the other hand, has considered sexual selection and evaluated it as a factor for symbolic communication. Since approximately two million years ago, human ancestors, like their primate relatives, competed and fought over acquiring females as sexual partners and reproducing with each other, but they were not particularly invested in childcare and raising offspring. Gradual behavioral changes for the protection of spouses and children led to greater efforts in establishing verbal communication. The growth of the brain size from that time transformed human sexual behavior from a multi-male, multi-female pattern to the current sexual dimorphism. Providing sustenance and protecting the family gave rise to discussions of concepts such as assurance and betrayal. The prerequisite for solving such a problem was language and speech.

Perhaps the most acceptable theory was first proposed by Baron Condillac[618] and Jean-Jacques Rousseau[601]. They believed that gestural and signaling language existed before spoken language, with Michael Corballis, a psychology professor at the University of Auckland, New Zealand, being its biggest advocate[89,619–621]. He believes that the most remarkable achievement of humans was the invention of language, which accompanied the process of culture construction along with the freeing of hands and bipedal walking. Genetic mutations like *FoxP*$_2$ have played a significant role in this event. The beginning of this process dates back to 100,000–200,000 years ago with the emergence of proto-language. Chimpanzees, infants, and toddlers can express many of their needs through gestures and pointing. The acceptance of bipedalism had a significant impact on the freedom of hands for sign language. Deaf individuals can also learn sign language. The use of icons, as visual language, is still relevant and often more effective and efficient in conveying concepts. Nevertheless, despite Corballis's commendable efforts, he fails to present compelling evidence for this cultural explosion. We still lack concrete and specific evidence for the transformation of gestural language physiology into the ability to produce sound and speech[622].

However, it is possible that the process of human speech ability may be a form of exaptation, which was previously proposed by Darwin as pre-adaptation. The term exaptation, first used by Stephen Gould and Elisabeth Vrba, refers to a transformation that was initially intended for something else but has undergone significant changes over time[47]. A classic example of exaptation is the presence of feathers in birds, which initially evolved for controlling body temperature but later became crucial for flight. Some argue that the structure of our face, that is arrangement of ears, nose, larynx, nasal shape, and mouth, is also a form of functional change for secondary activity known as speech[623]. Undoubtedly, the ability to speak is derived from coordination and synchronization (idiosyncrasy) among the body parts. While the larynx is primarily an organ for creating a respiratory connection between the mouth and lungs, it is functionally and visibly an organ for producing sound.

Returning to the theory of natural selection, it gained prominence with the publication of the famous article "*natural language and natural selection*" by Steven Pinker and Paul Bloom[624] in 1990. Pinker[610] extensively discussed this issue in his subsequent book, "*The language instinct*." He believed that there are multiple components involved in the formation of language, and it is better to discuss each one separately. They believed that for a complex biological structure to form, it must follow the path of natural selection. This path is a combination of reproductive success in different organisms accompanied by genetic and hereditary diversity, which is the primary driving force of evolution in these organisms. This process naturally occurs slowly and gradually. Sometimes genetic mutations have limited and minimal effects, and it is highly unlikely that this process is caused by a single gene mutation. Theory of mind, mental time travel, laryngeal structure, recursion, mind wandering, generativity, storytelling, memory, and conscious mind are likely to be among the most important factors.

14 Talking Body

RECURSION

In the early 13th century, Leonardo of Pisa, also known as Leonardo Bonacci, one of the most prominent Italian mathematicians of the Middle Ages from the Republic of Pisa, developed a computational method to aid merchants and capitalists in calculating profit and loss[625]. This method later became famous as the Fibonacci sequence. This sequence is formed by adding the two preceding numbers together, such as 1, 1, 2, 3, 5, 8, 13, 21, 34, … . Additionally, he also discovered the golden ratio (the ratio of length to width in rectangular shapes), which is represented by the Latin letter phi (ϕ). This ratio is obtained by dividing two consecutive numbers in the Fibonacci sequence or by adding two consecutive numbers and then dividing it by the larger number. That is defined mathematically and is claimed to have pleasingly harmonious qualities. In a logical term, recursive can mean computable or refer to the process of defining something in terms of something previously defined[626].

Recursion is, in fact, a type of computation that includes similar elements and structures within itself. In other words, recursion is a form of repetition that utilizes its own output for the next input and calls itself[627]. This type of repetition, with its generative capability, can continue indefinitely and utilize concepts of infinite order. However, unfortunately or fortunately, the lifespan of a human is limited. Recursion is a prominent characteristic of human cognition. Some scientists believe that language and mathematics' abilities are a shared capability. Different words in the English language are used to convey this concept, such as "repetition" and "reiteration," but the fundamental difference between them and recursion is the lack of utilizing their own output. In other words, repetition does not use previous information to create new information. Recursion, on the other hand, uses the addition of concepts in speech to construct its structure. In other words, it builds upon the foundations of speech and gains depth. In simple repetition, the output structure is entirely surface-level without any depth. Furthermore, recursion relies on placing an action or object within a previous action or object of the same kind[628]. Each substitution depends on the previous substitution being of the same kind. For example, the sentence "The ball of Mary's cousin's dog is in the garden" demonstrates how the meaning of each word depends on the preceding word. The same sentence in some languages like Persian could be made easily by adding sound "eh" (means "of" in English) to the end of each word and make a very lengthy sentence. Different groups interpret it in a way that is relevant to their own context. For instance, psychologists use it to refer to think about thinking, mental time travel, and theory of mind. The structure of recursion is also crucial in language development and expansion. For example, in English, phrases can be added to the right side of a sentence, while in Persian, they are added to the left side of a sentence. In the sentence, "Nathalie hopes that Mary knows that the school teacher teaches well," two additional phrases have been added at the end of the last sentence. In theory, these additions can continue infinitely, but in practice, limitations exist due to limited human memory capacity. In some languages like Piraha in the Amazon jungle, the structure of recursion simply does not exist[619,629].

Repetition exists in both human and animal life, although it can evoke different meanings or concepts. A famous example of practical application for these two concepts is the method of preparing and cutting a cake. Mixing the cake batter during its preparation is a form of repetition, but it doesn't matter from which point in the bowl we start mixing the batter. However, cutting each portion of the cake depends on the previous cutting point, as the previous cutting point must be determined. Therefore, the former is called repetition, while the latter is called recursion. However, in practice, recursion requires the existence and proper functioning of memory. This memory should remember what the previous output was in order to consider it for the next input. In fact, recursion is a set of

DOI: 10.1201/9781003538400-16

organized principles in all areas of mental activity. It appears that language has gradually evolved with the help of recursion and likely did not emerge suddenly[630]. It can be assumed that language has been shaped and evolved over thousands of years based on daily requirements and cultural interactions throughout various epochs of human history in the past millennia. However, in some languages, due to limitations in verbal communication, recursion may not exist or may not have emerged. Additionally, in these languages, recursive thinking may not have developed[631]. In other words, recursion may be part of a toolbox used to construct language but has not been utilized in all languages. The development of human linguistic ability could be the result of obsessive speaking and conversing or deliberate evolution in the thought system, and ultimately, it has been used to share thoughts[632,633].

Do animals also communicate? It seems that animals have the ability to establish vocal communication, but it cannot be called language. Vocalization is an important way of communication[634]. Auditory learning is observed in many animals, but for vocal learning, it is not sufficient alone and requires the ability to learn syntax. Human intelligence regarding syntactic structure enables us to establish communication between words or parts of words in receiving and producing speech. Probably, animal vocal communication is genetically fixed, controlled by emotional reactions, and is not voluntary[635]. Therefore, its organization occurs in subcortical regions of the brain, the brainstem, and particularly in the midbrain. For example, cats produce various sounds like hissing, growling, screaming, howling, and meowing, and although they establish vocal communication, none of them are considered speech. Damage to the midbrain in cats leads to their silence and inability to vocalize. Similar models can be observed in other animals. In the human model, laughter and coughing, for example, can be mentioned, which are actually a kind of inherent or instinctive vocalization not closely related to speaking. In other studies, on birds, it has been shown that although they vocalize during the learning of specific songs, they do not have the ability to learn new songs[636]. In contrast, human speech is entirely purposeful and intentional, and does not require the learning of new sounds. The ability of animals to vocalize can be derived from the need to defend their territory, find a mate, and mimicry to infiltrate the territory of conspecifics. Monkeys usually have three types of sounds to announce the presence of various predators, including snakes, eagles, and leopards but the primates' sounds depend greatly on the presence of the listener. These listeners are actually their close friends.

Numerous stories and naturally, several scientific studies about the capability of animal communication and speech have been written. The most famous story related to an uncastrated stallion called Clever Hans, who seemed to have the ability to respond to simple numerical calculations, express colors or letters with numbers by tapping his hoof on the ground as suggested by his trainer, Wilhelm von Osten (1838–1909). He believed that animals had a rational and mental learning capacity in the case of education. For example, when the horse was asked that if Tuesday is the eighth day of the month, what day will Friday be? The horse touches the ground 11 times. His fame was such that a New York Times reporter named Edward Heyn went to Berlin and later wrote a one-page article about him. The assessment of thirteen people commissioned by the Berlin Psychological Association failed to prove any deception. However, in 1907, a psychologist named Oscar Pfungst discovered that the horse was only reacting to the trainer's body language cues and did not have the ability to comprehend verbal language[637]. Hugh Lofting has introduced a short legged horse (cob) named Clever Nino with his master Blossom in his famous book "*Doctor Dolittle circus*" published a century ago, which has probably a conception of clever Hans[638]. Other studies, like the study by Keith and Catherine Hayes on a chimpanzee named Viki, showed that she could not learn more than three to four words. Further studies confirmed that the ability to understand words and sounds in animals is more limited than the ability to produce words.

Arguably, one of the most imperative contents of human language is its grammatical structure, which is recursive[639]. Although this type of grammatical structure formation process started million years ago, the past 1–2 million years have been a crucial stage of this change. It seems that the closest equivalent phenomenon to language in non-human primates is the use of hands rather than the vocal system. The former is intentional and goal-oriented, with the ability to learn, while the latter is simply considered as an emotional response, involuntary, fixed, and without changing shape. Other scientific investigations indicate that in training apes for establishing verbal communication, we have been much more successful using hand movements, gestures, and pointing, rather than using auditory and vocal methods. It appears that the body language of animals in their natural environment or in the wild has much fewer limitations compared to vocalization.

BODY LANGUAGE

Bees also communicate with each other. In 1919, the Austrian ethologist (a branch of zoology) named Karl von Frisch (1886–1982), in his studies on the behavior of female worker bees, observed a peculiar behavior when they returned from finding a food source. He termed this behavior the "bees dance." Later, studies showed that the round dance represents finding a food source at a close distance, and the waggle dance represents the distance, direction, and location of the food source in addition to the quality and nature of the food source. Von Frisch wrote his masterpiece about bees dance, "*The dance language and orientation of bees*" in 1967 and won the Nobel Prize in 1973 for his studies on the language of bees[640].

Baron Etienne Bonnot de Condillac, an 18th-century philosopher and clergyman, wrote a story about two abandoned children, a boy and a girl, who never learned any language and were wandering in the desert after a flood. Although he had a great interest in describing the process of language evolution, his position as a clergyman and the influence exerted by the church during that period made him cautious about directly expressing such a matter. Since he couldn't challenge the church's doctrine that language is a divine gift and couldn't base the ability of human speech on the innovative and blasphemous theory of evolution, he attempted to put it into writing in the form of a story or legend. He portrayed the story of a boy and a girl who were abandoned and used gestures and pointing to communicate, gradually adding sounds and voices to it. Initially, their communication was only through twisting body movements, understood only by their companion. However, their speaking ability gradually increased, communication became simpler and easier, and ultimately, language succeeded in coordinating with gestures and pointing. Although it seems to be a mere story, Condillac probably aimed to demonstrate how language as a human knowledge has evolved[618].

Jean-Jacques Rousseau[601], who didn't have the religious constraints of Condillac, wrote an analysis of the theory of gesture and pointing in language in 1782. About a century later, in 1872, Wilhelm Maximilian Wundt[641], the founder of the first experimental psychology laboratory and author of anthropological psychology, wrote a two-volume book on international sign language in Leipzig, Germany. Although he believed that sign language is the root of all languages worldwide, he thought that this language is the same for all deaf individuals around the world. Furthermore, he believed that this language lacks the ability to convey abstract concepts. However, it is now well established that sign language varies among different societies and possesses almost all the complexities of communication. Although McDonald Critchley presented dual theories about the evolution of language and its gestural roots, the most comprehensive theory in modern times was proposed by Gordon Hewes[642], an American anthropologist who obtained his degree from Berkeley, in 1973. It had been discovered before that teaching vocal language to apes is almost impossible, but communication with them through gestures and pointing is simpler and more feasible. It was later demonstrated that sign language is quite similar to spoken and vocal languages, and brain damage can lead to the loss of both abilities in humans. In 1991, Michael Corballis, a psychologist from New Zealand, revisited the theory of sign language in his famous book "*The lopsided ape*" and presented a well-grounded language theory in written form[89]. David Armstrong and William Stokoe[643] approached this perspective through sign language and pursued the concept in their book "*Gesture and the nature of language*" in 1995. In 2002, Corballis[621] compiled a comprehensive article entitled "*From mouth to hand*" and elaborated on the principles of transforming sign language into spoken speech.

While many verbal concepts have their roots in hand movements, such as understanding (grasp my point), emphasis and description on the subject (press a point), mention something (point to things), and getting stuck on ideas (seize upon ideas), other models have also been proposed for the evolutionary development of language. In Philip Lieberman's model[644], emphasis is placed on the function of rule-based clusters in the brain, while MacNeilage's model emphasizes the transition from oral movements used for eating to those used for speech[645]. Terrence Deacon's model focuses on the growth of the prefrontal cortex and the improvement of motor pathway control from the motor cortex to the larynx[616]. Sensory fibers travel from the amygdala to the anterior cingulate cortex and then to the PeriAqueductal Gray (PAG) matter. Uwe Jurgens[646] and Hans Kuypers[647] neuroanatomical studies have illustrated that motor fibers originating from the speech motor center (Broca's area) in the brain's cortex extend to the nucleus ambiguus in the brainstem, where motor neurons governing laryngeal movements reside. These pathways have been meticulously traced[623]. This connection does not exist directly in feline (cat) family or non-human primate species and may be seen as entirely indirect and polysynaptic. In chimpanzees, the connection to the nucleus ambiguus exists only for lip movements. Although damage to the motor cortex in human results in severe speech impairment, similar lesions in monkeys eliminate vocalizations. Sensory fibers also connect from the PAG to motor fibers in the ambiguus nucleus. In summary, it is analyzed that the presence of direct motor fibers to the brainstem plays a fundamental role in human speech abilities. Deacon believes that in the evolutionary process of motor fibers, the system of instinctual cries and vocalizations, such as laughter and crying, has lost the control of laryngeal movements to the motor fibers that originate from the brain's cortex in humans.

Mirror neurons first were discovered by chance in monkeys in 1992 and given its name four years later[648,649]. In the meantime, the discovery of mirror neurons for psychology by neuroscientists, especially Giacomo Rizzolatti in 1999 in Parma, Italy, which was akin to the discovery of DNA for cellular replication[650], greatly contributed to advancing this theory[651,652]. These neurons, located in area F5 (the fifth area of the frontal lobe in macaque monkeys, equivalent to the human motor cortex B44, B45), are stimulated by visual stimuli, such as observing the act of grasping food, and subsequently become active in performing the same action. In other words, those neurons are triggered by visual stimuli (see something) and possess the capability to perform actions. In fact, the motor response is the mirror image of sensory reception in those neurons. However, it should also be noted that observing an action must be simultaneous with observing an object for the mirror neurons to discharge or fire electrically, and be stimulated. This is often the case in nature; if an object does not exist, an image must be present in working memory. In other words, they transform visual information into knowledge of what others intend to do. These neurons are typically convergent and respond to a very specific stimulus, such as precise grasping between two fingers or a general stimulus like grasping something with a fist. They essentially possess a population code for a specific action. This code is not a simple yes/no response; rather each neuron has a level of confidence, and the higher the visual stimulation, the greater and stronger the discharge. Therefore, these neurons are figuratively referred to as "monkey sees, monkey does." This theory that mirror neurons have played a significant role in the evolutionary process of language was initially proposed by Giacomo Rizzolatti and Michael Arbib[651].

From an evolutionary point of view, at first mirror cells evolved to be able to control the feedback of our actions, which was necessary for learning. Their involvement in social interaction, gestural tasks, and various sensory-motor activities, which are vital for human vocalization, has been of secondary importance. Actually, mirror cells are specifically designed for certain functions[653]. Approximately 40% of neurons respond to visual stimuli for observing various actions such as looking, placing, reaching, and grasping. Fifteen percent of these neurons in the F5 region respond to audiovisual stimuli, such as tearing paper or cracking peanuts. One third of these neurons in this area is discharged by oral-facial stimulation and are called mouth mirror cells, which are activated by sucking, breaking, swallowing, and other digestive movements, and more than all, they are activated with gestures and oral movements to communicate verbally. These actions include lip smacking, sucking, protruding lips, lip grasping, chewing, mouth grasping, tongue protrusion, teeth grinding, and tongue touching. However, it should be noted that not all neurons in the F5 area are mirror neurons. Mirror neurons mostly located in the crown area and other neurons or canonical neurons that discharge only during the execution of a manual action surround the rest of the F5 area. The influence of these neurons on each other and their final output is not well-defined. Mirror neurons must have the capacity and ability to learn in order to recognize objects or to design movements that ultimately lead to the execution of the action. This phenomenon has likely arisen in the evolutionary process of the human brain due to what we call the plasticity of neurons. In other words, these neurons possess the property of being trainable, and conventional neurons in the same area may be involved in this process.

In summary, mirror cells are neurons that become active both by performing certain specific actions ad by continuously observing a group of actions. These neurons are called "potential mirror neurons" before being trained and "pseudo-mirror neurons" if they have the ability to respond and react to certain movements. Audiovisual mirror neurons can react simultaneously to sound and image. Sound causes neurons to be trained in a way that by hearing the sound, they mentally observe and execute actions such as cracking peanuts or tearing paper. In other words, hearing a sound can infer the consequence of an action[653].

However, the evolution and development of mirror neurons were not solely for the improvement of social behavior. It seems that simultaneously with the upsurge in unilateral and asymmetric performance of humans and the use of the dominant hand in this system, a feedback improvement process has evolved in the hand-object communication. This evolutionary process is based on a circuitry and flow that is effective for understanding and comprehension, essentially a practical combination of movement and goal achievement. Mirror neurons can represent and perform our expectations for reaching a specific goal. It is the expectation that determines the size and movements of the hand. Of course, this type of movement depends on the viewer's capacity and movement reserve, and the more it is, the more and more intense the reaction of the mirror neurons, and of course, its effect is reflected in the movement system of the observer[654].

The mirror system also enables us to have the ability to empathize with others, not solely because we comprehend and recognize those emotions, but because we felt them and undergone those experiences ourselves[653]. This issue relates to the concept of simulation theory, which suggests that these neurons have the ability to read the minds of others. In other words, they can simulate them in our own minds and activate the mental process that exists in someone else's mind within our own brain. However, the recognition of emotions is not necessarily synonymous with empathy, especially if we feel threatened by witnessing someone else's emotions. The ability to perceive visual and auditory details of others is also a factor in understanding empathy. Motivation and emotions create a blueprint of ourselves, which forms an encyclopedia-like collection in the mind, allowing us to interact with the surrounding world[653]. Another important aspect of mirror neurons is their role in understanding meaning of motor events. They not only comprehend but also their output dynamically activates a complex sensory-motor network, such as when a person thinks of doing something and ultimately carries it out.

Having the awareness and understanding of movement, the mirror neuron system has the ability to comprehend the minds of others[655]. In fact, the human brain can understand the type of action through sensory stimulation. This system has the ability to interpret an observed action in relation to a performed action. It is much more familiar and better stimulated by familiar movements such as lip reading or lip biting, and of course, it works with less intensity and weaker with certain movements like dog barking, and to some extent, it is controlled and regulated in a mirror-like system.

In many social situations, varying degrees of mirroring are necessary. However, the absence of complete mirroring processes can be associated with phenomena such as repetitive use of words, repetitive and purposeless actions, and sometimes autism. Although autistic patients can mimic facial expressions, they are practically devoid of empathy and feelings. Nevertheless, training and learning can enhance the mirror neuron's abilities for various actions. Overall, the functioning of mirror neurons involves self-action recognition, understanding others' actions, connection to the motor system, interaction and engagement with the individual's motor experience, and ultimately learning in the evolution and development of the mentioned aspects[653,655].

In the evolutionary process of vocalization, it appears that the essence of language has become ingrained within the neurons in the relevant brain area. This phenomenon may elucidate why language often accompanies sound, such as spoken language, and also be non-vocal and manual, such as sign language[656]. Mirror neurons are actually a larger part of a neural network, and in humans, they include motor (Broca) and sensory (Wernicke) areas of language, as well as the superior posterior (upper-back) region of the temporal lobe and some other regions. Rizzolatti and his colleagues[651] believe that the mirror system is actually a system for understanding and comprehending the actions of others, which can then perform those actions itself. This concept is the fundamental principle known as "the motor theory of language perception."

As mentioned, mirror neurons in monkeys are activated by observing movements, such as approaching an object, resulting in an electrical discharge in the neuron. However, in humans, merely mimicking that behavior, that is an action without the presence of an object, can also cause cellular discharge. This phenomenon paves the way that makes the concept of "referred action" understandable. This phenomenon is the main feature in understanding body language and is called "action displacement" due to "torso displacement"[657]. In other words, these neurons are activated not only by observing hand or mouth movements in the premotor cortex but also by reading phrases about movements of the limbs. Human hands, like monkey hands, have adapted extensively to life in trees, not only for grabbing branches and hanging onto them but also for picking fruits and even opening shells. Meanwhile, three-dimensional and color vision has significantly contributed to this phenomenon. It seems that although voluntary control over vocal production and speech has recently emerged, it has arisen precisely for this purpose. It appears that a portion of the motor system fibers (pyramidal tract) is used to control the movements of the subcortical speech centers so that humans can express what is in their minds. For monkeys, vocal production is primarily and solely based on emotional states, and they lack the ability to produce intentional sound and speech. Babbling is one of the similarities among humans, monkeys, and birds. Infants begin babbling around ten months of age without the need for feedback from others.

Babbling is likely an initial stage for controlling the vocalization apparatus and may be a prerequisite for complex imitation for vocalization, enabling control over the vocalization system. At this stage, by linking the auditory system with the motor control of the vocal system, a vocal imitator can create the necessary coherence and similarity between input and output for imitation. In fact, babbling allows for the adjustment and generation of a trajectory for imitating different signals for future life. New sounds are gradually added to old sounds and are then understood by hearing. This ability requires a strong memory and the ability to perceive its components, namely, syllables. Vocal learning ability is prominently seen in humans, and animals such as seals, dolphins, African elephants, bats, and some birds like passerines, hummingbirds, parrots, and songbirds have acquired this ability to some degree in their evolutionary process[658]. Non-human primates can overcome this limitation to some extent by controlling their larynx. From this, we can conclude that we, as humans, have the ability to understand the speech of others due to the way they express themselves, and the dominant sound pattern has little effect on it[659].

In the 1950s, Alvin Liberman and his colleagues[660,661] at the Haskins Laboratories in the United States attempted to create devices that could decode sounds, separate their elements, and then print them, in order to solve the problem of typist shortages. However, this endeavor did not yield successful results. Studies showed that basic units of speech sounds, such as the letter "b" in different words, have different acoustic properties. There is a distinction in understanding the letter "b" in words like "bunny" and "rabbit" compared to "cab." The feature that defines the vowels in words like "beet," "bat," and "boot" is not a random phenomenon, and it was created during the evolution of speaking. The position of a word in a sentence and the position of a letter within a word affect its articulation. For example, the letter "b" in "cab" in the sentence "call a cab" is generally not pronounced clearly. This phenomenon makes it challenging to program devices capable of understanding human speech. This observation applies particularly to explosive consonants, including the letters D, G, P, T, and K, which demonstrate clearer and more salient characteristics. In fact, many explosive consonants may appear to be pronounced similarly but have different auditory perceptions for human ears. In conclusion, we perceive sounds that are produced through similar methods involving the larynx, mouth, and lips as similar, regardless of their acoustic–physical properties. In other words, modern humans comprehend speech sounds based on how they are produced and seen, rather than based on a speech pattern. The well-known McGurk effect experiment by Harry McGurk[662,663] demonstrated that if you place a different sound on the movements of the mouth while articulating a sound, what is heard corresponds to what is seen. For instance, if the test subject pronounces the letter "B" but the sound of "P" is played over it, the listeners perceive the sound of "B" because that is what they see. This study shows the effect of lip and mouth movements on speech understanding, which represents a very strong illusion between what we see and what we hear. This study indicates the influence of lip and mouth movements on speech comprehension. Deaf individuals are highly skilled in lip reading. In normal people, lip reading is accompanied by the stimulation of central auditory neurons. Natural individuals also utilize lip reading to understand the words of their interlocutors when they are unable to hear sounds, such as in noisy environments or at a distance. Lip reading appears to be a speech modality dependent on lip movements. In other words, the mirror neuron system is designed in such a way that when a movement is observed, it tells or instructs how to execute it. This system is based on visual inputs and the production of sounds based on a performed action. In monkeys, mirror neurons respond to sounds resulting from certain actions, such as the tearing of paper, but unlike humans, they do not respond to sound production and vocalization. This ability, which is the merging of speech sounds with the functional output of mirror neurons, was added to the evolutionary repertoire over time, leading to the linguistic abilities of modern humans.

The subsequent point highlighted is that in non-human primates, vocalization is frequently spontaneous and driven by emotions, rather than being intentional and necessitating. In contrast, vocalization in the mirror system is based on clear intention and will allow for change, improvement, and learning. Animal studies in monkeys have shown that mirror neurons respond to actions such as tearing paper and breaking nuts, and these actions can be learned. However, these neurons do not respond to monkey sounds and calls. In other words, these neurons are specifically programmed to respond to sounds processed manually rather than vocally[664]. Although animals do not have the ability to speak, it seems that the mirror system has created a natural framework for language to ride on and undergo development and evolution. Additionally, in monkeys, mirror cells only respond to so-called transfer actions, such as reaching toward an object, and do not respond to non-transfer actions like pantomime or imitative movements. However, in humans, mirror cells respond to both types of actions. Responding to pantomime actions in the second group prepares the neural pathway to understand symbolic actions. Functional magnetic resonance imaging (fMRI) studies in the human premotor cortex have shown that cells in this area are not only activated by observing different movements but also by reading concepts associated with them. It is likely that mirror neurons have become interested in language somewhere in between.

Unlike its ape relatives, the early human walked on two feet, had free hands, and had the ability to gesture and display with their hands. Infants learn pointing from the age of one but start learning words from the age of two. It appears that finger pointing is a prerequisite for sign language[583,665]. The freedom of hand and body movement in three-dimensional space, along with the influence of language, allowed the early human to imitate the shapes, animals, and movements of objects with their hands and fingers in three-dimensional space. Gradually, they also acquired the ability to express emotional reactions through facial expressions.

Bonobos have more flexibility in displaying gestures compared to chimpanzees. They have a better ability to combine facial and manual communication signals to convey messages. Bonobos, gorillas, chimpanzees, and orangutans each have a large number of specific gestures for conveying messages, which can indicate the performance of one or several different actions[666,667]. In human language, in addition to a large number of imperative sentences, which are particularly useful in military environments, there are also a large number of descriptive sentences for sharing opinions and messages. Sign language and gestures also perform similar functions during childhood.

Early humans, although primarily walking on two legs, had not yet lost their old habits of living in trees. The use of tree fruits and occasionally finding a safe place to sleep away from the stress and danger of dangerous creatures were effective factors in tree-dwelling life. *Ardipithecus* (*Ardi*) separated from chimpanzees and bonobos about five million years ago. They had adapted to walking on two legs. Although their legs were still used for grabbing branches, at that time, *Ardi* was still considered a type of tree-dwelling creature. Bipedalism in that period was a type of limb position that allowed maneuvering in the forest canopy. Meanwhile, chimpanzees and gorillas, after the separation from *Ardi*, remained knuckle walkers, which hindered their progress toward hand usage. It can be considered that *Ardi* used their fingers as objects to demonstrate animal movements and simultaneously used facial expressions to convey emotional states. These methods are still used in ballet and other performances. In this process, gestures, similar to icons, were symbols resembling environmental and emotional conditions, enabling our ancestors to effectively communicate with each other. However, they were not yet proficient in walking on two legs, lacked sufficient skills, and were slow. They were referred to as "optional or facultative bipeds," meaning that walking on two legs was a choice, not a necessity or compulsion[668,669].

It seems that until the era of upright posture in humans (*H. erectus*), the culture of using facial expressions with emojis had not been established, and after that, approximately two million years ago, "obligate bipedalism" began. This process of predominantly free walking on two legs, along with an increase in brain size and the liberation of hands, brought about better and more extensive communication abilities for them. This phenomenon was probably a precursor to the emergence of grammar. In studies using fMRI, it has been shown that human mirror neurons respond to pantomime movements and symbols, such as the sign for silence, but do not respond to verbal commands. Pantomime is a form of free communication that is based on complex imitations and tries to attract the attention of the recipient by using a series of gestures similar to normal everyday behaviors. However, pantomime is an energy-consuming behavior, sometimes not very clear, obvious, or understandable, and in some cases, a single pantomime can express different and diverse concepts. Sign language is a natural successor to pantomime. Gesture movements involve the same brain areas that encompass spoken language. These areas are mostly associated with changing meanings or related symbols. These symbols include words, images, sounds, gestures, and objects. In sign language, some signs are conventional, and many of them are iconic. Even with structural damage to the brain due to various reasons such as strokes and inability to move facial muscles and/or limbs, many of the performative abilities do not disappear[670,671].

Sign language, unlike spoken language, is based on graphic patterns to represent the external world. Hands and arms create the surrounding world, and signals, in parallel and simultaneous form, convey concepts in a highly flexible and not overly dry manner. Gradually, sign language movements have been accompanied by spatial–temporal mapping between the gesture and its meaning. In many cases, sign languages use manual or bodily iconic indices[672]. For example, in Italian Sign Language, up to two-thirds of the language consists of manual signs. Similar numbers and values exist for British and American Sign Language. To increase communication speed, improve functionality, and overcome grammatical limitations, gradually conventional indices were used, becoming an informal and common phenomenon. Many gestures and signs occur simultaneously with speech, meaning that the modality of gesture and speech are semantically connected and occur synchronously in time, blending together. In fact, they originate from an integrated system, and the most prominent parts of speech and gesture usually manifest themselves together, either at the beginning of speech or in a prominent speech intonation. According to the theory of dynamic motor system, the system that activates the mouth and hands dynamically and flexibly influences each other, resulting in synchronization[673]. To acquire such manual abilities, children need at least to be one and a half years old.

According to the Swiss linguist Ferdinand de Saussure, the main characteristic and identifier of sign language is its conventional nature. However, why did language shift from gestural and manual references to speech? The F5 region of the brain in monkeys and its equivalent, B44, in humans are considered the motor areas associated with speech, requiring mouth movements and eating in humans. It seems that the hand grasping movements are coordinated with mouth movements. Scientific studies have shown a direct correlation between the degree of hand and mouth opening during simultaneous movements to grasp objects. Most animals communicate through sounds; they have the ability to produce sounds but cannot speak. It is likely that gesture and speech share a communicative phenomenon. In fact, there is a significant synchrony between gesture and speech. Speech itself is a type of gesture that involves movements of the lips, mouth, tongue, and even hands. We perceive hand gestures with the help of emitted light and our eyes, while we recognize vocal gestures through sound and our ears[674].

In addition to that, language is also dependent on memory. Visual memory, which works very well in humans, and auditory memory, which has improved significantly in terms of evolutionary processes for remembering auditory patterns associated with gestures or meaningful content. To achieve this improvement in auditory memory, specific anatomical changes in the evolutionary course of brain structure have been necessary. Two individuals named George Hickok and David Poeppel[675,676] proposed the theory of two pathways for speech perception. However, the description of the brain's anatomical structure comprising two fundamental pathways was first presented in a research study by Mishkin and Ungerleider[677,678] in 1982. The ventral pathway extends from the inferior (lower) temporal gyrus and insula to the inferior frontal gyrus and is responsible for speech comprehension[679]. It is likely that similar pathways exist in animals such as monkeys and dogs, which obey human commands. The dorsal (upper side) pathway, anatomically situated above the ventral pathway, gradually strengthened in humans and extends from the temporal lobe to the parietal and then to the inferior frontal regions. This pathway is primarily responsible for speech production[680]. In fact, this system has gone through the evolutionary process of transforming movement into voice communication; it is connected to the meaning and concept of verbs and helps to express and articulate words by using the pattern or scheme that the previous path draws. The ventral pathway has succeeded in producing meaning from sound. In *Austras,* which are similar to non-human primates, they mostly focused on the ventral pathway to comprehend vocalization and relate it to visual signals associated with mouth and facial gestures. The second component is not formed in animals such as monkeys and dogs, and it is likely a contributing factor to their inability to speak[681].

Actually, hand and face movements are understood in the visual cortex which is projected to both pathways. The dorsal (posterior or upper) pathway, which is mainly for spatial information and used to understand sensory–visual perception and convert it into action, is called the "*where system*" and the ventral (anterior or lower) path is called the "*what system*." The "where system" is related to habitual actions and is associated with hand movements that physically interact with objects or other beings, such as shooing flies away with hand movements. However, the second pathway, the "what system," is related to communicative actions and object recognition. In this case, the hands interact with other beings but carry the concept of communication, such as saying goodbye using hand movements. Humans have learned in their evolutionary process to transfer both visual and auditory information to the motor system. However, in non-human primates, only visual sensory inputs are transferred, which is why their ability to speak is very limited. The transfer of auditory inputs to the speech-related motor centers has evolved explicitly in intelligent humans. Paul Bach-y-Rita, American neuroscientist, in 1960s showed that it is possible to teach a blind person to see images with sound or tactile stimulation. In fact, he used to activate the above two paths with blind training. In other words, Braille alphabet is a form of training to see, using the sense of touch for the blind. These changes can be seen not only in the cerebral cortex, but also in the thalamus, which is the relay center for sensory messages[682,683].

Both the visual and auditory systems are involved in analyzing where and what the path is. The visual system is effective in analyzing by not only distinguishing certain matters but also in practical implementation based on visual findings. The auditory system is also important in understanding the spatial aspects of the path in three dimensions. The collection of visual and auditory pathways ultimately reaches the frontal lobe. Additionally, it has been identified that the anterior cingulate cortex in monkeys is involved in the initiation of voluntary vocalization. It appears that speech perception has become possible through the existence of a system with high resolution or accuracy for decoding auditory codes, accompanied by effective memory mechanisms and the ability for abstraction. Gradually, it has expanded and developed on the prefrontal cortex and established itself. The connection between posterior and anterior pathways to speech areas in humans is much better and more complete than in monkeys. The pathways in the human model are bidirectional and, concurrently using the visual, auditory, and motor cortex, not only to strengthen it but also to lateralize it greatly.

In the early 19th century, the concept of phrenologists (skull readers) was that the speech or language area was located behind the eyeball. For about 150 years, it was believed that the language structure in the brain has an output and an input center. The output center for spoken words was described in 1861 by Paul Broca, in the lower frontal cortex under the third convolution, in a patient named Louis Victor Leborgne[334,684]. About 15 years later, a 26-year-old German physician named Carl Wernicke (1848–1905) identified the input center for spoken words in the upper posterior part of the temporal lobe[685,686]. Damage to this area caused the individual to have the ability to speak, but the grammatical structure in their speech was severely impaired. The patient sometimes used self-made words, and occasionally, the severity of the disease was such that the patient's speech became incomprehensible. The patient's lack of understanding of others' speech led them to suddenly feel that the outside world was strange to them, and they might exhibit abnormal behavior. He termed this type of speech disorder aphasia, specifically identifying it as sensory aphasia, wherein the input system or words heard are incomprehensible. This stands in contrast to Broca's or motor aphasia, where understanding spoken words remains possible, yet the patient struggles with articulation and speech. He considered the second type of aphasia as a motor disorder in speech. Wernicke believed that a bundle named "fibrae propriae" running in the deep white matter of the brain connects the sensory and motor parts; and depending on the fibers that are damaged, the sentences the patient utters are incomprehensible, although they have the ability to understand spoken words[687].

Ten years later, in 1895, Ludwig Lichtheim[688] (1845–1928), a German physician, expanded the concept of speech disorders by connecting the sensory and motor aspects and developed a mechanistic understanding of brain function. Later, it was referred to as "conduction aphasia"[689,690]. Over the past century, the conceptual foundations of conduction aphasia have been established based on the idea that the sensory system is connected to the motor cortex through pathway called the arcuate fasciculus[691]. These fibers were first reported approximately 200 years ago by the German physician Johann Christian Reil (1759–1813). Just behind the Wernicke's area, there is the angular gyrus, which is directly connected to the visual (occipital) cortex, and its damage impairs a person's ability to write. Although the concept of the arcuate fibers is generally correct, there are still many details and ambiguities that have not been discovered. In the early 20th century, it became evident that a set of interconnected regions in the brain contribute to the higher cortical functions. Subsequently, the concept of split-brain became clearer; the left hemisphere predominates in language and analytical thinking, while the right hemisphere is responsible for emotional processing, prosody, and intonation. In the 1960s, Norman Geschwind[685] at Harvard University reintroduced the concept of communication pathways in specialized cortical centers to neuroscience. According to Fitch, initially, these pathways were believed to be involved in sound imitation but gradually their function changed, and they are now used for more complex tasks such as syntactic structure and semantic interpretation. However, there are other communication pathways that connect the visual, auditory, speech, and motor cortices to each other[623,692]. The main foundations of this theory are based on three concepts. First, the *phoneme*, which is the smallest meaningful sound unit that can distinguish two words (meaningful signs) from each other, such as "right" and "light" in English. Second, the *viseme*, which is the visual equivalent of the phoneme in the visual structure, although it has received less attention. However, lip movements for different words may be the same, and visemes can be similar. The third important unit is the *articuleme*, which is a type of movement unit that is produced and heard following the movements of the oral articulatory apparatus, phonation device, and its associated effect. Visual input carries the potential for guessing auditory input. Separating sounds, especially when auditory processing is difficult, can be very helpful. After that, the auditory information is transferred to the Broca's motor area (B44). This connection between visemes, phonemes, and articulemes is considered the cause of what is called the initial words (pre-words) or proto-lexicon and ultimately plays a role in the production of a type of initial language or proto-language. This type of language actually has essential elements of language structure in its evolutionary process but still has some deficiencies and has not been fully integrated into the linguistic system. This concept was first introduced by anthropologist Gordon Hewes[642] in 1973, besides that, he explained the concept of transformation of gestures and turning them into spoken language. However, Darwin had previously introduced the concept of musical language in 1871, without having semantic ability, which has recently gained attention from linguists. It is not clear whether this initial speech was about finding food or territorial range or even love. However, what is known is that this initial language has greatly assisted in the transfer of non-verbal general knowledge to meaningful communication.

Derek Bickerton, an English born linguist, defines proto-language as a type of verbal communication based on words without having grammatical structure. Unlike Darwin's theory, he believes that the mutation in information transfer through language is indicative of a change in type rather than degree, due to its high complexity. In comparison with other animals, he does not consider language merely as a means of communication but as a system for representing our mental thoughts and concepts through our sensory perception of the surrounding world[693,694]. This perspective has created a suitable cognitive domain for studying and researching language. Primates generally have the ability to produce proto-concepts, namely, nouns and verbs, which are the fundamental basis for forming sentences. In this perspective, language is a secondary and context-specific consequence for presenting concepts. Children under the age of two learn language in a stereotypical manner, starting with babbling and gradually attempting to convey concepts to the listener by putting together a set of words. This is essentially a form of proto-language. Even wolf-children like Genie and Kaspar, who have normal intelligence but have grown up in an environment deprived of language instruction, establish their own verbal communication in a similar way.

In the past, merchants, especially slave traders, used a collection of words called pidgins to communicate with sellers who didn't speak English language. It is a kind of inter-language or basic variety of language, structurally and lexically very restricted, and is not normally spoken[695] as a first language. This form of language never advanced beyond a pre-language stage in adults, but in children, it manifests as a complete language with the ability to generate grammar, known as creoles, and can be compared to other languages. Interestingly, all pidgins have emerged from speakers at various levels.

Broca's area, situated in the lower part of the frontal lobe, consists primarily of four regions, with the most notable ones being areas 44 and 45 as classified by Brodmann. Its active connection for controlling eye movements and limbs, its clear influence on sign language, and its communication pathway with motor neurons related to the larynx indicate its importance in speech production. In the human speech model, pantomimic gestures have likely transformed into oral imitation, which has further transformed into vocalized sounds, ultimately resulting in speech. However, the evolution of the dorsal–posterior pathway has caused a shift in gestural modality, and the use of the ventral pathway has facilitated quicker initiation of speech production after enabling auditory comprehension. The posterior pathway not only has undergone significant and fundamental evolution in humans but also plays a crucial role in the ability to adapt and learn vocalization. It simplifies repetition by mapping phonological perceptions to motor representations. A similar and homolog pathway exists in birds that sing and chirp. In humans, the posterior pathway has ultimately adapted for sound production and is responsible for converting sound into speech. In other words, facial and manual gestures are vocalized. In essence, the posterior system is a kind of mirror system that not only receives but also participates in the execution of actions. This action is effective in modulating and calibrating speech sounds, allowing us to articulate sounds as we hear them. This is probably why infants initially start speaking with babbling, and according to George Hickok, this behavior remains like tuning, where they continuously adjust and correct themselves based on the correct sound they hear[676,696]. A non-musician can only hear but cannot properly tune the instrument for the appropriate sound. Another structure that has developed in the human evolutionary path is related to the motor control of the larynx, which is primarily used for "learned vocalization." Damage to this region results in the loss of vocalization ability. A similar center is not observed in other primates. This pathway does not seem to be related to instinctive vocalizations such as crying, laughing, or shouting. The first pathway is a distinct neural pathway that is not seen in primate vocalizations and has evolved approximately seven times stronger and more advanced in humans. Various neural nuclei in the brainstem have also evolved for this purpose.

Sign language is not only dependent on hand movements but also on facial expressions. Head and facial movements can convey emotions and can therefore change the direction of a sentence. In sign language, the listener, that is viewer, focuses more on the facial expressions of the signer than their hand movements[697]. This ensures that the connection between the listener and the signer is not ambiguous and allows for better understanding. In other words, head and facial movements are highly effective in conveying the tone of speech. Sign language is essentially a powerful representation of a bridge between manual actions and language with an open structure. It should be noted that sign language is a type of language based on hand movements, but the gestures and postures are only accompanied by spoken language. There are reports indicating that some deaf individuals who have experienced a stroke in the speech motor (Broca's) area have difficulty and disruption in establishing communication using sign language, which is referred to as asymbolia. This phenomenon indicates the existence of a central control hub in the brain for interpersonal communication, encompassing both verbal and gestural forms. Gestures can be indicative, such as pointing with a finger toward an object, or they can be symbolic representations of three-dimensional spatial concepts or traditional cultural gestures, such as raising the index finger for asking a question or the thumb for showing positivity or approving something. Gestures are also observed in blind individuals who have been blind from birth. Gestures express our thoughts in a certain way. In many cases, different gestures are used to emphasize what has been said when speaking on the phone or in darkness, even without the listener being able to see the gestures.

According to Corballis, perhaps the initial part of the process of transforming hand movements into speech involved an increase in the use of facial movements. However, with improvement in speech ability, facial movements have become less prominent. Generally, language comprehension is not solely based on auditory information, and as mentioned before, visual input is highly influential in this matter. Infants probably have the ability to differentiate between their mother tongue and non-native language from around four months of age. At six months of age, they start producing repetitive syllables, known as babbling[620,621]. At the same time, they begin to orient their eyes toward the mouth when listening to speech. Around the first year of life, they start saying their first words, and during this time, the eyes of children are directed from the mouth toward the face and eyes of the speaker. However, the exact reason for this is not well understood. Is it for establishing and improving social communication or optimizing language skills? In fact, looking at the speaker's lips to initiate speech articulation movements is obliging and helps create a sensory-motor model for producing words. Vision also serves as a supplementary source of information for improving speech perception, especially in crowded places where the ability to hear sounds is reduced or when the speaker and listener are physically distant from each other. This phenomenon is also observed in elderly individuals who have experienced a reduction in their hearing abilities and rely on hearing aids, as well as congenitally deaf individuals. Furthermore, individuals with hearing loss from birth pay more attention to lip movements during conversations with others. Nonetheless, for listeners, language comprehension, other than their mother tongue, will be easier through the observation of lip movements of the speaker. The use of masks by speakers in the aforementioned situations can pose a significant challenge for listeners.

Gradually, human beings were able to add sound to their breathing process by vibrating vocal cords in the larynx, named "voice." In fact, it is the sound energy resulting from the airflow through the respiratory system from the larynx. The airflow, passing through a channel vibrating the vocal cords, is transformed into sound, which is the human sound source. The speed of vibration of the vocal cords determines the pitch of the sound, i.e., the fundamental frequency. The vibration of the sound in the throat, nasal cavity, and mouth causes a secondary frequency to be placed on it, which is called "formant frequency." This frequency is influenced by the lips, mouth, tongue, and cavities[698]. Actually, the formant frequency is the result of input sound energy emitted from the larynx and the output frequency of the filtering function of the vocal tract. So, by manipulating the size and position of the lips, tongue, jaws, and vellum and the length of vocal tract, we can make different harmonies. Therefore, there are essentially two types of frequencies: the fundamental frequency associated with the larynx and the formant frequency linked with the other parts of the vocal apparatus. The fundamental frequency determines the pitch, while the formant represents the timbre (precise distribution of relative strength of the harmonics of the fundamental frequency) of the sound. In the other words, timbre or tone is something that allows us to distinguish the unique sound qualities of different musical instruments[699]. The structure of the vocal tract in chimpanzees cannot produce vowels and sounds like humans because of the vellum position and tongue anatomy. In the course of evolution, humans crossed this barrier by changing the location of the vellum of the tongue.

In the production of sound, airflow during exhalation is essential for many living creatures and is produced in very few instances during inhalation. The inhalation process is an active and energy-consuming process necessary for life, while exhalation is a passive activity that occurs without energy consumption and is caused by the elasticity of the chest wall and related muscles. The most significant advantage of vocalization during exhalation is its energy efficiency, meaning it is almost entirely free and has been highly valuable in the evolutionary development of language. Therefore, one can speak throughout the day without consuming much energy. Of course, a person has to use energy for a very loud and operatic voice and movements of the jaw, tongue, and mouth. Crying in infants, barking in dogs, and in some cases, human laughter occurs during inhalation. In birds, frogs, dogs, crocodiles, and some mammals, there is a syrinx instead of the larynx, located in the chest cavity. Sound production or phonation occurs when air passes through the larynx and the vocal folds contract that is come close together due to their myo-elasticity. From a neurological standpoint, it is also passive without using too much energy[700]. In other words, we do not need electrical discharge of nerve fibers to contract and bring the vocal cords close together. The fundamental frequency, produced by the vibration of the vocal cords, can give rise to overtones, which are harmonics occurring at multiples of the fundamental frequency. Essentially, it acts like a filter, but the main pitch of the sound corresponds to the fundamental frequency. The separation of the sound source and the filtering system is a characteristic observed in vertebrates. However, in some wind instruments like the clarinet and trumpet, the filter and sound source are paired. Why didn't coupling occur in living organisms? The exact reason is not clear, but it is likely because they provide different, separate, and controllable information, enhancing the ability to articulate. The perception of the listener is related to the same pitch. These overtones may not be audible but are related to a source. The primary tonicity of the sound depends on the length of the vocal cords, the tension (degree of stretching) on them, and the tissue density. Generally, in males, the fundamental frequency is around 100 Hz, it means deeper voice (bass), while in females, it is around 200 Hz which shows a sharper tone or treble voice[644].

Hormonal changes during adolescence have an impact. This type of sexual dimorphism is a human characteristic and is not typically observed in animals. This phenomenon allows for the identification of the speaker's gender based on their voice. Increasing tension on the vocal cords leads to a lower pitch while augmenting the density of vocal cord tissue causes a reduction in frequency, resulting in a deeper voice. Elevated male hormones are accompanied by a decrease in tension and an increase in density of the vocal cords. Both of these factors reduce the fundamental frequency and make the voice deeper. Although there is no scientific evidence or established correlation between the fundamental frequency and body size, it is astonishing how people can accurately guess body size based on pitch and frequency modulation. Perhaps this misconception arises from a type of evolutionary inertia or an incorrect but learned experience.

When sound enters the supra-glottal space, the only fundamental factor affecting the formant frequency is the length of the vocal tract, which extends from the larynx to the lips. The longer the path, the lower the formant frequency, and the shorter the path, the higher the formant frequency, resulting in a sharper tone. The formant frequency has no effect on voice modulation, but it contributes to a more feminine, higher-pitched voice. This set collaborates with the lips, tongue, jaw, and soft palate, which are hardly connected to the organized work of twenty-five swallowing muscle. Five cranial nerves, which likely formed about 200 million years ago, assist these muscles. In fact, humans do not have any special organs for speaking. Lungs, respiratory tracts, mouth, and tongue are each made for a specific task of respiration but they have changed in the course of evolution. For example, the larynx is descended and the tongue, unlike many primates, which is wide and short, is relatively long, round, and perpendicular to the upper part of respiratory tract. The human vocalization system evolved from the ancestral aquatics and probably underwent multiple stages of exaptation. Another point to note is that the control of the upper vocalization pathway, including the skin of the nasal area and lips, is continuously linked in non-human primates, so their ability for facial and oral movements in animals like dogs and cats is limited, although their social interactions are facilitated by such movements. Over time, tongue movements and other speech organs were added, including the lips, soft palate, larynx, torso, sides, and root of the tongue. With the addition of tongue movements, facial movements diminished. With the changing vocalization pathway and improved control of language over new sounds such as vowels, consonants, and plosives, speech developed as a form of gesture and indication, not just sound[701].

Based on Corballis view on language evolution, speech is a kind of gesture that is half swallowed. Speech sounds, along with speech-related gestures and indications, are mapped in the brain. Although different gestures have their own symbolic responses, speech in the real world consists of sounds that rarely have a clear connection to what they represent. In other words, the arbitrariness or coloring of sounds is a personal sign of ownership in language. Due to this very characteristic, sign language is believed to be unrecognized as a real language, despite having a solid symbolic foundation. Sometimes words can be associated with objects. In a study, Wolfgang Kohler[702,703] presented two circles, one smooth and the other jagged, along with two new and unfamiliar names, "maluma" and "takete," to participants. The study revealed that 95% of individuals chose the first name for the smooth circle. That was actually observed by Dimitri Uznadze a Georgian psychologist and reported in 1924. Then Kohler[704] referred to Uznadze's experiment in a 1929 in his book "*Gestalt psychology*." Additionally, words can also be related to the shape of the mouth. For example, words like "you," "tu," "vous," or "me," "moh," and "noan" have mouth shapes that synchronize in different languages.

The sounds of letters also have an impact. For instance, labial consonants are often bilabial, and nasal sounds come out through the nose. Raising the tone and intensity of the voice is used to indicate height while reducing it is used to indicate a decrease in height. Sometimes the speed of an object is compared to the speed of speech, such as comparing it to the speed of a galloping horse. Some words are also formed based on their sound, such as "mew mew" or "quack quack," which is called "onomatopoeia." The theory of onomatopoeia was first introduced by Johann Gottfried von Herder (1744–1803), who studied under Immanuel Kant at the University of Königsberg. He presented the theory of onomatopoeia in his essay "*The Origin of Language*," which contradicted the Bible. His theory is now known as the "Bow-Wow" theory[600,705]. He attempted to find a logical and non-religious interpretation for the roots of words, but this theory did not explain how the connection between sound and meaning is established or how it is understood by listeners. However, he himself acknowledged this problem and knew that many names were not derived from animal sounds and were constructed completely careless to the animal sounds. He was forced to introduce the concept of "pre-language" as a bridge between human-like and modern human languages. Some scientists have searched for the roots of words in internal and instinctive cries of humans and refer to it as the "pooh-pooh" theory. The fundamental problem is that although there is a wide variety of sounds produced by mammals, they do not have a specific language or specific meaning. Furthermore, many patients who lose their ability to speak still retain the ability to laugh and cry, which are emotional-instinctive behaviors.

In this context, a devout Christian professor at the University of Oxford named Friedrich Max Muller (1823–1900), an eminent philologist, strongly emphasized the extensive connection between mental concepts, thoughts, and language. He believed that God has entrusted language to the human soul and he thought that the structure of language influences the speaker's worldview[706]. In other words, we speak in a way that reflects how we perceive the world. For example, the number of colors in different languages ranges from two to twelve, and in fact, in language, by limiting cognitive processes, influences our cultural reality; for example, we use gender-specific flower names for girls. Although Max Muller's theories are highly questionable, they still have a reputation among the general public. The survival of this theory is probably due to the validity of the belief that language affects thinking or certain types of thinking. In 1928, an English baronet named Richard Paget who specialized in speech and sign language, put forward this theory that sound symbolism is based on hand gestures. In his view, the movements of the tongue and lips likely imitate movements of other organs, such as hands and arms. He reiterated Darwin's claim that the distinction between gesture and speech is based on intensity, not type. However, the theory of gestural language learning was first introduced by Gordon Hewes in 1973 and later developed by other researchers[642].

The concept of symbol and schema was first introduced by eminent neurologists, Henry Head and Gordon Holmes in 1911, published in the *Brain* journal under the title "*The Body Schema*"[143,707–709]. As per their assertion, anything that contributes to conscious body movements is added as a model to ourselves. Later, Frederic Bartlett, British psychologist, introduced this term into psychology and considered it as an active organization of reactions or past experiences, in which the character of structural memory is of great importance. For the first time, he explained the concept of conventionalization[710] in psychology in his most famous book "*Remembering*" published in 1932. Both sign language and speech phenomena have elements of physical description and selected conventional symbols. This phenomenon of transformation and conversion from physical elements to speech phenomena is called conventionalization, and a gradually dominant descriptive system evolves into the second model. However, the new system maintains efficiency and ease of operation and is transmitted through culture, not power and body movements. It appears that emojis still play a role in growth in addition to language formation. Over time, gesture and sign language have lost their visual–iconic properties and become more conventional, while everyday speech has become much more conventionalized than sign language, making it incomprehensible to many people. This phenomenon has mainly occurred due to the conventionalization of language. Gesture and sign language may have evolved along with involuntary cries, but our ancestors gradually gained deliberate control and intention over these cries. These gestures were gradually refined, improved, and language gradually became dominant, leading to long-term coordination.

Gradually, the empty spaces between vocal gestures and hand gestures have been filled with facial movements. Facial expressions date back to the early stages of evolution to convey social signals and signs. The larger the body size, the more extensive and effective facial expressions become, such as orangutans and chimpanzees showing their teeth to enhance social cohesion or grimacing to indicate interest in grooming. Monkeys have different facial movements that are different in different groups, such as pursed lips and grinning. Although these movements may appear similar, they convey different meanings. Many facial manifestations represent happiness, joy, sadness, or anger.

These are emotional reactions that interfere with human speech, such as when you laugh or giggle. According to a hypothesis, facial expressions indicate the presence or absence of confidence. Many gamblers attempt to suppress these facial expressions while playing[621].

In primates, the motor cortex responsible for mouth and facial movements also involves varying degrees of voluntary control, that's why facial expressions of primates cannot necessarily be reliable. Facial movements are also a major part of sign language, so the listener looks at the face in addition to the hands. These intentional communications have evolved from hand-to-mouth and facial expressions. Head and facial movements form an essential part of our everyday conversations. For example, we nod our heads downward to indicate agreement, upward to indicate disagreement, wrinkle the face and forehead for disapproval, and raise our eyebrows to show surprise. Children establish connections with each other through playful facial expressions like sticking out their tongues or pulling faces. Additionally, non-volitional emotional behaviors like laughter and crying also rely on facial and oral movements to convey their messages[620,711].

As Corballis says, there is only one small step left from the face to the mouth. In accordance with him, speech is a kind of facial gesture that is half swallowed. Many of these movements occur in the mouth, but since they are not visible to the listener, sound has taken their place in the course of evolution. Therefore, humans have tried to perform vocal cord movements voluntarily in a way that the resulting sound is audible and intelligible[711]. The throat, tongue, soft palate, and lips are also involved, but it is still the lips that greatly contribute to auditory understanding. In other words, lip-reading has enabled people to hear what they see. As previously mentioned, the test of Harry McGurk and John MacDonald[662,663] supports this point. In their video test, a video of a person mouthing "ba-ba" is recorded, but the sound "pa-pa" is dubbed over it. However, most of the listeners perceive the sound they see, which is "ba-ba."

Laughter is a process that can be seen in some organisms to unite and stabilize the group. In monkeys, it can be observed during inhalation and exhalation, while in humans, it is primarily observed during exhalation. Perhaps laughter was a prerequisite for speech. Although genuine laughter is involuntary, voluntary control of laughter has been traced back to modern humans around three million years ago, marking the beginning of speech. The next stage involves the transfer of hand movements to the mouth. These two processes are strongly interdependent. In the motor cortex of the frontal lobe, neurons associated with hand movements are located alongside neurons associated with the mouth. A study by Maurizio Gentilucci[712] revealed that clenching the fists is accompanied by the activation of neurons responsible for opening the mouth. The more intense the clenching of the fists, the wider the mouth opens, and this phenomenon likely has a positive impact on speech articulation. In summary, as an object becomes larger, the pressure on the fingers to grasp it increases, causing the mouth to open wider and the produced sound to be louder for speech. This process of transforming and transferring movements from the body to the mouth is referred to as "miniaturization" similar to transforming large and bulky electronic devices like computers into smaller ones like cell phones[713]. Vocal language conserves a significant amount of energy, but sign language, due to the use of hands, requires more energy consumption. Speaking simultaneously occurs during the exhalation phase, which consumes significantly less energy compared to the inhalation phase. That's why most people talk incessantly. In the inhalation phase, the abdominal muscles, diaphragm, and chest are involved to create negative pressure in the lungs, while in the exhalation phase, these muscles are at rest and without tension, requiring no energy consumption. It serves as a good way to fill the swim bladder of a fish to enable it to swim effectively. However, this evolutionary change in humans has led to the development of human lung air sacs, and negative pressure in the pleura is vital. If water, other fluids, or air accumulate in the pleura, the lungs collapse, resulting in death[711].

After the hands were freed for walking on two feet, the liberation of hands from being used for sign language became part of the agenda. Here, the second revolution, the speech one, was taking place. The Devil had found a new job for them, which was creating tools, especially weapons for survival, writing to record spoken words, and drawing to preserve morale. The liberation of hands increased the ability to perform delicate movements, and overall, all the actions that had been effective in the technological revolution took place. The ability to speak and the blossoming of language also led to the liberation of the eyes. There was no longer a need to use the eyes as a receiving agent for learning and understanding sign language. Light, too, had no other practical application as a transmission agent. In other words, sign language was impossible to use in darkness. However, any obstacle in the path of light between the speaker and the listener hindered communication; and if, for any reason, the recipient was visually impaired, sign language was ineffective. Sound was sufficient for transmitting audio in darkness. Using the three-dimensional space created by the two ears made it possible to locate the speaker's abode accurately. By increasing the volume of sound without approaching the person, one could have an impact on the others. Nevertheless, at the same time, the values of sign language were in such a way that they made it difficult to answer the question of why sign language turned into spoken language. Sign language was well learned by early humans after the discovery of fire and had practical and fundamental applications in hunting. With this method, hunters could effectively communicate with each other before attacking the prey, including attention, precision, and understanding of the noise emitted by the prey. Furthermore, they could seamlessly coordinate tasks, allowing them to work with one hand while communicating with the other. Additionally, they could work with one hand and communicate with the other. However, the descendent of the larynx was accompanied by an increased risk of choking during swallowing. But according to Hughes, it seemed that the verbal capacity and vocabulary of modern humans gradually became limited by using symbols, emoticons, and sign language. The ability to learn, teach, and transfer cognitive and abstract concepts did not exist well with sign language. It seems that sign language had reached its full capacity at the end of the Paleolithic period, and perhaps this limitation of capacity was one of the very significant factors in transforming sign language into spoken one.

A broad vocabulary works well when we have a good system for organizing word files or listing words. It seems that phonemes in spoken language function similarly to alphabet letters in written language and enhance the ability to search for words. As specified by Arbib and Rizzolatti, the foundations of syntactic structure for spoken language have been established in the pre-language period[651,714]. Haldane believes that if there has been approximately a one percent improvement from sign language to spoken language in each generation, it would have taken about 4000 generations, or approximately 100,000 years, for a complete transition from sign to spoken language. Clearly, the speed of these changes has increased over time. Spoken language has had a significant impact on facilitating human communication and continues to do so. However, we still use sign language in our verbal communications. This process is effective for emphasis in conversations and better understanding of opinions, but when speaking on the phone or listening to the radio, only the sound remains. It is likely that in the evolutionary process of using sign language and limbs, it has become synchronous with our spoken language. Over the past 50,000 years, spoken language has gradually established its commanding position in interpersonal communication, but at the same time, this process has led to a unidirectional and linear nature of human communication, unlike sign language[604].

Respiratory issues also needed to be solved simultaneously. Perhaps humans transitioned from an aquatic to a coastal life. It was suggested that the control of respiratory movements during breath-holding and the control of lip and mouth movements began simultaneously in a long-term evolutionary process from this period. The larynx has also undergone its own evolutionary process. Studies by Victor Negus, an English surgeon during a 20-year period (1929–1949), have shown that the structure of the human larynx began with a sphincter ring. When fish dive into water, the closure of the sphincter ring prevents water from entering the airways. Frogs have two curtains in front of their larynx and close them when diving into water, producing sounds when expelling air periodically outside of the water[715,716]. Humans do not have the ability to close the pathway of the larynx.

The term "phonation" refers to the vocal source of acoustic energy. By modifying the rate, speed, and frequency of air, the created sound can be changed in meaning[717]. Many animals send signals related to the presence of a predator to their friends in this way. Humans can express their emotions by changing their voice frequency. For example, in tonal languages like Chinese, Thai, Igbo, or Vietnamese, the meaning of words changes by altering the frequency of the emitted air, expressing their emotions. The initial frequency of the sound produced by the larynx, known as the fundamental frequency, filters acoustic energy when exiting the upper airways of the larynx to produce maximum energy in a localized manner, which is called the formant. The formant plays a fundamental role in creating and determining voiced (vowels) and voiceless sounds and affects the resonance of the voice, speech rate, and information transfer. During rest, the interval between inhaling and exhaling is approximately two seconds. However, during speaking, the exhalation time is controlled by the contraction of the diaphragm, chest, and throat muscles. Therefore, unconsciously and involuntarily, the brain performs this task to ensure a consistent frequency of spoken words. Typically, uttered sentences have a constant frequency from start to finish, and for long sentences, a long exhalation and, of course, a long inhalation are necessary. Additionally, frequency changes in tonal languages like Mandarin Chinese are accompanied by changes in meaning and emotion. For example, the word "ma" in Chinese has four meanings. The most influential muscles in this programming and speech control are the respiratory muscles of the chest and abdomen. The diaphragm muscle has a few muscle fibers that can act as sensors, but they have less impact. All of these are indicators of evolution in our respiratory system.

At the same time, in infants, the tongue is flat and attached to the throat, so when sucking milk, the throat rises, and the nasal–pharyngeal airway connect with each other, creating a separate airway, allowing the child to simultaneously suckle and breathe[718]. In other words, at the beginning of human birth, like other mammals, they have two separate paths in the neck, one for breathing and the other for drinking. In fact, the tongue is anchored in the back of the mouth, and naturally, during swallowing, the epiglottis covers the larynx and swallowing occurs. Otherwise, during respiration, the epiglottis aligns with the nasal space and rests softly in alignment, forming the respiratory path consisting of the nasal cavities, larynx, pharynx that is a large cavity above larynx, and lungs. Naturally and obligatorily, infants breathe through their noses; their brains are programmed this way, and they generally do not have any experience of breathing through the mouth. This complex process allows them to develop their speech system until the age of eight. The elongation of the neck, changes in the shape of the skull and face, the horizontal positioning of the tongue in the upper part of the vocal apparatus, the vertical position of its root in the posterior and pharyngeal region, the descent of the larynx, and the growth of teeth are the most important phenomena in this process. Transforming the tongue into a horizontal part in the mouth and a vertical part in the throat is highly effective in producing vowels such as "u," "a," and "i." The process of correctly articulating vowels practically takes until the age of eight. In infants, the larynx is located at the top of the neck, and while breastfeeding, the infant can simultaneously breathe, thanks to the direct connection between the nostrils and the nasopharynx, while the swallowing pathway through the esophagus remains completely separate. However, as the larynx moves downward with age, it creates a common space in the oral and pharyngeal region, a strange and uncommon phenomenon in humans that distinguishes them clearly from other mammals[717]. Although the low position of the larynx is exceptional for humans, certain species such as deer, antelopes, and large cats like lions and tigers also have this feature. This morphological change has allowed the vocal tract to assume more shapes, resulting in a broader spectrum of produced sounds compared to other mammals, and its vocal effect is present in the presence of large-bodied animals. This speech-related change is also observed in the human larynx, which descends for the second time but only in adolescent boys, causing their voices to resemble the roaring of deer. This phenomenon significantly increases the risk of choking on food. However, this descent resulted in the positioning of the larynx in a vertical orientation relative to the mouth, which is crucial for the initiation and ability of speech. In this position, humans were able to produce or mimic a complete range of vowel sounds. Presumably, this situation initially enabled early humans to produce sounds similar to goats or sheep. Overall, the main function of the descended larynx is to increase the diversity of vowels and produce low-frequency formant sounds. Additionally, since the production of formant frequency depends on the length of the vocal tract, which is proportional to the size of the head, and the size of the head is proportional to the body size, one can infer precise indices of the size of a living organism based on its formant frequency. Animals' capability to detect formant frequencies allows them to identify the species and size of others, even in darkness. This phenomenon contributes to the development of specific evolutionary adaptations for animals.

In practice, there are three ways to reduce formant frequency and make the voice deeper; lengthen the nose, protrude the lips, and lower the larynx[118]. It has a special function for animals, especially adult males, to express the boundaries of their territory to other animals and prevent their encroachment. In other words, this frequency can indicate body size to other animals and be effective in preventing trespassing into their territory. Thus, the practical function of the descent of the larynx initially served to indicate the size and stature of an animal but gradually became utilized for sound production and acquired a linguistic function. Therefore, it can be concluded that the descent of the larynx in *Austras* initially served to indicate body size but gradually facilitated speech production and played a role in their evolutionary development. Another common misconception is that the descent of the larynx is related to the human ability to walk upright. However, many animals, such as gibbons, spider monkeys, and kangaroos, although they assume an upright position and occasionally move on two feet, do not experience a descent of the larynx. What seems to be more significant is the shortening and flattening of the face, which can be a good reason for the simultaneous descent of the larynx. Almost no bipedal animal, except humans, has a short and flat face[692].

The shape of the face is also considered important in this regard. Unlike other primates and even Neanderthals, the human face is smooth, thanks to the reduction in the size of the sphenoid (butterfly-shaped) bone in anatomically modern humans compared to Neanderthals. As a result, the size of the vertical and horizontal parts of the vocal tract becomes equal, enabling better and more diverse speech sounds[719]. It seems that in the theory of natural selection, violence is not necessarily present; instead, many creatures try to help each other in a way that benefits both of them. This phenomenon is called "recycling," which is a combination of modifying and redirecting an organ in a living organism to serve a new purpose. For example, the relatively incomplete design of the respiratory system in humans is a result of recycling the air sac in fish. The ability to use language for communication is also a form of recycling, although it has increased the risk of choking from food getting stuck in the throat[720].

The skull has taken on a rounder shape due to the enlargement of the temporal (auditory) and frontal (forehead) lobes. This phenomenon can facilitate the processing of speech. However, genetic mutations have played a significant role in this process[721]. Perhaps one of the most important mutations is related to the *Foxhead Box Protein* P_2 ($FoxP_2$) gene. Structural genes are responsible for producing proteins and enzymes, but the function of these genes is regulated by regulatory genes. Many of these genes are similar to each other and exist in different species. If a gene is missing in one species, a similar gene can be transferred from another species. For example, the *pax-6* gene is a regulatory gene for eye development and is found in mice, squid, humans, and flies. If this gene is missing in mice, it can be taken from flies and transferred to mice, and because they have a shared evolutionary history, it will have the same function in mice. The next question is how the blending of imitation and reference occurs with mirror neurons? In the early 1990s, doctors encountered a family known as the KE family in Great Britain, some of whom had a speech disorder from childhood that continued into adulthood. The main problem in using muscles, which was related to regular, fluent, and eloquent expression of words and having admirable speech. It was resulted in slow, difficult, contradictory, unclear, and sometimes incomprehensible speech. Initially, Hurst and colleagues[722] reported this speech disorder in three generations of a family. They had a severe speech disorder. However, its transmission in different generations indicated its genetic nature. Later, Mirna Gopnik, a linguistics professor at McGill University, proposed a theory of grammatical disorders resulting from genetic problems. Her theory[723,724] became known as the "grammar gene." In 2001, the gene related to the disease was identified in another unrelated patient, which was associated with the defective $FoxP_2$ protein on chromosome 7. Next studies have shown that the $FoxP_2$ gene differs between Neanderthals and modern humans. The primary function of this gene is the integration of the vocalization system with the mirror system. The function of both gene alleles is necessary, and the result is the expansion of speech abilities into a voluntary and learnable system. The presence of this gene in healthy individuals enables word generation and ultimately stimulates the Broca's speech area in the cerebral cortex. However, this phenomenon is not observed in patients with dysfunction in this gene. Disruption of the gene in the human model can lead to developmental verbal dyslexia, cognitive impairment, and orofacial motor disturbances.

The connection between *$FoxP_2$*, the KE family, and speech impairment was first identified by Cecilia Lai[725] and her colleagues in 2001. She believed that the mutation of the gene leads to disturbances in the anatomical structure and growth of the speech and language-related neural system during the embryonic period. The *$FoxP_2$* mutation likely occurred between 100,000 and 400,000 years ago in *sapiens* as two changes in the exon of chromosome 7. In birds, damage to this gene results in the loss of their ability to sing, while in mice, it causes a similar phenomenon with changes in motor learning. This gene is primarily observed in mammals. Wolfgang Enard's studies in 2002 demonstrated that the gene structure in humans differs from mice at three positions and from chimpanzees collectively at two positions. This change in humans likely occurred between 38,000 and 45,000 years ago, although some sources also refer to it happening 100,000 years ago[726]. While the freeing of hands for object manipulation and toolmaking was known since Darwin's time, this event allowed humans to point to objects and demonstrate them with hand movements while speaking. The ability to use language to describe toolmaking or talk about other matters while performing tasks or constructing tools led to the dominance of modern humans on Earth. Neanderthals disappeared, and the evolution of modernity strongly separated modern humans from their counterparts.

The key point is that a gene does not directly affect behavior and brain function; rather, this effect occurs through the production of different proteins. Secondly, a gene can play various roles in the brain and body, a phenomenon known as pleiotropy in technical terms. The gene *$FoxP_2$* is related to vertebrates from an evolutionary perspective and is involved in the development and function of many neural circuits in the cerebral cortex, basal ganglia, and cerebellum. Changes in two amino acids of this gene in the human model, when separated from the last common ancestor, likely contributed to the emergence of speech and language. This gene is observed in other animals, especially songbirds, and its disruption leads to the loss of their vocal imitation abilities. Damage to this gene in mice results in a lessening in their motor skills. Alterations or loss of nucleotide sequences of this gene in humans lead to a reduction in the timing and sequencing of oral-facial movements, impairing the fine-tuning of mouth movements necessary for language acquisition, specifically for speaking. Furthermore, *$FoxP_2$* regulates a gene network and affects neural growth and brain development, thus directly or indirectly influencing the plasticity of neuronal connections. These changes indirectly impact the coordination abilities of speech organs[727]. In the human and monkey models, *$FoxP_2$* differs by only two amino acids, while it differs by three amino acids in the mouse model. In the Neanderthal model, two key mutations have occurred compared to modern humans. Perhaps the mutation in *$FoxP_2$* can be viewed as the Chomskyan theory of a "big bang." It is likely that other genetic mutations have also occurred in this regard. In the past 20 years, numerous genes have been reported to be involved in speech disorders, but two famous cases are related to *Slit-Robo*, a cellular signaling protein for axon growth and their crossing to the other side with the help of *Robo* receptors on the cellular membrane. Another protein is related to the *$SRGAP_2$* gene, which has been duplicated twice in the human genome and its function is the pruning of dendrites in the prefrontal cortex. It seems that our ancestors had at least three major mutations in these genes and, at the same time, were equipped with the ability to learn complex sounds. These changes have led to remarkable control over respiratory and body movement. Certain specific genes have caused a reduction in the size of temporal muscles or have become vestigial. Lip smacking in ancient monkeys is considered as facial mimicry movements observed almost always in face-to-face interactions of monkeys[728].

The change in eye position in humans has also been very influential. Eye movement from the sides of the head and its placement in the front of the face has been a fundamental aide to language. Binocular vision and three-dimensional perception have emerged as a result. The combination of these factors has had a direct impact on the better understanding of limb, lip, and facial movements of the speaker. Additionally, the eyes have transitioned from a round shape with colored sclera seen in monkeys to an elongated egg-like shape with white sclera. Intense dark colored iris, among the whiteness of the eye, has made visual contact much better and clearer. Our practical experience has shown that sometimes communication and message transmission between two individuals, that is, speaker–listener, can be accomplished solely through direct gaze without words.

The transformation of fish ear cartilage into lower jaw is also indicative of a very important evolutionary phenomenon. The first jawless fish emerged approximately 450 million years ago during the Silurian geological period. Some fish found a new function for the ancestral fish's earliest ear. The muscles associated with the fish's ear became larger to be able to move it. This is how the movable jaw developed. The addition of teeth to the jaws led to the emergence of large and skilled predators. Teeth and jaws have been fundamental features in the evolution of vertebrates, especially mammals, for survival. The rest of them could perform their primary function, but one of them, which performed the specific task of capturing and processing food by slicing and chewing it, transformed into a jaw. Later on, the jaw, mouth, and teeth played a special role in speech. According to Peter MacNeilage, the oscillatory movements of the teeth, aided by the lips, tongue, and jaw, played a significant role in chewing and licking movements in mammals and transformed it into vowels and consonants. He believes that speech sounds have been formed from a framework called syllabic structure, and its content consists of vowels and consonants. He considers babbling as a dominant syllabic frame in which an infant tries to produce suitable syllables such as "ba ba" and "da da" to establish verbal communication using jaw movements for chewing, licking, biting, and swallowing[645,729]. Early primates gradually introduced these movements into their social communication domain. Approximately 370 million years ago, during the Devonian geological period, the first forests formed and covered the Earth's surface. Gradually, as lakes and water bodies dried up, animals transitioned from living underwater to breathing air on the land. Some of them had both gills and lungs, which they used according to the conditions and circumstances. Lungs were helpful in conditions of lack of oxygen in water, warm air, and shallow water but hunger made them turn from water to land and adapt to aridness instead of water.

The auditory system of mammals, with a slight difference in the middle ear, is similar to fish. The ear canal in fish is filled with a liquid that resembles water in terms of density, allowing sound to easily transmit to their ears in water. When we enter the air, this system no longer works because the sound bounces back when it hits the water in the closed chamber. To solve this problem in the course of evolution, the tympanic membrane and stapes bone were developed in the middle ear. Those ossicles were derived from the remnants of the hyoid bone. However, the quadrupeds didn't just listen to the sound of flying insects and the movements of their prey but they also needed to listen to each other and hear each other voices. Therefore, after evolution, ears that were sensitive to sounds transmitted through the air needed a sound production system: the larynx. Lungfish has a type of valve similar to a sphincter that separates the path of compressed air from the water. The passage of compressed air through this valve is accompanied by sounds such as screams, hisses, and groaning. The larynx system of lungfish is completely analogous and homologous to the human larynx. Since the laryngeal cartilage never fossilizes, it is not possible to determine a specific time for the emergence of a two-part larynx in quadrupeds. This process, which acts as a protective valve for the lungs, has remained essentially the same from the beginning with some minor changes. Liberman and his colleagues believe that the larynx descended very late in evolutionary history. This event likely occurred during the period of *sapiens* and before its descent, it probably had degrees of formant frequency production. Thus, it is possible that Neanderthals had a type of language encoded in the form of syllables. In other words, the speaker encoded the units of sound to create syllables, and the listener decoded them[118,730]. In summary, the descent of the larynx not only increased the human phonetic reserve but also improved the ability of encoding and decoding. This phenomenon results in a larger volume of information being transmitted to the auditory system.

The thickening of the hypoglossal (12th cranial) nerve and spinal cord in the cervical region has probably contributed to the motor abilities of respiratory muscles and the fundamental role of the tongue. This nerve is responsible for innervation of all, except one, intrinsic and extrinsic muscles related to the activity of the tongue. This surge in the ability to use language, which is critical for the production of spoken language, was occurred in H. ergaster/erectus (upright man). Prior to this biological period, the hypoglossal nerve was thinner and more similar to that of large primates. There are also other studies showed that there is not any correlation between the thickness of hypoglossal nerve and human speech[731]. This phenomenon, which began six to one million years ago, led to an increase in the ability to create airflow for vocalization in the larynx. Finally, let's remember that language is the final product of the conventionalization of the encoded speech syllables. It should be noted that this process separates voice communication from praxis which truly is motor action. The liberation of hands has allowed individuals to easily manipulate tools, as well as simultaneously have the ability to teach toolmaking to their friends and benefit from speech during the process. Technology has advanced concurrently with speech ability. Conventional words were used in language, leading to increased social dependency and improved interpersonal relationships. Self-verbalization has been the primary factor in human progress.

Overall, a common language has been the fundamental basis of a nation. Thus, each social group strives to keep itself distant from the pressure of foreign cultures and languages. Unlike sign language, the lack of symbolic elements in spoken language has reduced its vulnerability to penetration by other cultures. Furthermore, the diversity of sound-based speech systems has led to language acquiring the character of a fortress, which is very difficult to breach. This phenomenon has caused human groups to remain distant from each other while forming different cultures. Although sign language has a longer lifespan, it has less diversity. For example, in World War II, the Japanese broke the American military information codes due to their mastery of the English language. Americans were forced to communicate with each other in the Native American language called Navajo for a period of time. For the Japanese, this language was incomprehensible and sounded like gibberish[732]. It seems that if all human beings communicated with each other using sign language, the concept of a nation would be significantly diminished, and perhaps we would have a more peaceful world. Nevertheless, indigenous languages are valuable for preserving identity, and international languages help facilitate communication effectively[733]. The belief that indigenous languages are primitive, inferior, and sub-standard is incorrect. Different languages have a positive impact on art, culture, tourism, economic prosperity, creating cultural diversity and a new life. Moreover, believers in a single language like Vietnamese, Cambodian, Rwandan, and Burundian have suffered greatly in civil wars. The latter two countries are unique in Africa due to their shared language. Perhaps the belief and conviction that the existence of a single language leads to global peace and tranquility is merely a myth.

However, sign language is still used and holds great value. First and foremost, sign language greatly aids the hearing-impaired and prevents their severe limitation in today's modern society. Secondly, the existence of symbolic elements for communicating with other individuals, whether they share the same or a different language, or aiding in expressing concepts more fully, is effective. Charles Peirce, an American philosopher sometimes knows as the father of pragmatism, believes that there are three forms of signs for conveying meaning. First of all, emoticons or icons, which are visual symbols. Visual language is a type of language in which people use graphical elements instead of words such as hieroglyphics in which images convey information[734]. Secondly, there are symbols like animal sounds, roars, gestures, and logos. Thirdly, there are indices, such as pirate flags, where the consequence of encountering them is evident[735]. Animal roars or gestures can have different forms but convey the same meanings. For example, in monkeys, lowering the head or showing the buttocks is a sign of approval. Meanwhile, some vocalizations in monkeys are genetic and do not require presence among peers for learning. For example, a rough and harsh voice in gorillas is a sign of dominance, control, and chest-beating to display power. The use of emojis on cell phones is still widely used to respond or comment on various topics. However, the use of gestures can be a valuable addition to emphasize the meaning of a sentence. For instance, in the sentence "He went to...," we can clearly indicate his direction of going by pointing our finger, thus completing the sentence. This phenomenon demonstrates that modern language has not completely separated from manual signs and signals. As Ludwig Wittgenstein, Austrian philosopher said, it is not the word "red" that represents the color red, but rather the reference to the red object that does so. Ultimately, sign language provides effective assistance to those who do not want their voices to be heard. The use of gestures and pointing by soldiers during wartime is very common for self-preservation. In many other cases, a simple gesture without words effectively conveys the meaning.

In conformity with the theory of Tomasello, the learning of religious or social rituals and obligations can occur through an individual learning process. In other words, when Entity A performs behavior X, Entity B shows behavior Y in response to it. Afterward, Entity B predicts behavior Y in response to action X. In the human model, when parents open their hands to hug a child, the child moves toward them. Gradually, the child learns to move toward them by opening their hands and eventually pointing toward their embrace. Such ritual-based behaviors, whether religious or non-religious, have led to mutual understanding of other's movements. As the number of individuals increases, this two-way communication transforms into social teaching and learning. These movements are then added to the repertoire of human gestures. In this regard, the use of pointing with the index finger is specific to humans, and only chimpanzees that have been in captivity from the beginning can learn this gesture. Some of these behaviors are only understandable and meaningful to modern humans who have the capacity to interpret, understand, and respond. For example, when a cat stands in front of a closed door and meows to go outside, this behavior can only be understood, comprehended, and responded to by humans who have the ability to interpret it[736,737].

For many organisms, separating the social from the biological environment is almost impossible, but it differs when it comes to primates. They possess various types of traditions, experiences, and tribal customs. For example, the use of tools has different patterns in monkey societies. From here, it is clear that culture is important in biological evolution, which is similar to the behavior of modern humans as hunter–gatherer, farmer, and urban dweller. Another mechanism is called "Baldwinian" evolution, where individuals in a society learn a required skill through social interactions, gradually establishing it within the community[49,50]. This idea demonstrates that all skills are not necessarily biological, such as navigating the internet or virtual space. Therefore, in many cases, initial learning is necessary for further learning and can be considered the primary stimulus. Other factors include precise visual attention, increased working memory capacity, which gradually increase and strengthen in subsequent generations. This issue is also important when it comes to the emergence of language. For instance, there must be some form of a primary language that later transforms into a complete language for communication. A complete language is one that allows the creation of new words and the construction of meaningful and specific sentences by combining them. The presence of grammar is the main factor in understanding sentences and inferring concepts from them.

But was this change in our ancestors from early language to full language a merely biological change or just some kind of invention with a broad social consensus without a clear impact on the genetic blueprint of the brain? Archaeological studies from the past 250,000 years do not show signs of human culture such as burial customs and art, but they do demonstrate abundant evidence of experimental toolmaking and the use of fire. It seems that this change occurred in the past 70,000 years due to the creation of language. It is not possible to accurately answer the question of whether the human brain at that time had evolved enough to be receptive to a complete language from a primary language. For example, our brain is prepared for many tasks in the modern world, such as playing computer games, which were created and introduced to the market in recent decades without the human brain having prior knowledge of their existence. Overall, it appears that the neurobiological evolution of human language began approximately 100,000 years ago and continued for several thousand years after the transformation from a primary language to a complete language, which took a long time. The emergence of the Indo-European language occurred around 6000 years ago, and the ability to write was added to it around 5000 years ago. Consequently, understanding the emergence of grammatical structure is a fundamental question. Therefore, the next step in language construction was the formation of grammar. The question is why and how grammar became part of language, and what its purpose is, and how it can function entirely. In the meantime, the role of three phenomena—memory, time, and mind—and their importance in shaping the structure of language must be clarified.

15 The Little Horse

LANGUAGE FRAMEWORK

In *Genesis*, it is stated that Hebrew's god created language for humanity. While knowledgeable in language he did not know the names. Adam was tasked with assigning distinct names to birds, insects, quadrupeds, and other creatures. However, the gift of Greek deity, Prometheus, to man was fire. Both of them were vital elements and so important for the cultural evolution of the humankind[738]. It seems that the selection of names was purely based on chance, and even onomatopoeia was not used to name them. Plato also discusses the dialogue of Socrates with two others, Cratylus and Hermogenes, in his book "*Dialogue*." He also believes that the names of living beings were chosen by chance and coincidence.

Our inner thoughts as humans consist of mental concepts that arise from our interpretation and analysis of the external world. These phenomena include observing objects, behaviors, actions such as playing and jumping, and understanding different feelings. However, the key point is how we can convey them to others in the absence of their tangible and objective presence of those phenomena. These symbols can take the form of sign language, different sounds, and even written texts. These symbols can provide humans with the ability to refer to an object or concept that is not physically present. For concepts and assumptions to be understood by both the speaker and the listener, it is necessary for each symbol to have the same meaning in their minds. This process is essential not only for past events and incidents but also for future phenomena.

It seems that the number of concepts existing in the mind practically corresponds to the number of words in our vocabulary. For example, there is an apparent difference between men and women in terms of the spectrum of colors and their naming. Perhaps other animals have fewer concepts in their minds. These symbols take shape in the form of conventions, symbols, words, gestures, etc. A good English college student typically knows up to 50,000 words, and this number can reach 70,000 for individuals with higher education. Terms that are formed based on convention and agreement rather than physical similarity are called "conventional agreements." We call these symbols words in spoken language. Although other things like prepositions are also called words and they are added to the text. To use symbols, they must be learned because they do not have much visual similarity to the corresponding concepts[739]. It seems that this phenomenon exists not only in humans but also in other animals, with the difference that they do not have the ability to express words. They can transfer some concepts such as the presence of a predator or prey to each other through certain behaviors or different sounds. This, itself, indicates Darwin's motto that "the change in the evolutionary process is in the degree of the ability to perform a task, not its type"[711,740]. Words convey concrete and abstract concepts and enter people's minds in various forms.

Cerebral hemispheres are each engaged in their own tasks[741]. The left frontal lobe is central to speech production, while the right temporo-parietal junction plays a fundamental role in visual–spatial processing and speech prosody. The right hemisphere is involved in the continuous expression of events and the establishment of their interdependence. It also plays a role in the prosody and melodic aspects of speech. The right hemisphere also preserves these characteristics for sign language. The left hemisphere plays a significant role in the sequential processing of speech (speech cohesion), but it cannot utilize "space" as a linguistic feature associated with the right hemisphere. Therefore, it relies on "time" for this task. Sign language utilizes the right hemisphere for encoding language and demonstrates better performance in expressing the shape, size, visual range of an object, its spatial position, and certain movement features within the visual range. However, the left hemisphere excels in data processing, specifically in analyzing high-frequency spatial movement patterns with fine and rapid hand and facial movements within the visual range. The right

DOI: 10.1201/9781003538400-17

hemisphere predominantly operates in general and large-scale movements, global recognition of faces, establishing social interactions, decoding emotional expectations of individuals, and comprehending comprehensive and holistic actions, exhibiting greater capabilities.

How much of the events that occur throughout our lives do humans remember? Milan Kundera, the Czech novelist, in his novel "*Ignorance*," strongly insists that we do not remember much. He describes the relationship between memory and nostalgia[742]. Of course, measuring the amount of memory in our brain is very difficult, but it will certainly not be very high. Let's remember that our episodic memory in the hippocampus has limited capacity, although it can virtually store many things in mind. This limited and virtual capacity can only be filled with unlimited words, which, when combined, express various concepts in the form of sentences. The grammatical structure of language emerged when it was necessary to express our mental concepts as words in sentences[743]. In this case, grammatical structure is required to change the meaning of words by altering their order in sentences. This grammatical structure is usually present in many languages, such as Persian, with the order of Verb-Object-Subject (VOS). In some languages, like English, it is Subject-Verb-Object (SVO), and at least in four languages spoken by tribes in Venezuela (Warao)[744], Brazil (Nadeb or Kaburi), and Indonesia (Tobati or Yotafa), which is an Austronesian language, and Western Australia (Wik-Ngathan), which is a Wik kind of language, the grammatical structure of all is OSV[574]. However, all three grammatical structures can exist in six thousand different languages. Sign language, through having a comprehensive element, places less importance on word order. To form the grammatical structure of a language, there was a need for the emergence of a large number of new words. Although they often did not have meaning on their own, they collaborated with main words in constructing sentences, and many of them were derived from these very words called "function words."

Another important point is the generativity of grammatical structure in language. The generative nature of the grammatical structure is due to the presence of two parts of components and elements such as verbs, nouns, and prepositions, and the rules governing their arrangement to form a sentence. Sentences may be grammatically correct but lack meaning. For example, instead of saying "The cat ate the meat," we can say "The meat ate the cat," but the second sentence does not have a thru meaning. Sometimes sentences can be so long that understanding them requires re-reading or repeated readings, reducing their practicality in conversation and speech. However, grammatical structure itself is autonomous in the sense that it is one of several levels involved in the formation and meaningfulness of sentences. This means that phonemes must be correct and arranged in a way that is expressible. Single morphemes and words must be correct in form, and only then does the grammatical structure take shape. This structure helps not only in constructing complete sentences but also in constructing open-ended sentences. After the emergence of grammatical structure, it acquires semantic meaning. In fact, generative grammar arises due to the autonomy in syntactic structure, so the analysis of a sentence includes their syntactic and semantic parts.

Perhaps the phenomenon of generativity is related to our unlimited nature and our spatial experiences. For all empirical purposes, time and space have expanded infinitely and are not exclusively for human. In fact, for the first time, Paul Deane proposed the idea that language structure is derived from our "spatial perception inheritance." This idea is related to the function of the inferior parietal lobule (IPL). Although IPL is specifically involved in spatial perception, the left IPL is involved with the structure of language comprehension. Imaging studies have shown that the left IPL's metabolic activity increases significantly when grasping an object with the right hand. It appears that the function of the left IPL has aligned with the speech centers[745].

Now, we return to the importance of the hippocampus. The right hippocampus aids in our spatial understanding, navigation, and orientation in the surrounding world, while the left side reconstructs episodes of old memory in the past for us. The unidirectional function of the hippocampus is likely related to our speech abilities and can also help us construct future imaginations. As John O'Keefe states, many prepositions in English refer to time and space, such as "against, at, among, around, between, along, from, in, and to." Scientific studies have shown that damage to the right hemisphere of the brain leads to significant impairment in memory (amnesia) regarding concepts that are visually spatially encoded, while damage to the left hemisphere affects episodes that are linguistically encoded[746–748].

The existence of mechanical phenomena can be influential in the formation of linguistic structures. Deacon believes that the construction of objects and tools possesses the property of recursive embedding. The main periods of toolmaking in humans include the Oldowan and the Acheulean phases, approximately 2.5 million and 700,000 years ago, respectively. In the second phase, humans learned to attach a handle to a blade. Gradually, their ability to connect pieces improved. Constructing such tools, like a hand axe, has been a cognitive leap forward, resulting in the expansion of human symbolism and imagery. At the same time, there was a noticeable increase in the flexibility of hand and arm movements. In addition to tool construction, it has a hierarchical property and represents smaller goals that are integrated into the main objective. The main consequence of the various stages of tool construction is the creation of a temporal disjunction between different actions, similar to language, where all words must be prepared in the mind and arranged in sequence before being uttered[616]. In a broader perspective, language itself can be a tool or a type of technology created by humans, not necessarily a form of thought. The construction of tools, language, and other engaging phenomena such as music and painting can be considered as human responses to the post-arboreal era. Ultimately, it can be concluded that adaptation is a two-way process involving changes in the living organism and its surrounding environment.

THE LITTLE HORSE

In 1956, Charles Scoville, a neurosurgeon based in New Haven, Connecticut, opted to conduct surgery on a 27-year-old man who suffered from uncontrollable seizures. The surgery was successful, and the bilateral hippocampus and entorhinal cortex of the patient, Henry Molaison, were removed. While the patient's seizures improved to some extent with the available medications at that time, Henry experienced two major problems. The first problem was episodic memory loss, meaning he couldn't recall past events and incidents[749]. The second problem, which was as important as the first one, was the inability to form maps and plan for the future. However, his working memory, which is required for daily tasks, remained intact[750]. During his interviews with Suzanne Corkin, a clinical psychologist, Molaison did not remember meeting her multiple times for interviews. Even if Corkin left the room for a few minutes during the interviews, upon her return, Molaison didn't recall her absence. Henry generally had no recollection of what he did in various hours of the previous day[751]. Nevertheless, his working memory, such as remembering phone numbers, engaging in conversations with others, riding a bicycle, etc., was considered normal. Although his semantic memory, which refers to general knowledge and concepts, was also relatively good, he had no idea about the future. When asked what he planned to do for studying or leisure the next day, he usually couldn't provide a precise answer. He often deferred planning for future events. He not only forgot the past but also had forgotten his future[752,753].

As Eric Kandel[476,754] puts it, memory is like glue. It connects all our moments, and without it, our conscious and unconscious lives turn into small, separate, and meaningless fragments of time; cognitive abilities, learning, and children's intellectual growth are disrupted, and life becomes empty and meaningless. The nature of memory, what it consists of, is a very big question that we still don't have a definitive answer to. Ramón y Cajal by presenting the hypothesis of the existence of neurons as the basic cellular structure of the nervous system in 1890, believed that memory is a kind of change in the structure neurons and their synaptic connections (protoplasmic kisses as he said) between them. It took shape as concepts and repeatedly processed in the mind[755]. The theory of synaptic plasticity, proposed by Ramon y Cajal, suggests the ability to modify the strength of synaptic connections through their repeated use. Of course, the term synapse was first proposed three years later by Charles Sherrington, an English neurophysiologist. In his next book "*Degeneration and regeneration of the nervous system*" in 1913, Ramon y Cajal, pointed out that learning is associated with an upsurge in the number of synapses. Another group, based on the theories of Karl Lashley and William Adey, introduced the concept of bioelectric fields, which result from the integration of activity from many neurons[756]. Later, studies in the seventies by Tim Bliss and Terje Lomo[274] showed that the process of learning and memory following repeated stimulation, that is long-term potentiation, of neurons in the hippocampus is accompanied by an increase in the number of biochemical receptors and surface spines in the membrane of neuronal dendrites. The retrieval process is also associated with the reactivation of the same neural network. However, memory does not necessarily work as a voice recorder. Memory could be changed, fragmented, and added new pieces to it. In other words, memory reconstructs itself, and as a result, the possibility of error is high. According to Fredric Bartlett, professor of psychology at Cambridge University, in his famous book "*Remembering*," memory is a mixture of information that is built on the basis of our previous knowledge[710].

Recent studies conducted in the past 60 years by Kandel and colleagues have shown that synaptic strengthening occurs through modulation in the release of chemical transmitters from presynaptic neurons[475]. The construction of memory is not limited to specific parts of the brain; instead, all brain components are involved in various forms. The structure of the nervous system utilizes global mapping, a dynamic phenomenon where the continuous and repetitive activity of the motor system brings about constant changes. Synaptic changes in extensive areas of the global mapping in the brain constitute the basis of memory. Memory is a process of continuous space–time re-categorization, which by its nature requires constant motor activity in order to be able to repeat a task, such as writing with a pen, which ultimately manifests as unconscious performance. Based on Gerald Edelman's interpretation, memory is like the glaciers in the Alps, melting and freezing again in different seasons, constantly changing shape while always being present[757].

Humans generally do not remember memories from before the age of three to four. The first memories that remain in our minds are from around the age of five and onward. Humans have several main forms of memory, which are divided into two basic types. The first form is explicit or declarative memory, which represents conscious facts that we have memorized through experience and can consciously recall later. It essentially demonstrates itself when an event that is mentally significant occurs. This form of memory relies heavily on the structure of the hippocampus. The second form is implicit or unconscious memory, also known as procedural memory, which, although we have learned something, we execute it on a daily basis in a completely unconscious and automatic manner. This memory is evolutionarily the most primitive form of memory and has little connection with the hippocampus. It is primarily associated with basal ganglia, amygdala, and cerebellum[758]. Learning with this type of memory is more challenging but is stable, enduring, and operates quickly and efficiently. Examples include everyday activities such as driving, riding a bicycle, and writing, where we don't pay much attention to the process, and when we focus too much on them, we may not perform them correctly. It is not highly localized within the brain's structure, and it is triggered and engaged through different parts of the brain, such as the cerebellum, the basal ganglia, the amygdaloid complex, and other reflex centers in the brainstem[759].

Explicit memory is divided into two categories of long-term and short-term memories. Working memory typically lasts for a few seconds or minutes and is generally forgotten after performing the related task, such as seeing a phone number, memorizing it for the short-term period, and immediately forgetting it after dialing[760]; however, long-term memory remains in the mind due to repeated exposure. Working memory is strongly dependent on consciousness because the content of attention (internal attention) remains in a fixed and stable state until the next action takes place[301]. Long-term memory is also divided into two parts: episodic memory and semantic memory. Working memory is primarily associated with the frontal lobe; however, episodic memory is related to the hippocampus, and executive memory, such as cycling, is associated with the cerebellum. Semantic memory refers to our understanding, awareness, and knowledge of a specific phenomenon. Although we have learned it at some point, we do not recall the exact time we learned it. For example, we know the capital city of our country, but we don't know when we learned it. Almost all of our acquired knowledge is related to semantic memory. Episodic memory is about events that we remember when they occurred. In other words, we have a specific time frame for them in our mind. For example, we know what we had for breakfast this morning. The timing of this event is significant in our mind[761,762].

The primary sensory cortices and associated areas in the brain cortex receive and process sensory information. Meanwhile, the frontal areas encode motivations, attention, and working memory. While the retrieval of long-term memory initially relies on a healthy hippocampal structure, it does not require it in the long run, and damage to it does not affect the content of memory[305]. Due to the association of the memory pathway with the limbic system, memory encoding can be influenced by sensory content associated with it. However, functional memory impairments can be accompanied by behavioral disorders such as increased sexual arousal, overeating (hyperphagia), and excessive use of lips and mouth to understand the surrounding environment. Occasionally, semantic memory can acquire an episodic form, such as solving a specific problem during school time when other students were unable to solve it. It then moves away from episodic form and is registered in semantic memory. Additionally, episodic memory can sometimes merge with semantic memory. In other words, a part of the desired phenomenon is evaluated based on episodic memory, while another part is evaluated based on semantic memory, and this is called autobiographical memory. The awareness of our children can be dependent on their everyday stories or stories told by their parents to the extent that these stories become consolidated in their minds. As episodic memory function increases between the ages of two and eight, semantic memory performance decreases during this period. Learning a language requires both forms of memory. Procedural memory is used for learning grammar, while explicit memory is used for lexical knowledge[750,759].

After the publication of "*On the origin of species*" by Charles Darwin, Reverend Samuel Wilberforce (1805–1873), Bishop of Oxford, in a debate with Thomas Huxley attempted to refute the theory of Darwin. Sir Richard Owen (1804–1892), a scientist and the superintendent of the natural history department of the British Museum, assisted him in this matter[763]. Although they were defeated in the debate, Owen later became the subject of ridicule and anger from Darwin due to his statement about the existence of a "horse in the brain" later called the hippocampus, with specific functions and activities. We can have vivid images of the world around us, reconstruct the future in our minds, and navigate through the imaginary world created by our minds. All of this requires creating an environment within our minds that simultaneously incorporates our feelings and emotions from the external world. The presence of place cells and grid cells in the hippocampus provides a metric framework for our spatial perception. Specifically, in animals, there are head direction cells and border cells that determine the physical boundaries of the environment. The electrical discharge in place cells is such that smaller and more distant images are formed in the anterior part of the hippocampus, while larger and closer images can be seen in the posterior part. In other words, we can determine the location of a living being based on the discharge of its place cells in the hippocampus. Scenes are coherent spaces containing objects that we have the potential to interact with. If we add the dimension of time to them, they can be considered as episodes or events. In fact, in the human mind, places do not exist without the concept of time, and what we perceive as time is necessary for constructing mental episodes. According to physicists, time does not exist on its own and emerges synchronously with the creation of space. With the help of the hippocampus, we can distinguish past episodes from the future or create images of the future while remembering the past. Visual perception is crucial for the formation of space–time (spatial–temporal) layouts. In animals that have not evolved as much as intelligent humans, the use of olfactory sense has replaced visual perception. Therefore, recalling episodes of mental images, with or without navigation, requires the functioning of the hippocampus, and bilateral damage to it is particularly associated with severe impairments in these tasks, especially in the anterior region[764]. The most significant function of the hippocampus is related to the encoding and replaying of specific events and incidents, as well as constructing individual representations for creating a mental image of the future in the anterior part of the hippocampus. Although the posterior region is responsible for recalling the past, it can also record and encode a phenomenon imagined for the future in the region associated with the past. As a result, we remember a phenomenon that did not happen. Another function is recording different locations in the spatial structure of our minds. The neurons there work like a GPS for vehicles. These neurons, known as place cells, design maps of the places where an individual has been in different scales in the mind. The scale of these maps is smaller in the posterior part of the hippocampus and larger in the anterior part, indicating the ability for magnification and zooming in the posterior region. The encoding process occurs simultaneously. As a result, the process of recording and recalling places is associated with their corresponding time, creating a spatial–temporal layout in our minds. In a study, it has been shown that the hippocampus of London taxi drivers, who had a better mastery of the city map, was larger than their less experienced counterparts[746,753].

Continuous stimulation of the cells in this region leads to strengthening their connections with higher-level cells and ultimately provides long-term protection for them. It appears that information is initially registered in this area and then in other parts of the brain. The hippocampus may play an influential role in finding them. In other words, mental time travel (MTT) occurs in a multidimensional space–time. The most well-known cells studied in the hippocampus are place cells. These are pyramidal-shaped cells that become electrically active when laboratory mice are placed in a certain environment. This region is referred to as the "firing place field." Discovering these cells has contributed to the theory of constructing a cognitive map of the environment. In fact, clear and salient spatial cues strongly stimulate and control place cells. These cells seem to be heterogeneous neurons with different functional properties. Additionally, those are stimulated by personal movements, environmental boundaries, objects, smells, and the orientation of individuals' heads, and this ensemble creates its spatial map in the hippocampus, which can provide a basis for learning and cognition[765,766].

In experiments conducted on mice in laboratory settings, it has been observed that place cells are active during rest and sleep, and they can even be activated by other cells during wakefulness to find a new path. This phenomenon can be similar to mental time travel. Overall, the hippocampus has two fundamental functions. First, it has the ability to construct a cognitive map of a route that is formed prior to performing a behavior, aiming to find goals. Second, it houses an episodic memory system in which mental time travel is engaged. In other words, the hippocampus's function is to design a virtual and unreal action in both time and space[253,762]. Humans, like many other animals on Earth, move in space, making the concept of space and location fundamental and foundational. This phenomenon, both physical and mental, does not seem unlikely or extraordinary that mental time travel has evolved through the process of redesigning or pre-designing spatial movements and has become consolidated in the brain/hippocampus. If we go back 66 million years ago to find our common ancestor with the rat, we realize that despite significant advances and mental abilities of the human species, remembering and designing in space within our real world has remained constant and preserved its fundamental form. It can be speculated or concluded that mental time travel was one of the earliest cognitive branches of intelligent humans that has grown and evolved. This process is crucial for all mobile organisms in space place, as they need to know where they have been, where they are, and where they will be. Perhaps we can conclude that the process of space place is indeed time. We often remember places with the times we have been there. To create mental time travel as an episodic mental–visual phenomenon, we need a driving force and raw material to retain information before it is erased from the mind.

Mapping different spaces and locations in the mind at different times (present, past, and future) requires the "updating" of hippocampal information. When a living being moves, there is a continuous need to redraw the map in the mind. This process, which is fast and flexible, occurs between the entorhinal cortex, the hippocampus, and the neocortex. These cells are arranged in a grid pattern. The generated modules (fields) function differently depending on the geometric layout of the environment and the position relative to the individual. These grids have different spatial scales and create a three-dimensional image in the mind. In the posterior parts, the larger and clearer view of the components of the region is observed, while in the anterior parts, it mainly processes the general aspects of the image in a more distant view[747,765,766]. In studies conducted on mice in laboratory, it has been found that the posterior part of the hippocampus processes an area of about one meter, while the anterior part processes an area of about ten meters. This allows organisms to have the ability to zoom in for details or refocus to see the overall picture in different times and locations. It is quite similar to reading details of a book page by page, but after finishing reading, only a general image of the information remains in the mind, which is, of course, due to the function of the hippocampus. This is exactly like letters in language, with a very limited number of letters, an infinite number of sentences can be constructed, as Chomsky said: discrete infinity.

Endel Tulving believes that remembering is a form of self-awareness in which a person projects themselves into the past to re-experience the episodic events that have occurred. He called it autonoetic consciousness[767]. Knowing and being aware does not shift consciousness (e.g., being aware of the boiling point of water), but autonoetic awareness has the property of recursion and can bring the individual's previous experience into present awareness. That's why all adults know that boiling water is hot and has a burning property for the skin. This phenomenon is similar to embedding one phrase within another, or a sentence within another sentence. Of course, placing one event within the depth of another can be seen as virtual pieces in episodic memory[619,768]. For example, I remember thinking about something that happened a few days ago. In the above statement, there are three time periods. First, I remember it now (the present time); second, what I was thinking about yesterday, and third, an event that occurred a few days before that. In other words, the pieces of episodic memory are intertwined recursively to create a new concept. In the case of amnesia, with Henry Molaison being a classic example, episodic memory is impaired, while semantic memory remains intact. However, episodic memory relies on semantic memory. For example, you cannot remember going to the hospital if you don't know the meaning of the word "hospital." The concept of the word "hospital" is related to semantic memory, but going to that place at a specific time is attributed to episodic memory. In other words, episodic memory is constructed upon semantic memory, and without semantic memory, episodic memory does not function. Perhaps that's why we don't remember anything from the first three years of our lives.

Alexander Luria, a Russian neuropsychologist, introduces a person named Solomon Shereshevsky[422] in his famous book "*The mind of a mnemonist*"[421]. Solomon possesses an extraordinary memory and remembers many small and significant events in great detail that occurred in the 1940s in the Soviet Union (present-day Russia). Similar cases have also been reported, such as Daniel Tammet, who was able to learn to speak Icelandic language within a week[769]. Lu Chao had the ability to recite pi number to over 67,000 decimal places, but all of these individuals had low intelligence. It seems that excessive abilities in a specific area can lead to weaknesses in abstract thinking, inference, and a sense of humor. As mentioned, an educated English-speaking adult knows about 50,000 words, which themselves are combinations of the concepts of who, where, when, whom, and what. It should be noted that a portion of our memory can be deceptive. In fact, part of it is constructed by our knowledge and awareness, while the other part, which includes detailed specifics, is formed or completed through mental imagery in our minds and is referred to as "paramnesia." Human memory has the quality of reconstructive hammering, which is a kind of iterative process that changes and can be accompanied by significant and fundamental errors in real situations, such as giving testimony in court or in clinical medicine. In this regard, the famous story of hypnotist Hippolyte Bernheim (1840–1919), a French neurologist in 1884, about inducing an artificial sleep (hypnotism) in a girl to elicit a false memory of molestation, is reported[770,771]. He insinuated or suggested to his patient, Marie, that she saw an elderly bachelor molesting and raping a young girl through a keyhole. Three days later she remembered the name of rapist and the girl, too. In another report, Elizabeth Loftus, an American psychologist, recalls her mother's death at the age of fourteen, even though she was asleep at the time of the incident. Moreover, many people claim to have seen the famous cartoon character Bugs Bunny at Disneyland, and even one-third of them say they have shaken hands with him, despite the fact that Bugs Bunny is a creation of Warner Bros. Pictures animation studios. Such false memories have led to the imprisonment of many individuals in the 1990s and, in some cases, even to death, occasionally by electrocution in the electric chair[772].

These phenomena are referred to as "false memory" and can be interpreted as a type of association memory illusion. They can also be observed in bilingual individuals, although they are more prevalent in the dominant language[773].

From an evolutionary perspective, memory is a form of adaptation to the environment, allowing living organisms to create greater stability in their lives through behavioral changes. Procedural memory, as mentioned before, is a form of memory used for learned tasks without the need for conscious thinking, such as riding a bike or driving, primarily involving the functioning of the cerebellum and basal ganglia. On the other hand, working memory is related to performing tasks using reasoning and is associated with the prefrontal cortex. Part of this behavioral adaptation is called "instinct," which is developed in early organisms and transferred genetically[774]. However, genetic transmission alone is not sufficient; environmental conditions must also be present. Instinct can manifest as complex behaviors that are not easily justifiable to us, such as seasonal migration of birds, food hoarding and calling in squirrels, or dam construction in beavers. Although these behaviors are seasonal, they can be influenced or even disrupted by environmental factors, especially in human performance. In intelligent beings like humans, various instincts, such as emotional reactions, sexual tendencies, and even the inclination to disappear, can manifest in different forms.

Following the studies by Eric Kandel and Scott Small at Columbia University, it has been shown that mice do not develop Alzheimer's disease[775,776] but can experience age-related memory impairment (ARMI). The results indicate that proteins produced by single-stranded RNA molecules called transcripts have a positive impact on mouse memory. Destruction of relevant genes, especially a type of gene called *RbAp48*, strongly causes age-related memory disorders in them. This disease is primarily associated with damage to the dentate gyrus in the hippocampus, while Alzheimer's disease is characterized by damage to the entorhinal cortex in the brain. The *RbAp48* gene produces nuclear proteins, which means proteins that enter the cell nucleus. These proteins are part of a protein complex called *CREB*, which is involved in the conversion of short-term memory into long-term memory. Additionally, the hormone osteocalcin, produced by bones, increases the activity of the *RbAp48* gene and is responsible for promoting the production of *CREB* protein[475,476]. Thus, it can be concluded that exercise in older ages also plays a significant role in preventing memory disorders, reflecting the Persian proverb "a healthy mind is in a healthy body."

Sign language, like other human languages, has its own phonology, morphology, and specific syntactic structures, and it is a symbolic abstract system that can express all human thoughts. However, its fundamental difference from spoken languages lies in the method of transmission, which uses hand and facial movements instead of auditory perception through the ears. Just as there are multiple spoken languages, different sign languages are observed in different nations.

Verbal and visuo-spatial domains in working memory are associated with spoken language and sign language. The verbal loop is related to spoken language, while the visuo-spatial loop is related to sign language. There is substantial scientific evidence that suggests that increased and enhanced use of sign language improves visuo-spatial cognition in individuals, such as visual mental imagery, mental rotation for understanding spatial structures, understanding other spatial structures, and face recognition[777]. It appears that individuals who are deaf have greater proficiency and mastery in these aspects compared to normal individuals. Pointing and signing requires spatial mapping of objects in the scene. On the other hand, the perspective of a referred scene can permanently change, making highly important, the ability to present that perspective to a congenitally deaf individual.

In sign language, lexical items are formed by a limited set of hand movements, handshapes, facial expressions, and spatial locations. In Chinese sign language, the characters are formed based on a limited number of radicals and correct shapes that are connected spatially and visually in the structure of writing. In many languages like English and Persian, there is a correspondence between sound (phonology) and orthography, a property that does not hold true in Chinese[778]. Chinese characters are mostly meaning-based rather than sound-based, without a direct one-to-one relationship between sound and symbol. Therefore, their pronunciation is based on what is stored in memory. Similarity exists in reading Arabic numbers, where the number 2 is pronounced as "two" when standing alone, but as "twen" when combined with zero (20). Such differences can be observed in many languages, including Persian. Thus, Chinese or Japanese readers, compared to readers of English or Persian, need to remember more visual shapes in their minds and memories[779]. Typically, a proficient Chinese reader should memorize between two to twenty thousand visual characters, some of which are composed of multiple components. This is considerably more challenging than learning English, which has a total of 26 letters, or Persian, which has 32 letters, with a limited number of compound vowels (diphthongs). For this reason, it has been determined in numerous studies that speakers of languages like Chinese and Japanese have better abilities in perceiving and comprehending space, shape, form, and mathematics compared to their English-speaking counterparts, and they perform better in these areas.

In other words, the ability to learn lexical (alphabet-based) languages like Persian, Arabic, and English is related to their strategy and ability in phonological coding, whereas the ability in tonal (sound-based) languages, with an estimate of more than 1.5 billion speakers, like Chinese, Japanese, and Igbo (Nigerian ethnic group) is dependent on their visual memory capacity[780]. Overall, good auditory coding for alphabet-based languages and visual coding for tonal languages are of great importance. Additionally, language mapping plays a role in tonal languages at the morphometric level and in lexical languages at the phonemic and phonetic levels. In fact, it can be considered that the output of speech and articulation and their input in alphabet-based languages are based on phonemes, which is unnatural or at least highly abstract, and the actual form of language should be based on visual coding in sign languages. Sign languages and visually based scripts enhance visual memory. It seems that the stimulation of visual perception in the surrounding environment has been influential in the evolutionary development of human intelligence, leading to the emergence of such forms of morphology in the brain. Meanwhile, it appears that teaching sign language, even characters of languages like Chinese or Japanese, to children in elementary schools can effectively enhance their visual-spatial cognition. The Khoisan languages in South Africa use click sounds, such as the "nuch-nuch" sound, to express words. Other click sounds include Bantu, Nguni, Sotho, and Kwangali in southern Africa and Botswana. Initially, researchers thought that these sounds were similar to the sounds of animals like birds and ostriches, but later they realized that there is no set of animal sounds that can be similar to the Kho language[781,782]. According to an analysis, this language has over 48 different click sounds.

Learning in the evolutionary process of intelligent humans is a type of adaptation to cope with environmental events. This adaptation can be for chance-based events that have been or could be possible throughout human life. Another famous model of learning is classical conditioning, which was first introduced by the Russian physiologist Ivan Petrovich Pavlov (1849–1936). In 1904, when he was studying the digestion of food in dogs, accidentally noticed that the dog's saliva secretion coincides with the ringing of a bell even when no food is present. Behavior is formed and remains with its consequences and results. These are essentially physiological phenomena that occur involuntarily in response to a previously neutral stimulus. However, the most famous study in this area was conducted by John Watson and Rosalie Rayner Watson, his second wife, in 1920s[783]. Watson was a psychologist and founder of behaviorism[784] but their notorious, unsightly and unethical study involved the conditioning of a child, Little Albert, to fear a white rat by pairing the sound of a hammer hitting a metal rod with the sight of a monkey, dog, and rabbit. This study became a landmark in behavioral psychology. Such studies are no longer conducted due to ethical concerns. Another form of conditioning is operant conditioning, which is a relatively voluntary behavioral phenomenon in response to stimuli based on rewards and punishments. It appears that many human behaviors are based on conditioning and learning. The learning of language by intelligent humans is likely a form of this conditioning as well.

Both forms of conditioning gradually occur through natural stimuli without the need for consciousness; they are not deliberate and are stored in the unconscious and/or non-conscious memory. On the other hand, the prominent and conscious memory of humans can demonstrate more flexibility in adaptation to the environment as they are not immediately influenced by the surrounding environment. It seems that language and the ability to speak are constructed based on two influential factors: memory and learning in the evolutionary process. Semantic memory, which operates as a large storage of facts and words, encompasses lexicon or vocabulary, some grammatical rules, and the pragmatics of language; all of which are related to semantic memory. However, a significant portion of grammar and its execution is associated with non-conscious memory. That is why humans usually do not consciously think about when speaking their native language and naturally express words[785]. However, when it comes to speaking other languages, especially when learned at an older age, they have to heavily contemplate it. Episodic memory functions as a mental diary for personal experiences, enabling us to remember various phenomena at different times. Episodic memory is likely unique to humans and allows us to regulate our behavior based on specific episodes in the past. It allows us to visualize future episodes in our minds, plan for the future, and compare different scenarios. The recursive process, which is one of the fundamental principles of language, requires the use of both episodic and semantic memory.

From another perspective, language is a collection of environmental stimuli and external performance. In other words, it is a cycle of receiving and performing, like opening and closing a door or turning and nodding one's head[713,786]. The construction of sentences is intended for communication, so grammar, syntactic structure, vocabulary, meaning, and cognition all intertwine and merge in a path. Syntactic structure in different languages have significant similarities, and the differences between languages mainly lie in lexicon and phonemes, which result from cultural evolution in different languages and ultimately serve common communicative goals. Achieving this goal is the result of adaptability and the opening of a hierarchical structure that can execute a set of actions to reach its goal. In essence, although syntactic structure is entirely abstract and mental in form, its existence is crucial for constructing purposeful sentences. Meanwhile, the listener is required to open multiple words in their mind, decrypt them, and search for their meanings in their mind. This inferential and obligatory task is accomplished with the help of the narrative memory and cognitive abilities of the individual.

16 Time Flows

MENTAL TIME TRAVEL

Numerous animal behaviors, such as finding or locating food or seeking a mate, are instinctive. However, the concept of mental time travel (MTT) encompasses nearly all facets of human existence. Examining and evaluating MTT in animals is not easily possible because it involves language and human characteristics. Animals primarily live in the present moment, but humans like to express their past experiences and future aspirations using their memory. Memory is not merely a representation of the past, but rather a reflection of dynamic changes in the brain that allow for the repetition of an action. These dynamic changes involve alterations in the number and shape of synapses between neurons. The existence of episodic memory and MTT, alongside a strong desire to narrate our life stories, likely marks the onset of the formation of time.

Humans are as interested in storytelling as they are in sexual relationships[787]. This phenomenon is observed in almost all human beings every day. Narration constitutes a significant part of human thinking, and storytelling can be considered a form of thinking. Based on Katherine Nelson[788,789] research, storytelling in childhood transforms episodic memory into autobiographical memory, eventually leading to personal history and a broader understanding of the concept of "self." Stories can be processed in the past or the future, and they can be timeless or transition between the past and the future. John Niles refers to this type of human as "*Homo narrans*," that is, storytelling human[790]. Aristotle believes that poetry is superior to historiography because it is related to events that did not happen and expand the mind. Stories generally form the basis of our social world, and while they can somewhat detach us from reality, they enable us to have a better understanding of others. The roots of stories are likely found in play. That is why a drama in English is referred to as a "play." Additionally, Brian Bond, a British military historian, believes that mammals and birds also learn how to hunt or fight through play. However, their behavior during play and learning is distinct from their behavior in real conflict. The best quality of play is the existence of "rules." Many children follow rules during play, and play, in essence, occurs in reality[791].

The first stories were told and gestured around the fire. Human control over fire, which dates back to around one million years ago, and regular control of it, around 400,000 years ago, had a significant impact in this regard. Fire, besides creating warmth and light, driving away wild animals, cooking food, making it delicious, enhancing its nutritional value through better absorption, reducing the price of hunted food, and the ability to share it, turned storytelling into an integral part of early human life[242].

The aforementioned factors have led to the local fire becoming a means of communication within a social group or commune, increasing social activity, fostering interest among individuals, and facilitating their acquaintance with one another. However, lengthening the day with the light derived from fire has caused a shift in the circadian rhythm of sleep–wake cycles in the body. The consequences and benefits of this rhythm change have persisted for modern humans to this day and seem to intensify or worsen over time. Inventions such as the light bulb, television, and LCD/LED displays have resulted in individuals no longer listening to or seeing stories in their solitude. Pantomime in Rome and Greece, ballet in Italy, respect for fire and worship in Zoroastrian temples, in ancient Persia, and Kathakali, a dance accompanied by music, in India, have been added to the development of storytelling. Additionally, an interest in immortality has existed in epic stories such as Gilgamesh, but, like many stories, the protagonist ultimately realizes that disease led to death, which is a sorrowful and inevitable end to life[792]. This fascination with storytelling in daily life causes humans to be abundantly inclined toward backbiting. Gossip is likely inseparable from human speech abilities.

DOI: 10.1201/9781003538400-18

While other animals like mice, dogs, and chimpanzees may have the ability to engage in MTT, the human model is much richer, prominent, and intriguing. Furthermore, humans have the ability to share their thoughts verbally. As John Niles puts it, modern humans are speaking–storytelling beings[790]. Stories are usually expressed after games, and games in animals, especially hunters, involve intelligence and are not merely about play. Games occur in the present, while their stories harken back to the past, requiring additional elements. Firstly, the phrase "Once upon a time" is necessary, which takes the story to the past or the future. Secondly, stories possess a narrative quality[629]. They often have a relatively complex structure, typically unfolding sequentially, and events unfold over time, culminating in the sharing of the mental and verbal aspects of the story. The narrative form of stories has changed with socio-cultural progress and surreal stories, stories where time is not a factor, have entered human literature. It seems that stories are generally a mixture of real experiences and imaginary fantasies derived from activities and games[711].

Hunting has been a crucial factor for the survival of early humans. Understanding the local environment and region has been highly important for hunting, and as the Earth expanded and the need for hunting grew, sharing information among peers became necessary. Storytelling has existed in the form of conveying knowledge to children for over two million years, starting from the Pleistocene era, and has played a fundamental role in human evolution. Initially, storytelling was likely performed through pantomime. This behavior is considered a form of sharing human knowledge at that time and has had a significant impact on human survival. Moreover, stories have not merely served the purpose of knowledge transfer; they also include introducing imaginations, feelings, emotions and designs, plot development, invention, spatial visualization and mental imagery, and creating cultural beliefs and ideas[793].

In traditional societies, storytelling was primarily a masculine activity. Storytellers were regarded as hunters and respected individuals in the community. This phenomenon served as a one-way road to influence and acquire power. The ability of eloquent speech has been the most important attribute for entering the power game. However, it does not seem to have changed much in contemporary societies[692]. Men are louder and more vocal, attempting to impose obedience through increased intensity and volume of their voices, seeking dominance in the discourse arena. Overall, eloquence is a prominent attribute that has a significant influence on power acquisition, even in current societies. While many eloquent speakers have nothing substantial to say, they strive to use language to gain more power. Their social performance is like peacocks displaying feathers and crowns of kings on their heads. However, women benefit from their speaking power in friendly conversations, prosody in their speech tone, and gossip. These conversations aim for friendship, companionship, and solidarity, not merely power acquisition. Perhaps this type of speech could be interpreted as a rebellious behavior of women against forceful and boastful men[711].

The high sales of tabloid newspapers and the interest in watching films and reading stories stem from this very human attribute. Among these, storytelling and professional story writing will be a valuable job in the modern world. Stories based on human conditions and relationships, detective stories, crime stories, action stories, and stories that depict mysterious and ambiguous human relationships are popular on a global scale. The grammatical structure of language has likely evolved through episodic memory and has created a significant source of information about human relationships, initially aided by episodic and later by semantic memory. By sharing this information and creating symbols for future planning, language structure was utilized to express semantic concepts, but its foundation lies in episodic memory, mental time travel, and the sense of the passage of time[740]. Semantic memory, which expanded with the help of episodic memory, together were able to effectively contribute to the formation and evolution of language by reinforcing or relaying events and everyday occurrences. Gradually, the addition of symbols that were stored and categorized as distinguishable groups within the mind led to the creation of a combination of conventional signs for communication. Many conventional symbols in sign language gradually became customary and traditional, registering as mental images similar to what we observe in everyday life and sometimes carrying more conceptual value than hundreds of words. However, in some cases, drawing an image to indicate an action can be challenging, while expressing a word or sentence symbolically is easier and more energy-efficient. These symbols act as labels, and our minds can guess or create them, resulting, for example, in the formation of mental images when different animal species are mentioned.

Due to human interest in speed, efficiency, and optimal energy consumption, frequently used words with high usage frequency are shortened. As George Zipf, an American linguist, said, lazy people have made these words[794,795]. Tense construction in many languages is created by changing verbs, using constraints for emphasis and greater precision, and employing different words to refer to locations. Gradually, the concept of time has been added to the language to express more accuracy. Recursion in the words helped us to write today a topic what we were talking about yesterday about the incident that happened in the distant years. During evolution, our complex mind has greatly contributed to the ability of our recursive speaking simultaneously. Therefore, we use books, writings, films, plays, and even popular soap operas for time travel[796].

Because the understanding of space and location predates time, temporal concepts are forced to have spatial interpretations. For example, to indicate the past time, the spatial concept of back is used, and for the future time, the word front is used. However, this behavior is not consistent across all societies. In the Aymara tribe, considered a precursor of Inca culture, in the Andes mountains (now in Bolivia and Peru), the concepts of forward and backward in time are used in reverse[797]. In the Chinese language, the future is up, and the past is down. But in the Amazonian Piraha tribe, time is only the present, with some concepts and words related to far and near, thus influencing their thought process[798]. For them, when someone dies, they go slightly out of memory. There is no past for them to reflect upon, and there is no future to build upon. This same thought process occurs in their minds when merchants arrive by boat; when they leave, everything is over, and nothing else exists. There is only a boundary between experience and non-experience; no spectrum exists[799].

The grammatical structure of language has been formed based on our understanding of the flow of time and the ability of our minds to engage in time travel. Without the existence of time, language might have grown in different paths. The Piraha tribe, due to the absence of verbs with past and future tenses, uses methods such as whistling, singing, and humming. Their speech tone has great richness, and they can tell lies and jokes, inseparable elements of being human. But the most prominent and valuable asset of language is its generative capacity[611,623]. We can comprehend, understand, create, say, and write words and sentences we have never seen or heard before. Language depends on rules that can be combined in a way that carries meaning and understanding. In fact, it is the combination of words, not their individual elements, that allows for the emergence of meaningful episodes. Grammar plays a crucial and beneficial role in constructing these combinations and making sentences meaningful. The most important of these words are nouns, which often serve as subjects or objects, and then, the presence of the object becomes significant. As it was said most languages, like Persian, use the order of subject, object, and verb in sentences, while in some languages like English, the combination of subject, verb, and object is usually used. Only four languages in the world use the verb, subject, and object combination in declarative sentences. The basic order in sign language is the same as English language; while the concept I mentioned is valid, it is not always expressed in that manner[629,711].

In general, episodic memory and combinatorial rules have made possible the communicable connection between people and the addition of time, especially future time, has been able to transform stories and myths into human speech. Myths can occur in the past or future, and of course, these stories take on an adaptive form and allow humans to go beyond their experiences. In other words, they have led to the drawing and sharing of our experiences as humans, so that we can shape ourselves in a probable future. The grammatical structure, by creating verbs for the future and past, has been very effective, real, operative, and efficient in shaping and bringing about religion. Not being able to speak for the future or the past makes it impossible for the religions to give us the gift of the future, and they also cannot warn us about the wrong deeds we have done in the past. Without language, religion could not describe or explain the world before birth, realm of death, after death, and the afterlife. In other words, the grammatical structure of the language makes it theorizable and expressible so that it can substitute, squeeze, and push it in the minds of believers.

These stories have created social cohesion, traditionalism, and conformity to the existing laws in large sections of society and continue to do so. Divine religious books are compatible with such a style and prepare humans for the occurrence of such events. These books are primarily moral stories and do not describe a specific ideology or political doctrine. These stories generally revolve around invisible and immortal gods, supernatural forces, and contain laws and beliefs that have been deeply rooted in people's minds over time. These stories clearly involve magical and conjuring acts. For Jesus Christ, at least 37 miracles are written in four Gospels. Belief in miracles reduces anxiety, upsurges hope for the future and lessens fear of death. The belief in divine punishment itself is an effective factor in reducing social misbehavior and achieving salvation in the promised world for a community that has not yet learned lawfulness. The belief based on the existence of an all-powerful and omnipotent God who created the earth, heaven, and hell is perhaps the greatest trick ever performed in human history. The belief in monotheism, a god who leads you to heaven, is based on good behavior with a desirable morality in an ephemeral and lawless world, and of course, this requires the existence of a god who has similar thoughts feelings and emotions as humans. According to Robin Dunbar, British anthropologist, the structure of religion is based on a recursive thinking because it is a form of social activity based on shared knowledge and beliefs. The foundation of these shared beliefs is based on having awareness of the thoughts, interests, and beliefs of others. In keeping with him, "I imagine that you think that I believe there are gods who intend to influence our opinions because they apprehend and understand our wishes and desires"[800]. This type of recursive speech or thinking is of the fifth-degree type. In Dunbar's opinion, although religion does not have compatibility and is not agree with the theory of evolution and is in contradiction with it, there can be a scientific explanation for religion. Religion can be a form of neuro-psychologically adaptive behavior as long as it has the ability to promote adaptive behaviors in the real world. David Lewis-Williams, a South African archaeologist, supposed that religion was able to drive the Neolithic revolution from 12,000 to 17,000 years ago. During this period, humans shifted from being hunter–gatherers to domesticating animals and tried to create suitable habitats for their lives. By sacrificing animals and sometimes humans, they ensured that power remained in the hands of the rulers. This kind of thinking resulted in religion as a form of cultural adaptation, created to maintain peace and order in human settlements that were constantly accompanied by population growth[801]. Religion persists in its role of fostering urban civilization and communal activities that gathering people. In addition to storytelling, religion has had continuous and effective cooperation, especially in the development of art, music, and architecture, particularly in the creation and development of worship sites. It could be concluded that science is also constructed from human religious demands for answering the questions of where we come from and where we are going. However, religion, like language, has a dark side. Both cause believers and speakers of different religions and languages to alienate themselves from others. The formation of nations is based on language, and the formation of communities is based on religion. People's minds have become separated from each other based on language and religion[800]. This has led to wars and bloodshed between believers of different religions. In general, these stories cultivate moral qualities and help develop emotions and instill the concept of good and evil, right and wrong, in the mind, as in the end of the stories, good individuals are almost always successful and victorious. However, the concept of good is heavily dependent on linguistic or religious affinity. The foundations of these stories coincide with crime novels, which are accompanied by numerous secrets and mysteries.

Mystery stories are essentially literary works that originated from the West, unlike religious literature, which is exclusively associated with the East. The most important feature of mysteries is its depth of characterization and memorability along with its ideological conservative in defense of the social status quo[802]. In the mid-19th century, authors such as Edgar Allan Poe (1809–1849), with his reclusive detective "*The murders in the rue morgue*" published in *Graham* magazine in 1841, Auguste Dupin; and Wilkie Collins, English novelist especially known for "*The woman in white*" (1859) and "*The moonstone*" (1868), both mystery stories, began creating mystery books. Then, individuals like Arthur Conan Doyle, the creator of Sherlock Holmes, continued the tradition. The pleasure of writing these stories was so great for Arthur Conan Doyle that after the death of his legendary character, Sherlock Holmes, in the "*The final problem*" in 1893; he brought him back to life again in the book "*The return of Sherlock Holms*" in the following year, which was published as late as 1905. After that, prominent authors like Agatha Christie, the creator of Hercule Poirot and Miss Marple, Ian Fleming, the creator of James Bond and the Belgian writer Georges Simenon with his French vernacular character of Commissaire Jules Maigret entered this field.

Other stories teach us what goes on in the minds of others. We understand their motivations and actions. These books are of great importance for social understanding and the formation of common goals. However, myths delve into something beyond human capabilities. Children stories involve various beings such as angels, talking animals, and extraterrestrial creatures, and perhaps that is why myths like *Harry Potter* capture so much attention. They may serve as stimulants for adaptive and future-oriented behaviors[803]. They influence beliefs about life after death and provide solace for the fear of death, aiding in its subsequent oblivion. Divine books are more based on storytelling rather than being rooted in specific doctrines.

Although many animals have the ability for MTT, the extent of mental travel created by humans is vast and boundless, and its horizon is uncharted. In all likelihood, this vast realm of limitless imagination has led us to build big cities, create various tools, and employ complex machines. Language itself has evolved from storytelling to technical and mathematical capabilities for communication. Meanwhile, the special feature of human MTT has played an influential role in the formation of language grammar. It should be noted that language structure is not solely the result of MTT. Words are stored in semantic memory, and their grammatical rules are so ingrained in implicit memory that their retrieval does not require the slightest thought or contemplation. Although the language is initially formed based on episodic memory, it gradually sinks into semantic and implicit memory so that it creates a safe boundary for itself. However, it is worth noting that although language was initially used to express and share experiences in episodic memory and to express related stories. It has been utilized for other purposes such as teaching and educational function. Those usages have shown themselves more with the advent of toolmaking period.

Humanity learned writing around 5000 years ago, although unlike the ability to speak possessed by almost all people worldwide, many are still unable to write[804]. Franz Boas (1858–1942), German-American anthropologist, in his book "*The mind of primitive man*," mentions the fundamental difference between civilized and primitive humans as the ability to write[805]. Paul Broca also believed that literacy has been influential in the growth of the frontal lobe and human civilization[806]. Modern anthropologists no longer explicitly adhere to this categorization and do not divide humans into civilized and primitive groups. However, they emphasize the impact of literacy on the structural and functional development of the brain and the acquisition of complex cognitive skills, which require a significant amount of time for learning. This demonstrates how neuroplasticity gradually transforms literacy, a hardware system hardwired based on the human genome. Writing is a human invention that involves focusing the eyes on words (foveation) and utilizing the left occipital cortex for perceiving the shape, size, location, and font of letters or words with functional changes in the brain. The plasticity of the brain's structure helps gradually establish the visual pattern of letters in the motor system, enabling a child to learn writing. Of course, learning certain letters such as p, b, q, and d in the English language may be more challenging due to their spatial orientation. There are similar examples in other languages. Handwriting brings power with itself. Scientific research shows that the creation of syllabary (calligraphy) was first done about 3000 years before Christ in Eelam, Mesopotamia, now located at southwest of Iran, by the Sumerians, and then separately by the Chinese (1300 BC) and then by the natives of Mexico (600 BC) recreated. The three basic strategies that were initially used for writing included the use of voice, syllables, and words. Letters are used to make the alphabet, however, the sound that goes with it can be different, which is called a phoneme. For example, the English language has 26 letters and forty phonemes. The second strategy is to use logograms instead of words. Chinese Mandarin and Japanese Kanji languages use this feature. The hieroglyphic language of the ancient Egyptians was also like this. The third strategy is to use hyphens in words, which are called syllabaries. The principle of this phenomenon is the use of a voiced letter after a silent letter (like the word fa-mi-ly). The creation of handwriting the Sumerians was aimed at increasing agricultural production in the Fertile Crescent for auditing and accounting. According to Claude Levi-Strauss, French anthropologist, the invention of writing was for the enslavement of other people. The syllabary did not have much use for the hunter–gatherer man[807]. In the Sumerian language, they used three different signs. Logograms functioned with pictorial structure in language. Phonetic signs that were involved in spelling syllables, letters, grammatical elements and parts of words, and finally determinatives that are not pronounced but destroy ambiguities and inconsistencies in writing. The construction and creation of calligraphy in different regions of the earth by different peoples is not necessarily an independent invention, but it can be due to something we call it the "idea diffusion." In other words, the idea is taken from other scriptings, but it is made in a local style. Sequoyah also known as George Gist, an illiterate person of the Cherokee tribe of Indians in Arkansas used the pattern of white people's language and was able to create a writing model called the Cherokee syllabary for their language in 1820[808]. Similar work has been done in Iran for the Persian language branched from Aramaic and something similar for Chinese, Egyptian and Korean languages. The Korean Han'gul language is an adaptation of the Chinese and Mongolian languages. Of course, it should be kept in mind that the initial syllabaries were complicated and incomplete and created ambiguity.

Paul Broca also believed that literacy was effective in the development of the frontal lobe and human civilization at the same time. Today, anthropologists do not believe in the above division and do not divide humans into two groups, civilized and primitive, but they correctly insist on how literacy affects the structural and functional development of the brain and ultimately the acquisition of complex cognitive skills; skills that require a very long time to learn. This shows how plasticity in the nervous system causes literacy to gradually transform a hardwired system based on the human genome. Writing is a human invention that is accompanied by functional changes in the brain due to the need to focus the eyes on words (foveation) and use the left visual cortex (occipital) to understand the shape, size, location, and font of letters or words. The plastic structure of the brain helps the visual pattern of the letters to be developed in the motor system and the child gradually learns to write. As it was mentioned earlier, learning some letters like p, b, q, and d in English will be more difficult because of their spatial orientation.

Stories have connected people to each other and have been influential in shaping culture. Every culture has stories of valor and greatness that strengthen the sense of shared heritage. In modern times, this is achieved through writing and the use of words, but in our pre-verbal and pre-written era, gestures were used. The Piraha tribe in the Amazon is linguistically impoverished compared to classical languages[798]. They have a smaller vocabulary, lack explicit verbs for the past and future, and indirectly refer to the future and past. It is likely that at some point in time, they have forgotten their past and lost their sense of history.

In the course of language learning, in many languages, rhyme and meter were initially used, but with the emergence of writing ability, the intensity of this matter decreased, although it is still used for teaching children through poetry and rhyme. However, epic stories have maintained their poetic quality even in the current era. The first epic book, "*The epic of Gilgamesh*", was written about 4000 years ago in Nineveh, Iraq. The composition of "*Iliad*" and "*Odyssey*" by the Greek poet Homer dates back to approximately 800 years ago, "*Metamorphoses*" by Roman poet Ovid, "*Ayadgar-i Zariran*" which is a Zoroastrian middle Persian heroic poem, and Ferdowsi, the Persian poet, composed "*Shahnameh*" about a 1000 years ago. Other books in this field include Dante's "*Divine comedy*," G. Chaucer's "*The Canterbury tales*," J. Milton's "*Paradise lost*," and Lord Byron's "*Don Juan*," which belong to the 19th century.

WANDERING MIND

Daniel Everett, who ventured to the Maici River in the Amazon rainforest as a religious missionary to study and investigate the language of the Piraha tribe, found that it was impossible to impart or introduce religion to the tribe members. It seemed that the form and grammatical structure of the Piraha language were such that the concept of afterlife in another world did not exist for them. It is not known for sure if Everett later joined the atheist group for this reason or if there was another reason in the back of his mind[798,809].

The concept of language is dependent on the structure of memory. Different linguistic groups and structures have been observed in different languages, and they are understood in their own specific ways. For example, in ancient Hebrew, there is no tense, and temporal constraints are used to indicate the completion or continuation of a sentence. In Fijian, which is an Austronesian language of Polynesia, there are markers and specific signs for the past and present tense, but their use is entirely optional, and there is no verb like "to be" which appears a strange defect to the learners[810]. In Xu, a language spoken by African natives, there is no marker for sentence timing, but temporal constraints such as now, yesterday, later, and finally, exist. Sinhalese, a language belonging to the Indo-Aryan group in Sri Lanka, has only past and non-past tenses. Many languages do not have specific grammatical features. These examples show that cognitive linguistic structures are actually based on historical products that have themselves been influenced by cultural choices.

Episodic memory has its own unique characteristics and can identify occurrences at specific times. It may not be very precise, but it can provide a general idea. This allows us to have a general understanding of the concept of time. Episodic memory allows us to go back in time and retrieve and review past experiences. At the same time, it enables us to have the ability to travel to the future. Thomas Suddendorf in 1985 coined the term "Mental Time Travel" (MTT) to describe the concept of mind wandering over time[811,812]. This phenomenon is a specific cognitive capacity of humans anatomically related to the right prefrontal cortex. Episodic memory, along with MTT, allows us to construct a vocabulary for the future. Natural selection must undoubtedly help organisms not only survive in the present but also in the future. Perhaps one of the very important reasons why intelligent humans have been able to reproduce their population significantly is their ability to perceive and evaluate the future. Meta-cognition refers to having access to our own minds. In other words, it is the ability of an organism to know what it knows and what it does not know. The foundations of human meta-cognition begin around the age of three. Meta-cognition is an essential component of our cognitive, social, and communicative lives. However, it is not known for certain whether meta-cognition in humans and other animals fundamentally differs or not. But one thing is clear: human meta-cognitive behavior is dependent on prominent conscious memory and mental time travel.

A resting brain is also working, like an idling car. Contrary to many animals, humans often think about things that are not directly related to their immediate environment[813]. They ponder over past events, possible future occurrences, and phenomena that may never happen. This kind of stimulus independent thought, also known as mind wandering or mental travel over time, is actually the default state of human thinking. Evolutionarily, mind wandering has been found to be beneficial for learning, reasoning, and planning for the future, but it comes with a heavy emotional cost[814,815]. Many religious or ideological systems have been built around the notion that living in the present moment and not dwelling on the yet-to-come is futile. However, is living in the present moment and enjoying what exists truly compatible with the evolutionary trajectory of intelligent humans? Maybe we can consider mind wandering as a kind of cognitive control failure. This phenomenon particularly distracts the mind from the task at hand during focused activities or studying, resulting in wasted time, which humans pay a price for. Working memory likely has a direct relationship with mental activity and MTT, and it can be continually and permanently trained. Mental time travel not only helps us plan for future tasks but also keeps individuals' attention oscillating among various streams of thoughts, including current sensory information, future planning information, and retrieved information from long-term memory. In other words, thoughts flow and switch from one to the other, allowing us to have appropriate behaviors for multiple goals. This phenomenon prevents the mind from getting exhausted by performing a single task continuously and has a positive impact on the learning process.

In a research study, it was shown that while reading Leo Tolstoy's "*War and Peace*" for 45 minutes, on average, each person experienced 5.4 episodes of mind wandering. Recent studies have shown that 46% of individuals engage in mental travel during the day [629,711]. These thoughts are usually self-directed or goal-directed. Thoughts related to the past are mostly (up to 70% of cases) focused on the individual's own past, while thoughts related to the future are nearly equally distributed between future goals and self-focus. Other studies have shown that blood flow in the brain increases regionally during these mental activities. When the brain is not focused on a specific task, it starts traveling to the future or the past, consciously or unconsciously. However, it is not necessarily appropriate to consider mind wandering as a negative and undesirable phenomenon. Mind wandering is primarily associated with the prefrontal and parietal lobes in the brain, as well as a pre-established default network that facilitates shifting from the present moment to non-present time. Its typical form occurs during daydreaming, while a more complete form is observed in dreams during sleep. These dreams are usually related to past issues and can sometimes be disturbing. During sleep, the motor system is inactive, and since it requires memory related to previous matters, it is closely connected to episodic and semantic memories. However, many events observed in dreams are not real and are merely constructs of our wandering mind. These constructed unreal phenomena can enter our beliefs in wakefulness and appear real[816].

The ability to visualize the future by thinking about it begins around the age of three to four with the help of episodic memory. Patients with episodic memory disorders, such as Henry Molaison, not only struggle to remember the past effectively but also lack the ability to mentally travel to the future and plan for themselves[751]. Clive Wearing, a musician who suffered from a brain infection called herpes encephalitis at the age of 46, stated, "I am effectively dead, except for clinging to the present." He practically cannot recall anything except for a short period and also lacks the ability to travel to the future. Actually, he developed chronic antegrade and retrograde amnesia[817].

Mental time travel appears to be a unique human characteristic. Although evaluating such travels in other animals is challenging or even impossible because they can't speak, some studies have shown that a small number of animals can mentally store a short period of time for future evaluation and decision-making. It is likely that these animals have the ability to perceive the future based on their instinctive behaviors, such as obtaining food, migrating from one point to another, and building nests. Some of these behaviors are also learned in their environment. Wolfgang Köhler's study demonstrated in his book "*The mentality of apes*"[818] that chimpanzees do not have a good understanding of the future or the past[819,820]. It is possible that the last common ancestor between humans and chimpanzees diverged with the discovery of the ability to engage in mental time travel. A bird called Clark's nutcracker has a habit of storing seeds it finds for future consumption. It also has the ability to find them after hiding them. This indicates that this bird can plan for the future and retrieve its food reserves[821,822]. For scientists, having a memory that can answer questions about when, where, and what signifies having sufficient evidence to prove the existence of at least episodic memory for mental time travel to the past. However, the results of these studies have faced opposition in scientific communities, and counter-examples have also been reported. The main problem in animal studies is the inability to establish effective communication due to the lack of language. In other words, evolution in mental time travel may have occurred simultaneously with the evolution of language and speech abilities, and temporal travel of the mind may have been a prerequisite for learning language.

Based on Steven Pinker's theory, language is fundamentally designed to be able to tell who did what, to whom, when, and why, based on questions constructed from what, how, where, when, and who. These words precisely define the concept and importance of episodic memory[609,610]. These concepts can also be effectively applied to the future. For example, who will do what for whom in the future? And when we explore the future, conditional sentences and phrases will also become valuable. For instance, if it rains (another phenomenon will occur), the trees will become lush, or if it doesn't rain, we will enjoy sunny weather. In other words, language allows the sharing of our personal temporal episodes, and our abilities to construct a clear and visible future. That's why we are compelled to specify elements of episodes such as individuals, objects, actions, qualities, and occurrence times in our speech.

For this reason, different cultures have used various verb tenses to solve or modify temporal issues. These languages use past, present, and future verb forms to indicate temporal differences in events. Some languages, like Chinese, use constraints, changes in tone, or other markers instead of verbs to indicate this aspect. In very simple and primitive languages like Piraha in Brazil, spoken by around two hundred people, they simply use relative time expressed through two pseudo-time morphemes, indicating whether an event happened in the present or non-present time. According to Daniel Everett, the Piraha primarily and fundamentally live in the present[809]. Additionally, referring to time is done through the understanding of place or recalling it with the help of the hippocampus. For this reason, as previously mentioned, many temporal concepts and constraints are constructed based on spatial concepts. It seems that in the evolutionary process, the right hippocampus has specialized in spatial understanding, while the left hippocampus has specialized in temporal conception. It should be noted that as humans, we are required to express spatial–temporal events in a sequence and temporal flow. This leads to the emergence of nesting and embedding structures in sentences. The emergence of modern humans has been accompanied by an increase in the duration of infancy until adulthood, which is referred to as childhood, where additional stages of growth are added. Childhood is estimated to range from 2.5 to 7 years of age, which is approximately the same period as the MTT and the acquisition of syntactic language structures. It seems that in the evolutionary process, this period is unique to humans. Initially, sign language was more effective in expressing spatial concepts, but the addition of time has created limitations for it, and gradually, spoken language has proven to have better functionality.

In general, there is a consensus that Chomsky's theory, which considered grammar as the product of an inherent and universal human syntactic structure, is not correct or ground-breaking[584]. Instead, the final result is the product of human culture, intelligence, and genius, which have emerged within the limited framework of the human brain. Therefore, language is not only an expression of our speaking ability but also a reflection of the structure of human thought, encompassing basic and smaller elements such as everyday events, different situations, objects, materials, places, goals, and times. In other words, the ability to generate language is an indicator of the emergence of human cognitive processes.

Planning for the future is a prominent human characteristic. It should be noted that the association with episodic and semantic memory helps us plan for the future. Initially, planning was for daily activities. What is our next meal plan? Where do we want to travel? What is our retirement plan? Planning is the main property of our thinking. Suddendorf and Corballis[811] expanded the concept of MTT in 1997. Based on their opinion, this property arises at the end of the evolutionary period of humans. According to them, other animals' behavior is based on their current circumstances and motivations, lacking the ability to plan and predict for the future[812,813]. This theory is based on the hypothesis proposed by Bischof-Kohler. However, there are no strong reasons to prove that toolmaking by animals was a spontaneous phenomenon rather than an urgent need[823,824]. For example, crows use sharp-edged leaves from Pandonas trees as tools for feeding. They insert the sharp end into holes containing worms and watch over it with their beaks. This behavior is probably pre-planned and learned. Orangutans and bonobos usually keep objects they will need in the next 14 hours. Their MTT ability is likely limited to this extent. Chimpanzees keep tools similar to anvils and hammers for breaking food seeds. However, this behavior cannot be definitively attributed to their MTT ability.

The characteristic of recursion exists in MTT, and with its help, one can imagine an event that has occurred in the present time in the future or the past[768]. Events are time-dependent for us, and although MTT can represent the past or the future, we live in the present time, and what happens in the present has an impact on what we predict for the future. We live with time. We try to forget the negative events of the past and envision a better future in our minds. We live with clocks, calendars, and specific days. We journey millions, even billions of years into the past, all the way to the Big Bang, yet we still contemplate what existed or how it was before that.

Many of our behaviors are instinctual. These instinctual behaviors, which form the basis of evolution, are for our own survival or that of our offspring. These behaviors have evolved toward a better future and do not require MTT. Another part of our behavior is learned from parents and society, but MTT is a capacity that has evolved through natural selection, and its adaptation has been faster and more flexible than genetic probabilities in a complex world. MTT can lead us into an uncertain future. Although many organisms, including humans, are dependent on seasonal or day–night changes, these changes are probably due to sunlight and their internal circadian rhythm. However, as humans, we have classified time entirely in a linear manner according to intention and will. Seconds form minutes, minutes form hours, and so on. Although our real life is between birth and death, since humans have understood the concept of time, they have asked themselves to what extent we can stretch time metaphorically. Physicists estimate the beginning of the universe to be around 14 billion years ago. This is a very long time, and even with the help of MTT, it seems very difficult and impossible to travel to its end. Children usually have a basic ability to construct probable future events in their minds from the age of four onward, and they generally do not recall events before that. It seems that MTT, speech ability, and the ability to construct the future all grow and evolve simultaneously in our minds.

We have divided time into the past, present, and future. The pervasive concept of time has a high cost for us: the intangible cost of guilt from past carelessness, inappropriate behaviors, and past mistakes. Anxiety about what will happen in the future, and ultimately the worst and most dreadful of them all with a very distressing consequence: death. The ability to comprehend death, as the most intimate and perhaps the most terrifying human experience, along with the existence of time and MTT, is a unique human characteristic. Many religions believe in concepts such as life after death (immortality, eternity), the resurrection of the dead (incarnation), and divine wisdom (enlightenment), sometimes in multiple stages. According to the Indian philosophy, the moment of death is a chance for ultimate change called "Moksha." Alike other religions they believe that death like birth is a passage to another world[825]. However, this phenomenon leads people to believe that time continues after death, and they perceive the continuation of time as synonymous with the continuation of life. Humans cannot experience their own death; it is "the other" death that leads them to contemplate, and anxiety is the characteristic of the feeling of approaching death. One of the important functions of religions is to prove this mental belief that life after death will continue in places like paradise, hell, Nirvana, or Valhalla[89]. This mindset provides solace from the cold and sheer darkness of death. This type of thinking actually forms the foundation of faith. Faith causes believers to take any action and even sacrifice their lives for the sake of their beliefs and thoughts, due to the reward they believe will belong to them in the future. Belief in life after death has been observed in various customs and burial practices since ancient times. Placing objects and belongings in the graves of individuals indicates the existence of a belief in life after death. The oldest record of such behavior is related to the Qafzeh Cave in Israel, where a deer skull was buried along with the body of a child. It seems that this type of awareness of the beginning of human life has begun with the emergence of intelligent species[826–828].

Such behaviors have also been observed in Neanderthals, although these conducts may be purely behavioral rather than spiritual habits. According to some researchers, MTT leads to the creation of the concept of self-awareness, which is stretched over a time span. According to Hazel Markus and Paula Nurius, we can have "probable selves" by representing ourselves in the past and future[829]. Based on William James' perspective, every individual can have three "Me's: a past self, a present self, and a future self created through social circumstances." These selves can raise the possibility that the evolution of language has been accompanied by the evolution of MTT. As it was said before, time is designed specifically and uniquely to tell who did what to whom and why, where, when, from whom, and with what. This collection and combination that Steven Pinker added to the criteria of "when, where, what" create these qualitative phenomena as episodic memory counters. These concepts can create future tense verbs when combined with conditional speech. For example, if the weather is good, we will go on a trip. Conditioning can also create concepts related to the past, although it is not necessarily required.

New theories have emphasized the importance of the hippocampus in the formation of spatial concepts in episodic memory. Referring to a location may be the functional basis of the hippocampus. The hippocampus may analogously encode temporal codes along with space and time, meaning that temporal–spatial (time–space) episodes are represented in memory. For example, in the English language structure, a large number of words refer to both spatial and temporal concepts like: around, about, opposite, along, among, between, and against, although some of them, like since and until, have purely temporal usage. Some even believe that the right hippocampus is primarily involved in the retrieval of spatial information, while the left hippocampus is involved in the recovery of temporal details, facts, and coordination with the left hemisphere for speech. The most important parallel phenomenon between language and MTT is the phenomenon of generation. Language can linearize two-dimensional temporal–spatial concepts and express events only with a structured timeline. Language of gesture, in this regard, has a greater ability to express the dual characteristics of time and space in an event and can simultaneously express spatial concepts separately. Language has become so intertwined with episodic memory that sometimes it is unclear whether what is said is necessarily a real phenomenon stored in the mind or a story constructed and elaborated in our minds and stored in the hippocampus[830,831].

MIND READING

The concept of being able to decipher the minds of others or decode their thoughts has captivated intelligent beings since antiquity. The genesis of devices known as lie detectors often stems from this very notion. But can't we read each other's thoughts? Can't we be aware of another person's mind? And understand the thoughts of our friends? Can't we enter the minds of individuals through their eyes? Sometimes, when our friends ask us for money and we are unwilling to help them, we make up an excuse to not give them money, and of course, they clearly know that the reason we provide and what we say is different from what we have in mind, and we know that they know. In simple terms, we are aware of each other's minds. There are numerous daily examples of our awareness of the minds of our friends and colleagues that arise from frequent contact with them. The ability to understand what goes on in other people's minds is called theory of mind, which itself is a phenomenon of being intertwined and recursive.

Actually, this phenomenon is something enters into the mind of others. Reading the mind is a characteristic to be possessed, not to be done, and it is strongly manifested in humans, although it is observed to some degree in animals such as chimps. Mind reading forms the basis and the heart of the cognitive process for social communication. It seems that language plays an essential role in the development of theory of mind and its lack of development has a significant impact on shaping social relationships. Damage in such a process can be associated with diseases such as autism and schizophrenia. If the parents are well communicating in sign language and can teach this ability to their children, there is no disorder in the process of mind reading in congenitally deaf people[832].

Piping plover, a sparrow-sized shorebird in North America, falsely pretends to be injured in order to keep the predator away from its nest and to save the lives of her offspring, so that the greedy predator will follow her. In other words, she is lying on purpose[833]. Similar behaviors can be observed in other animals as they defend their offspring against predators. Understanding and being aware of the minds of others is a concept that humans have been interested in for a long time. The formation of the Supernatural Society in London, consisting of scientists, philosophers, doctors, and writers, aimed to determine the existence of supernatural forces such as spirits, levitation, trance states, and communication with spirits. Of course, there were also performers like James Randi, a Canadian-American stage magician, and Uri Geller, an Israeli-British illusionist, who claimed to bend spoons using mental power[834]. Many ordinary people resort to concepts such as subconscious awareness, remote control, or communication with the deceased until a scientific and clear explanation is provided to them regarding these matters. Concepts like the idea of dualism, which means the separation of the soul from the body, extended by René Descartes, are additional causes. He believed that after death, the soul separates from the body and flies into another world, and of course, we humans tend to believe and be inclined to such beliefs. It seems that humans are creatures made of flesh and blood that shape their thoughts, actions, and thinking in their brains. The belief in dualism and the separation of the soul from the body may be the result of humans' ability to mentally travel in time during their lifetime.

The ability to understand and comprehend the thoughts of others generally begins around the age of one and gradually develops fully within the first ten years of life. Great apes can understand the intentions of others but do not have the ability to understand others' false beliefs. Jean Piaget examined children's ability to evaluate what they themselves see compared to what the interviewer sees through the "Three Mountains Task," which was developed by him. The test was developed to check the child's thinking for egocentrism[835] that is inability to differentiate between self and the other. A model of three mountains with different shapes and sizes is placed in front of the child, and they are asked to describe what they see. Then, the child switches places with the interviewer, and they are asked again what the interviewer sees. Younger children do not have the ability to understand that what they see is different from what the interviewer sees. In fact, reading the minds of others is difficult for them. Theory of mind, also known as mind reading, refers to the daily ability to attribute separate and independent thoughts to oneself and others for the purpose of predicting and describing behavior. It seems that we infer and interpret the behavior of others based on what they say, but in reality, our behavioral experiences are involved in interpreting it. In addition to this method, there are at least two other classical methods for understanding the thoughts and emotions of others. The first is access to and analysis of the emotions of others, and the second is the ability to understand their movements or intentions to perform an action. Unlike empathy and socio-emotional situations, which are related to the insula and cingulate gyrus, theory of mind, and mind reading abilities are dependent on the prefrontal cortex, temporo-parietal junction, and temporal pole.

From a practical standpoint, theory of mind is classified into different levels and stages. Level zero indicates that there are no beliefs, interests, or goals in the mind and is usually applicable to non-living beings such as rocks and tools. Then, at the first level, a living being only has its own beliefs but does not concern itself with other beings of its kind. For example, ants individually collect food and do not directly interact with their surroundings or fellow ants. The second level represents a living being (me) that knows that another being also has beliefs and beliefs of its own. At the third level, a deeper understanding is acquired. "Does he know that I know?" For example, does he know that I know where the sweets are hidden? The "Sally-Anne Test," which was developed to measure peoples' social cognitive ability, can be used to investigate these phenomena.

The Sally-Anne Test, which was initially designed by Simon Baron-Cohen to evaluate autistic children, practically examines theory of mind[836]. The test is conducted by introducing Sally and Anne, both dolls, to the child initially. Sally puts a small marble into her basket and then leaves the room. After Sally's departure, Anne takes the marble out of the basket and puts it into her box. When Sally returns, the crucial question is asked to the child: "Where will Sally look for her marble?" The correct answer to this question indicates that the child can respond based on others' perspectives. In the initial study by Simon Baron-Cohen, approximately 20% of autistic children gave the correct answer to the question. The correct answer demonstrates that the child can understand that Sally has her own beliefs and thoughts, even though they do not align with reality. This ability typically begins around 4–5 years old and fully develops around 8–10 years old. This test has the ability to evaluate children in understanding inferential and false belief tasks. The human brain is highly capable of understanding how others think. Almost all of us have experienced this ability. Some of it is inherently innate, while another part is culturally learned. Shared emotions that others see, hear, or feel, as well as inference and belief-sharing through storytelling, stem from this concept. The strong and effective mechanisms through which we humans can evaluate others' beliefs occur due to the presence of a specific part of human personality known as "social senses." This phenomenon is one of the cognitive preconditions of human beings for the evolution of their social structures, which are acquired or taught from childhood.

It seems that the survival ability of humans during the Pleistocene era was not necessarily due to physical abilities but rather to intelligence and social collaboration. Language allows us to share inner thoughts. Of course, the concept of thinking must also be clarified. Words are deeply intertwined with our mental processes. Immanuel Kant, a prominent German philosopher, regarded thinking as synonymous with speaking, and according to Merlin Donald, a Canadian psychologist, when the brain begins to construct a coded verbal world in the structure of the mind, simultaneously with the evolutionary changes in the prefrontal cortex; the result of this process is thinking and words[837], actually thinking needs words. It appears that these two are inseparable from each other. Overall, thinking is a phenomenon that emerges in the human subconscious, and its outward manifestation is language. However, Steven Pinker believes that language is a useless environment for thinking because words can have various meanings. In his view, language is a means of communication, not a form of thinking[610]. However, language alone has a significant impact on many thoughts. This phenomenon can arise from awareness of others' opinions, reading or hearing their opinions, or through education and learning, joining in. However, as a language-sharing system, it is highly adaptable. Humans adjust their behavior in many cases based on the speech of others. Children easily stay away from danger through learning and education from their parents. Others learn various topics and concepts through stories and written texts. Thinking in children is predominantly narrated through storytelling, referred to as "narrative thinking." Although stories are described to children in their mother language, they remain spatially in their minds as events unfold.

In fact, theory of mind also has the property of mutual awareness, for example, I know that you know that I know. The problem precisely starts from that point "when I know that you know." If I know how you think about me, then I can behave in a way that deceives you, and deception starts from here, and the prerequisite for deception is talking and lying[629,711]. In other words, *sapiens* started the speech with lies. The capacity for reading minds varies among different individuals. A group of autistic patients, with a great ability in this area and of course the ability to understand deductive mathematical concepts, has been labeled as Asperger's syndrome. However, mind reading has led people to evaluate others based on what goes on in their minds rather than solely on their physical presence. For example, seeing a stranger only in a dark and rainy night compared to seeing the same person in a bright sunny day has a different interpretation for us. Humans like to apply such interpretations to animate objects or things, such as concepts like feminine or masculine fragrance, and feminine or masculine colors. Additionally, similar cases are seen in children's cartoons such as Winnie-The-Pooh, Donald Duck, Pinocchio[838], and others. Deception, lies, and trickery in cartoon characters resemble more the ability of human mind reading than what actually happens in reality. One of the fundamental debates among scientists is whether animals have the ability to read minds or not. Some believe that Darwin's theory of evolution, which states that the difference in human characteristics from other animals in evolutionary terms is merely a matter of degree and not kind, is fundamentally incorrect. Michael Tomasello believes that chimpanzees have the ability to understand goals, purposes, and intentions, but they lack awareness and the ability to understand the beliefs and needs of others[736]. When a chimpanzee points to food with its hand, it means it wants it, but it is not aware of our understanding and comprehension of its behavior. Humans and our ancestors have had and have a deeper understanding of this phenomenon. This level and degree of understanding are likely prerequisites for language. This gradual emergence of complexity in social behaviors is known as "social intelligence."

One of the characteristics of human language is its indeterminacy. This means that words in sentences can have multiple meanings especially in proverbs, metaphors, and mottos. For example, in English, "time is brain" implies that quick action is necessary for treating brain diseases, but we express it in a very simple way: "time is brain." There are other expressions in different languages. Scott-Phillips[839] writes in his book "*Speaking our minds*" that this phenomenon is an inherent and unavoidable property of human language. While this property makes language uncertain and difficult to understand, it can sometimes lead to misunderstandings but also allows for flexibility and innovation[840,841]. This form of language, which is both expressive and inferential, is called "descriptive–inferential," meaning it both describes and analyzes itself. For example, when you lift your empty cups and gesture to the waiter in a restaurant, he understands that he should refill them. Another example is during exams, where simply raising your hand to the examiner indicates that you have a question about the exam process.

Our minds try to discover various deceiving methods and act as an organism to ensure survival. However, pretenders try to deceive the minds of others through their behavior and actions. This has led opportunistic individuals to try to exploit the talents of this miracle for their own benefit. This ongoing process continues until someone succeeds. The concept of lying emerges from here, and a deceptive person comes into existence. All of this is based on the sequence of being intertwined and gives rise to an arms race in understanding. The ability to acquire something based on recursive thinking is called "Machiavellian intelligence"[518,519]. The depth of recursive thinking ability in humans allows them to read the minds of other living beings better than other creatures. As Daniel Dennett says, we interact with other people based on what we think is going on in their minds, not based on their external manifestations and apparent behaviors. This collection of other people's minds includes their beliefs, desires, hopes, fears, etc. The spiral of deception and trickery increases the collective intelligence of a species, which is highly effective in creating the next generation[842]. Mental time travel in other people's minds creates a platform that is specifically and universally relevant to humans.

Recursive thinking and mind reading demonstrate different levels and degrees of intention and will. The following examples are simple sentences that demonstrate the property of recursive thinking. Level zero: Only instinctive and reflexive reactions; Level one: Nathalie wants to go out with her friend. Level two: Albert knows that Nathalie wants to go out with her friend. Level three: Nathalie believes that Albert knows that she wants to go out with her friend. Level four: Albert believes that Nathalie knows that he (Albert) knows that they (Nathalie and her friend) want to go out. And this story continues in the same way.

The ability to read minds has allowed human beings to attribute living personalities to inanimate objects. The remaining names of stars and planets are derived from this type of thinking. This phenomenon has played a significant role in storytelling and mythology for children. Perhaps the highest ability and triumph of humans have been in creating an imaginary character without a physical body known as God. It seems that recursive thinking has greatly contributed to shaping this mindset. A fundamental change in the cognitive niche (cognitive changes in animals resulting from environmental changes) of our modern human ancestors has enabled their survival and endurance through increasing population and social cooperation, and this population growth has also increased social complexity[843]. Social relationships resulting from familial and kinship connections have grown faster than their numbers, thus leading to the formation of more complex structures. This increase in numbers has been accompanied by the formation of families and tribes, but according to Robin Dunbar, human communities were accustomed to functional groups of 150 individuals, which is called "Dunbar's number"[511]. The structure of these groups has also grown horizontally, forming different subgroups with varying specializations. In such structured societies, the formation and development of theory of mind is crucial for stability and effective cooperation. The most important function of theory of mind during that period was hunting for survival. Increased cooperation and better collaboration led to quicker and better mental connections among individuals. The enhancement of this cooperation and social intelligence gradually enabled humans to surpass environmental challenges, but a greater danger loomed over them, which was their own existence. Complex and sometimes unconventional behaviors of individuals in a society with a complex structure posed a much greater threat than problems such as climate change, food scarcity, diseases, and predatory animals.

It seems that among animals, dogs have a good understanding of the concept of gesture and pointing[844]. They possess the skill and ability to mysteriously comprehend what goes on in their owner's mind. They can effectively understand gestures and pointing to identify which container the food is hidden in. Although dogs have evolved from wolves, their abilities far surpass those of wolves, possibly due to domestication. Brian Hare, professor of evolutionary anthropology at Duke University, believes that dogs gradually developed a liking for human leftovers and some were selectively bred to resist attackers[845]. Another animal that has domesticated to some extent without human intervention is the bonobos. Chimpanzees are known to be very aggressive, attacking females and infants, while bonobos are more peaceful and as mentioned, resolve their issues through sexual services. Many bonobos were on the verge of extinction due to their calm behavior in the Congo region until protected areas were established for them. Interestingly, in the domestication process, brain size has reduced, but their practical abilities have improved. The brains of dogs are smaller than those of wolves, bonobos have smaller brains than chimpanzees, and modern humans have smaller brains than Neanderthals, yet their cognitive abilities have advanced.

Deception in chimpanzees manifests itself as a tactical trickery or bluffing. This kind of behavior, which is a form of thinking and can be observed in obtaining food among chimpanzees, is a situation based on intention and will. In humans, this phenomenon is displayed as lying. Machiavelli believed that lying and deception toward a liar and deceiver brings double pleasure. However, this constant cycle of deception leads to an upsurge in the resilience and critical thinking of society. Modern and densely populated societies make it easier for deceivers and liars to remain anonymous among numerous and unknown individuals in society. Accomplishing this in smaller communities is much more challenging. In other words, intertwined social life has led to an increase in the collection of deception, thus enhancing social intelligence and the ability to read minds.

Indeed, in animals, there are not elaborate reasons for the benefit of mind reading. Overall, it appears that theory of mind is relevant to animals with abundant social lives. Social life, especially its human model, requires clear communication. This communication is not primarily between two bodies but between two mental states. Our mental function occurs on a daily basis in the form of conversations, which are calibrated and regulated based on theory of mind. The founder of the concept of theory of mind, Herbert Paul Grice, a British philosopher of language, argues that genuine language requires the speaker to have an intention chain of beliefs that are expressed by the speaker and lead to the modification and understanding of that belief in the listener's mind[846]. The expressed sentences contain different thinking processes and implicit intentions, which are dependent on specialized modules related to theory of mind. These modules are inherently created, forming a pragmatic or utilitarian set that operates automatically, just like maintaining balance when walking. Ascribing different mental states such as intention, knowledge, beliefs, and so forth to selves and the others needs having a noble, worthy, and processed theory of mind.

We are constantly constrained in our interpretation of ongoing conversations based on our memory-recorded awareness and cultural habits. Conversations are arranged and distinguished in real-time and progress forward. In other words, our mind continuously and moment by moment adjusts itself with a thinking process, reducing our need for linguistic comprehension. This causes us to become entrenched in the intended meaning and simply grasp the desired concept. We may forget other environmental events, which overall can lead to the emergence of unclear and irrelevant information in sentences and discourse. As a result, all participants in a conversation arrive at a shared mental understanding of what is being discussed. In essence, language is nothing but a meeting of minds, and conversation, a shared flow of thoughts[847]. In conversation, the speaker implicitly assumes entry into the listener's stream of thought and adjusts it in a way that the listener comprehends it. However, it is possible for the speaker to misinterpret the listener's stream of thought or vice versa, resulting in one of the two individuals making a mistake, and the listener remains unaware. In fact, language is a kind of selective exercise and selectivity in which information is shared as much as it is instructive for mutual understanding. This phenomenon poses difficulties in speech and conceptual communication for individuals with autism who have fundamental challenges in theory of mind.

At the same time, it must be acknowledged that gossip, backbiting, chats, and daily talks are so intellectually insignificant that they make humans seem like newcomers who have just learned a trick and are eager to show it off. On the other hand, satire, proverb, and metaphor are excellent examples of the concept of theory of mind, where the listener is required to grasp the hidden meaning in humor and metaphor. Theory of mind allows us to express and share unspoken thoughts accurately and comprehensively through the flexibility of language. This, in turn, enables the true message in language to be displayed with greater clarity, prominence, and importance. However, linguistic impairment in autistic patients is primarily related to pragmatism, which refers to the language's adaptation to social concepts and the themes of the real world. The use of language in these individuals is more about gathering information rather than sharing it, but understanding proper language is not only related to information gathering; it also includes theory of mind and sharing them[848].

It should also be noted that the roots of gesture and pointing in language initially served merely as pointing[621]. In monkeys, who do not possess a high level of mental capacity, more than 96% of their communication with humans is in the form of directive concepts or orders. However, among humans, these concepts exist in both directive and descriptive forms, starting from around fourteen months of age, and theory of mind progressively develops and completes by the age of four. In fact, theory of mind is a prerequisite for language before its formation and development in the mind. It seems that the process of natural selection since the Pleistocene period, from 2.6 million years ago to about 12,000 years ago, and the transformation of humans from arboreal creatures to terrestrial beings, turning them into omnivorous living beings with the ability to speak and think, was highly effective. However, the main question is why this event occurred only in the intelligent species (*sapiens*) among approximately twenty *Homo* species.

References

1. Ashwell K. Neurobiology of monotremes: Brain evolution in our distant. Collingwood, Australia: CSIRO Publishing, 2013.
2. Alvarez LW, Alvarez W, Asaro F, Michel HV. Extraterrestrial cause for the cretaceous-tertiary extinction. Science 1980;208(4448):1095–108.
3. Alvarez LW. Alvarez: Adventures of a Physicist. New York: Basic Books, Inc., 1987.
4. Sanderson K. Did dust from the Chicxulub asteroid impact kill the dinosaurs? Nature 2023;622(7984).
5. Parks J. Asteroid dust found at Chicxulub Crater confirms cause of dinosaurs' extinction. Astronomy. 2023. (https://www.astronomy.com/science/asteroid-dust-found-at-chicxulub-crater-confirms-cause-of-dinosaurs-extinction/).
6. Rose KD. The earliest primates. Evol Anthropol 1994;58(2):159–70.
7. Gruber T, Clay Z. A comparison between bonobos and chimpanzees: A review and update. Evol Anthropol 2016;25(5):239–52.
8. Simpson G. The beginning of the age of mammals. Biol Rev 1937;12(1):1–46.
9. Western D, Ssemakula J. Life history patterns in birds and mammals and their evolutionary interpretation. Oecologia 1982;54(3):281–90.
10. Goldby F, Gamble HJ. The reptilian cerebral hemispheres. Biol Rev 1957;32(4):383–420.
11. Northcutt RG. Variation in reptilian brains and cognition. Brain Behav Evol 2013;82(1):45–54.
12. Herculano-Houzel S, Kaas JH, de Oliveira-Souza R. Corticalization of motor control in humans is a consequence of brain scaling in primate evolution. J Comp Neurol 2016;524(3):448–55.
13. Felton AM, Felton A, Lindenmayer DB, Foley WJ. Nutritional goals of wild primates. Functional Ecology 2009; 23:70–8.
14. Sousa AMM, Meyer KA, Santpere G, et al. Evolution of the human nervous system function, structure, and development. Cell 2017;170(2):226–47.
15. Ribas de Pouplana L, Delaye L, Becerra A, Lazcano A. The nature of the last common ancestor: The genetic code and the origin of life. Boston, MA: Springer; 2004:34–47.
16. Goodall J. Chimps: Atheneum. New York, NY: Byron Preiss Book; 1989.
17. Wood B. Origin and evolution of the genus Homo. Nature 1992;355(6363):783–90.
18. Tobias PV. Raymond Arthur Dart (1893–1988). Nature 1989;337(6204):211.
19. Oxnard CE. The place of the australopithecines in human evolution: Grounds for doubt? Nature 1975;258(5534):389–95.
20. Leakey MD, Hay RL, Curtis GH, et al. Fossil Hominids from the Laetolil Beds. Nature 1976;262 (5568):460–6.
21. McHenry HM, Berger LR. Body proportions of Australopithecus afarensis and A. africanus and the origin of the genus Homo. J Hum Evol 1998;35(1):1–22.
22. Lockwood CA. Sexual dimorphism in the face of Australopithecus africanus. Am J Phys Anthropol 1999;108(1):97–127.
23. DeSilva J, McNutt E, Benoit J, Zipfel B. One small step: A review of Plio-Pleistocene hominin foot evolution. Am J Phys Anthropol 2019;168(Suppl 67):63–140.
24. Kimbel WH, Villmoare B. From Australopithecus to Homo: The transition that wasn't. Philos Trans R Soc Lond B Biol Sci 2016;371(1698):20150248.
25. Grant BR. Evolution in Darwin's finches: A review of a study on Isla Daphne Major in the Galapagos archipelago. Zoology (Jena) 2003;106(4):255–9.
26. Kutschera U. A comparative analysis of the Darwin-Wallace papers and the development of the concept of natural selection. Theo Biosciences 2003;122:343–59.
27. Beddall BG. Wallace, Darwin, and the theory of natural selection: A study in the development of ideas and attitudes. J Hist Biology 1968;1(2):261–323.
28. Fehr E, Fischbacher U. The nature of human altruism. Nature 2003;425(6960):785–91.
29. Combe G. The constitution of man. New York, NY: Cambridge Uni. Press, 2009/1828.
30. Lazar JW. Nineteenth- and twentieth-century brain maps relating to locations and constructions of brain functions. J Hist Neurosci 2022;31(2–3):368–93.
31. Wade NJ. The Darwins and Wells: From revolution to evolution. J Hist Neurosci 2010;19(2):85–104.
32. Bolhuis JJ, Everaert M. Birdsong, speech, and language: Exploring the evolution of mind and brain. Cambridge, MA: MIT Press, 2013.

33. Shermer M. In Darwin's shadow: The life and science of Alfred Russel Wallace. New York: Oxford Uni. Press, 2002.
34. Wallace AR. On the tendency of varieties to depart indefinitely from the original type. In: Smith C, ed. Alfred Russel Wallace classic writings. Bowling Green, KY: Western Kentucky Uni.; 2009/1858.
35. Boyer DM, Yapuncich GS, Chester SB, et al. Hands of early primates. Am J Physic Anthropol 2013; 152:33–78.
36. Smith BH, Tompkins RL. Toward a life history of Hominidae. Annu Rev Anthropol 1995;24:257–79.
37. Sarich VM, Wilson AC. Rates of albumin evolution in primates. Proc Natl Acad Sci U S A 1967; 58(1):142–8.
38. Sarich VM. Generation time and albumin evolution. Biochem Genet 1972;7(3):205–12.
39. Suntsova MV, Buzdin AA. Differences between human and chimpanzee genomes and their implications in gene expression, protein functions and biochemical properties of the two species. BMC Genomics 2020;21(Suppl 7):535.
40. Davis AC, Arnocky S. Darwin versus Wallace: Esthetic evolution and preferential mate choice. Front Psychol 2022;13:862385.
41. Belyayev DK, Trut LN. Behaviour and reproductive function of animals. II. Correlated changes under breeding for tameness. Bull Moscow Soc Natur Biol Series 1964;69(5):5–14.
42. Mukhamedshina IA, Kharlamova AV, Trut LN. [Does the selection of fox for tame and aggressive behavior changes their ability to focus attention, and training the motor skills?] (In Russian). Zh Vyssh Nerv Deiat Im I P Pavlova 2014;64(5):521–30.
43. Wilkins AS, Wrangham RW, Fitch WT. The "domestication syndrome" in mammals: A unified explanation based on neural crest cell behavior and genetics. Genetics 2014;197(3):795–808.
44. Theofanopoulou C, Gastaldon S, O'Rourke T, et al. Self-domestication in Homo sapiens: Insights from comparative genomics. PLoS One 2017;12(10):e0185306.
45. Wagner GP. The logical structure of irreversible systems transformations: A theorem concerning Dollo's law and chaotic movement. J Theor Biol 1982;96(3):337–46.
46. Whiting MF, Bradler S, Maxwell T. Loss and recovery of wings in stick insects. Nature 2003;421 (6920):264–7.
47. Gould SJ, Vrba ES. Exaptation: A missing term in the science of form. Paleobiology 1982;8(1):4–15.
48. Baldwin JM. A new factor in evolution. Am Nat 1896;30:441–51.
49. Sznajder B, Sabelis MW, Egas M. How adaptive learning affects evolution: Reviewing theory on the Baldwin effect. Evol Biol 2012;39(3):301–10.
50. Scheiner SM. The Baldwin effect: Neglected and misunderstood. Am Nat 2014;184(4):ii–iii.
51. Hauser M. Our chimpanzee mind. Nature 2005;437(7055):60–3.
52. Tooby J, DeVore I. The reconstruction of Hominid evolution through strategic modeling. In: Kinsey WG, ed. The evolution of human behavior: Primate Models. Albany, NY: SUNY Press; 1987.
53. Pinker S. The cognitive niche: Coevolution of intelligence, sociality, and language. Proc Natl Acad Sci U S A 2010;107(Suppl 2):8993–9.
54. Clark A. Language, embodiment, and the cognitive niche. Trends Cogn Sci 2006;10(8):370–4.
55. Leakey LS. New finds at Olduvai Gorge. Nature 1961;189:649–50.
56. Leakey LS, Leakey MD. Recent discoveries of fossil Hominids in Tanganyika: At Olduvai and near lake Natron. Nature 1964;202:5–7.
57. Leakey LS, Tobias PV, Napier JR. A new species of the genus Homo from Olduvai Gorge. Nature 1964;202:7–9.
58. Leakey RE. A review of the evidence for our African origins. Ethn Dis 1991;1(1):8–20.
59. Leakey RE. Further evidence of Lower Pleistocene Hominids from East Rudolf, North Kenya, 1973. Nature 1974;248(5450):653–6.
60. Howell FC. Homo habilis in detail. Science 1991;253(5025):1294–5.
61. Tobias PV. Homo habilis and Homo erectus: From the Oldowan men to the Acheulian practitioners. Anthropologie 1980;18(2/3):115–9.
62. Haeusler M, McHenry HM. Body proportions of Homo habilis reviewed. J Hum Evol 2004;46(4):433–65.
63. Wood B. Human evolution: Fifty years after Homo habilis. Nature 2014;508(7494):31–3.
64. Prang TC. New analyses of the Ardipithecus ramidus foot provide additional evidence of its African ape-like affinities: A reply. J Hum Evol 2022;164:103135.
65. Meldrum DJ, Hilton CE. From biped to strider: The emergence of modern human walking, running, and resource transport. New York: Kluwer Academic/Plenum, 2004.
66. Lovejoy CO. Evolution of human walking. Sci Am 1988;259(5):118–25.

67. Lovejoy CO. The natural history of human gait and posture. Part 1. Spine and pelvis. Gait Posture 2005;21(1):95–112.
68. Gruss LT, Schmitt D. The evolution of the human pelvis: Changing adaptations to bipedalism, obstetrics and thermoregulation. Philos Trans R Soc Lond B Biol Sci 2015;370(1663):20140063.
69. Rosenberg KR, DeSilva JM. Evolution of the human pelvis. Anat Rec 2017;300(5):789–97.
70. Lewis CL, Laudicina NM, Khuu A, Loverro KL. The human pelvis: Variation in structure and function during gait. Anat Rec (Hoboken) 2017;300(4):633–42.
71. McNutt EJ, Zipfel B, DeSilva JM. The evolution of the human foot. Evol Anthropol 2018;27(5): 197–217.
72. Wang W, Abboud RJ, Gunther MM, Crompton RH. Analysis of joint force and torque for the human and non-human ape foot during bipedal walking with implications for the evolution of the foot. J Anat 2014;225(2):152–66.
73. Lovejoy CO. The natural history of human gait and posture. Part 2. Hip and thigh. Gait Posture 2005;21(1):113–24.
74. Helms JA, Schneider RA. Cranial skeletal biology. Nature 2003;423(6937):326–31.
75. Russo GA, Kirk EC. Another look at the foramen magnum in bipedal mammals. J Hum Evol 2017; 105:24–40.
76. Turner KL, Abernethy EF, Conner LM, et al. Abiotic and biotic factors modulate carrion fate and vertebrate scavenging communities. Ecology 2017;98(9):2413–24.
77. Mackowetzky K, Yoon KH, Mackowetzky EJ, Waskiewicz AJ. Development and evolution of the vestibular apparatuses of the inner ear. J Anat 2021;239(4):801–28.
78. Thelen E, Fisher DM, Ridley-Johnson R, Griffin NJ. Effects of body build and arousal on newborn infant stepping. Dev Psychobiol 1982;15(5):447–53.
79. Little MA. Evolutionary strategies for body size. Front Endocrinol (Lausanne) 2020;11:107.
80. Pritchard VE, Malone SA, Burgoyne K, et al. The relationship between handedness and language ability in children. Wellcome Open Res 2019;4:30.
81. Mesquita RM, Catavitello G, Willems PA, Dewolf AH. Modification of the locomotor pattern when deviating from the characteristic heel-to-toe rolling pattern during walking. Eur J Appl Physiol 2023; 123(7):1455–67.
82. Turner W. On M. Dubois' description of remains recently found in Java, named by him Pithecanthropus erectus: With remarks on so-called transitional forms between Apes and Man. J Anat Physiol 1895; 29(Pt 3):424–45.
83. Marsh OC. The Ape-Man from the Tertiary of Java. Science 1896;3(74):789–93.
84. Stone R. Meeting briefs: Java Man's first tools. Science 2006;312(5772):361.
85. Ciochon RL, Huffman OF. Java Man. In: Smith C, ed. Encyclopedia of global archaeology. New York: Springer-Verlag; 2014.
86. Morriss-Kay GM. The evolution of human artistic creativity. J Anat 2010;216(2):158–76.
87. Marston AT. The Swanscombe skull. J Royal Anthropol Institute Great Britain Ireland 1937;67:339–406.
88. Pfeiffer JE. The emergence of man. New York: Harper & Row, 1969.
89. Corballis MC. The lopsided ape: Evolution of the generative mind. New York: Oxford Uni. Press, 1993.
90. Ready E. Neandertal man the hunter: A history of Neandertal subsistence. Explor Anthropol 2010; 10(1):58–80.
91. Wong K. The mysterious downfall of the Neandertals. Sci Am 2009. (https://www.scientificamerican.com/article/the-mysterious-downfall/).
92. Hershkovitz I, Weber GW, Quam R, et al. The earliest modern humans outside Africa. Science 2018; 359(6374):456–9.
93. Kerner G, Neehus AL, Philippot Q, et al. Genetic adaptation to pathogens and increased risk of inflammatory disorders in post-Neolithic Europe. Cell Genom 2023;3(2):100248.
94. Harper KN, Armelagos GJ. Genomics, the origins of agriculture, and our changing microbe-scape: Time to revisit some old tales and tell some new ones. Am J Phys Anthropol 2013;152(Suppl 57):135–52.
95. Lawton G. Death: The evolution of funerals. New Scientist, 2012.
96. Howell FC. European and northwest Affrican middle pleistocenehominids. Cur Anthropol 1960; 1:195–232.
97. Howell FC. Man's evolutionary past. Soc Biol 1972;19(2):128–35.
98. Shipman P. The Invaders: How humans and their dogs drove Neanderthals to extinction. Cambridge, MA: Harvard Uni. Press, 2017.
99. Rightmire GP. The dispersal of Homo erectus from Africa and the emergence of more modern humans. J Anthropol Res 1991;47(2):177–91.

100. Mori F, Nakajima K, Tachibana A, et al. Reactive and anticipatory control of posture and bipedal locomotion in a nonhuman primate. Prog Brain Res 2004;143:191–8.
101. Aiello L, Dean C. An introduction to human evolutionary anatomy. San Diego, CA: Academic Press, 1990.
102. Russo GA, Kirk EC. Foramen magnum position in bipedal mammals. J Hum Evol 2013;65(5):656–70.
103. Dao AH, Netsky MG. Human tails and pseudotails. Hum Pathol 1984;15(5):449–53.
104. Coppens Y. East side story: The origin of humankind. Sci Am 1994;270(5):88–95.
105. Whittle MW. Clinical gait analysis: A review. Hum Mov Sci 1996;15(3):369–87.
106. Kharb A, Saini V, Jain YK, Dhiman S. A review of gait cycle and its parameters. Int J Computa Engin Manage (IJCEM) 2011;13:78–83.
107. Cieri RL, Churchill SE, Franciscus RG, et al. Craniofacial feminization, social tolerance, and the origins of behavioral modernity. Cur Anthropol 2014; 55(4):419–43.
108. Hewes GW. Hominid bipedalism: Independent evidence for the food-carrying theory. Science 1964; 146(3642):416–8.
109. Steudel K. The work and energetic cost of locomotion: The effects of limb mass distribution in quadrupeds. J Exp Biol 1990;154(1):273–85.
110. Watson JD, Crick FH. Molecular structure of nucleic acids: A structure for deoxyribose nucleic acid. Nature 1953;171(4356):737–8.
111. Watson JD, Crick FH. Genetical implications of the structure of deoxyribonucleic acid. Nature 1953; 171(4361):964–7.
112. Klug A. Rosalind Franklin and the discovery of the structure of DNA. Nature 1968;219(5156):808–10.
113. Luboga SA, Wood BA. Position and orientation of the foramen magnum in higher primates. Am J Phys Anthropol 1990;81(1):67–76.
114. Morgan E. The descent of woman. London, UK: Souvenir Press, 1972.
115. Morgan E. The scars of evolution. London, UK: Souvenir Press, 1990.
116. Cunnane SC. The aquatic ape theory reconsidered. Med Hypotheses 1980;6(1):49–58.
117. Morgan E. The aquatic ape. New York, NY: Stein and Day, 1982.
118. Fitch WT, Reby D. The descended larynx is not uniquely human. Proc Biol Sci 2001;268(1477):1669–75.
119. Langos D, Kulik L, Mundry R, Widdig A. The impact of paternity on male-infant association in a primate with low paternity certainty. Mol Ecol 2013;22(13):3638–51.
120. Sheldon BC. Relating paternity to paternal care. Philos Trans R Soc Lond B Biol Sci 2002;357(1419):341–50.
121. Finlayson C. The improbable primate: How water shaped human evolution? Oxford, UK: Oxford Uni. Press, 2014.
122. Dixson A. The evolution of neuroendocrine mechanisms regulating sexual behaviour in female primates. Reprod Fertil Dev 2001;13(7–8):599–607.
123. Marzoli D, Havlíček J, Roberts SC. Human mating strategies: From past causes to present consequences. Wiley Interdiscip Rev Cogn Sci 2017; 9(2): e1456. doi:10.1002/wcs.1456.
124. Walker KK, Walker CS, Goodall J, Pusey AE. Maturation is prolonged and variable in female chimpanzees. J Hum Evol 2018;114:131–40.
125. Thornhill R, Gangestad SW. The evolution of human sexuality. Trends Ecol Evol 1996;11(2):98–102.
126. Gray PB. Evolution and human sexuality. Am J Phys Anthropol 2013;152(Suppl 57):94–118.
127. Anholt RRH, O'Grady P, Wolfner MF, Harbison ST. Evolution of reproductive behavior. Genetics 2020;214(1):49–73.
128. Mautz BS, Wong BB, Peters RA, Jennions MD. Penis size interacts with body shape and height to influence male attractiveness. Proc Natl Acad Sci U S A 2013;110(17):6925–30.
129. Fine C, Dupre J, Joel D. Sex-linked behavior: Evolution, stability, and variability. Trends Cogn Sci 2017;21(9):666–73.
130. Bolte S, Neufeld J, Marschik PB, et al. Sex and gender in neurodevelopmental conditions. Nat Rev Neurol 2023;19(3):136–59.
131. Bartels A, Zeki S. The neural basis of romantic love. Neuroreport 2000;11(17):3829–34.
132. Walum H, Young LJ. The neural mechanisms and circuitry of the pair bond. Nat Rev Neurosci 2018; 19(11):643–54.
133. Neylan TC, Kluver H, Bucy PC. Temporal lobe and behavior: Kluver and Bucy's classic preliminary analysis of functions of the temporal lobes in monkeys; 1939. J Neuropsychiatry Clin Neurosci 1997; 9(4):606–20.
134. Nahm FK. Heinrich Kluver and the temporal lobe syndrome. J Hist Neurosci 1997;6(2):193–208.
135. Cataldo DM, Migliano AB, Vinicius L. Speech, stone tool-making and the evolution of language. PLoS One 2018;13(1):e0191071.

136. Liang X, Sun L, Liao X, et al. Structural connectome architecture shapes the maturation of cortical morphology from childhood to adolescence. Nat Commun 2024;15(1):784.
137. Roth G, Dicke U. Evolution of the brain and intelligence in primates. Prog Brain Res 2012;195:413–30.
138. de la Torre I, Mora R. Oldowan technological behaviour at HWK EE (Olduvai Gorge, Tanzania). J Hum Evol 2018;120:236–73.
139. Zaidner Y, Weinstein-Evron M. The emergence of the Levallois technology in the Levant: A view from the Early Middle Paleolithic site of Misliya Cave, Israel. J Hum Evol 2020;144:102785.
140. Adler DS, Wilkinson KN, Blockley S, et al. Early Levallois technology and the Lower to Middle Paleolithic transition in the Southern Caucasus. Science 2014;345(6204):1609–13.
141. Cascalheira J, Bicho N. On the chronological structure of the Solutrean in southern Iberia. PLoS One 2015;10(9):e0137308.
142. Westley K, Dix J. The Solutrean Atlantic hypothesis: A view from the Ocean. J North Atlantic 2008; 1:85–98.
143. Head H, Holms G. Sensory disturbances from cerebral lesions. Brain 1911;34(2–3):102–254.
144. Fostowicz-Frelik L, Tseng ZJ. The mammalian skull: Development, structure and function. Philos Trans R Soc Lond (B Biol Sci) 2023;378(1880):1–5.
145. Kyomen S, Murillo-Rincón AP, Kaucká M. Evolutionary mechanisms modulating the mammalian skull development. Philos Trans R Soc Lond (B Biol Sci) 2023;378(1880):1–14.
146. Glansdorff N, Xu Y, Labedan B. The last universal common ancestor: Emergence, constitution and genetic legacy of an elusive forerunner. Biol Direct 2008;3:29.
147. Evans KM, Waltz B, Tagliacollo V, et al. Why the short face? Developmental disintegration of the neurocranium drives convergent evolution in neotropical electric fishes. Ecol Evol 2017;7(6):1783–801.
148. Di Ieva A, Bruner E, Haider T, et al. Skull base embryology: A multidisciplinary review. Childs Nerv Syst 2014;30(6):991–1000.
149. Bastir M, Rosas A, O'Higgins P. Craniofacial levels and the morphological maturation of the human skull. J Anat 2006;209(5):637–54.
150. Chao MV. Periphery: How your nervous system predicts and protects against disease. Cambridge, MA: Harvard Uni. Press, 2023.
151. Piaget J, Inhelder B. The psychology of the child. New York: Hachette Book Group Inc, 1969/2018.
152. Stefani R. Some consequences of the evolution of the human pelvis. Arch Ital Anat Embriol 1984; 89(2):131–40.
153. Pavlicev M, Romero R, Mitteroecker P. Evolution of the human pelvis and obstructed labor: New explanations of an old obstetrical dilemma. Am J Obstet Gynecol 2020;222(1):3–16.
154. Bolk L. Origin of racial characteristics in man. Am J Biolog Anthropol 1929;13(1):1–28.
155. Verhulst J. Louis Bolk revisited. II-Retardation, hypermorphosis and body proportions of humans. Med Hypotheses 1993;41(2):100–14.
156. Kirkwood TB, Austad SN. Why do we age? Nature 2000;408(6809):233–8.
157. Toussaint O, Remacle J, Clark BF, et al. Biology of ageing. Bioessays 2000;22(10):954–6.
158. Schumacher B, Pothof J, Vijg J, Hoeijmakers JHJ. The central role of DNA damage in the ageing process. Nature 2021;592(7856):695–703.
159. Preuss TM. The human brain: Rewired and running hot. Ann N Y Acad Sci 2011;1225(Suppl 1):E182–91.
160. Orchard ER, Rutherford HJV, Holmes AJ, Jamadar SD. Matrescence: Lifetime impact of motherhood on cognition and the brain. Trends Cogn Sci 2023;27(3):302–16.
161. Lovejoy CO. The origin of man. Science 1981;211(4480):341–50.
162. Servin-Barthet C, Martinez-Garcia M, Pretus C, et al. The transition to motherhood: Linking hormones, brain and behaviour. Nat Rev Neurosci 2023;24(10):605–19.
163. Kirkwood TB, Rose MR. Evolution of senescence: Late survival sacrificed for reproduction. Philos Trans R Soc Lond B Biol Sci 1991;332(1262):15–24.
164. Kirkwood TB. Evolution of ageing. Nature 1977;270(5635):301–4.
165. Hawkes K. Human longevity: The grandmother effect. Nature 2004;428(6979):128–9.
166. Hawkes K, Smith KR. Brief communication: Evaluating grandmother effects. Am J Phys Anthropol 2009;140(1):173–6.
167. Liberman A. Word origins and how we know them: Etymology for everyone. London, UK: Oxford Uni. Press, 2009.
168. Patke A, Young MW, Axelrod S. Molecular mechanisms and physiological importance of circadian rhythms. Nat Rev Mol Cell Biol 2020;21(2):67–84.
169. Sehgal A. Physiology flies with time. Cell 2017;171(6):1232–5.
170. Roenneberg T, Merrow M. The circadian clock and human health. Curr Biol 2016;26(10):R432–43.

171. Panda S. The arrival of circadian medicine. Nat Rev Endocrinol 2019;15(2):67–9.
172. Luykx JJ, Boks MP, Terwindt AP, et al. The involvement of GSK3beta in bipolar disorder: Integrating evidence from multiple types of genetic studies. Eur Neuropsychopharmacol 2010;20(6):357–68.
173. Siegel JM. Sleep function: An evolutionary perspective. Lancet Neurol 2022;21(10):937–46.
174. Franken P, Dijk DJ. Sleep and circadian rhythmicity as entangled processes serving homeostasis. Nat Rev Neurosci 2024;25(1):43–59.
175. Wolf BO, McKechnie AE, Schmitt CJ, et al. Extreme and variable torpor among high-elevation Andean hummingbird species. Biol Lett 2020;16(9):20200428.
176. Siclari F, Valli K, Arnulf I. Dreams and nightmares in healthy adults and in patients with sleep and neurological disorders. Lancet Neurol 2020;19(10):849–59.
177. Borbely A. The two-process model of sleep regulation: Beginnings and outlook. J Sleep Res 2022; 31(4):e13598.
178. Solms M. Neurobiology and the neurological basis of dreaming. Sleep Disorders-Handb Clin Neurol 2011;98:519–44.
179. Blum H, Kafka JS. History of psychoanalysis: Freud and dream contributing to and celebrating the centenary of "The interpretation of dreams" as the origin of the psychoanalytic method. Int J Psychoanal 2002;83(Pt 2):483–6.
180. Leibin V. The interpretation of dreams as part of Freud's psychobiography. Psychoanal Rev 2003; 90(6):811–28.
181. Ito H, Kakishima S, Uehara T, et al. Evolution of periodicity in periodical cicadas. Sci Rep 2015;5:14094.
182. Aserinsky E, Kleitman N. Regularly occurring periods of eye motility, and concomitant phenomena, during sleep. Science 1953;118(3062):273–4.
183. Dement W. The occurrence of low voltage, fast, electroencephalogram patterns during behavioral sleep in the cat. Electroencephalogr Clin Neurophysiol 1958;10(2):291–6.
184. Mota CMD, Madden CJ. Neural circuits of long-term thermoregulatory adaptations to cold temperatures and metabolic demands. Nat Rev Neurosci 2024;25(3):143–58.
185. Roth TC 2nd, Lesku JA, Amlaner CJ, Lima SL. A phylogenetic analysis of the correlates of sleep in birds. J Sleep Res 2006;15(4):395–402.
186. Tan CL, Knight ZA. Regulation of body temperature by the nervous system. Neuron 2018;98(1):31–48.
187. Ungurean G, Barrillot B, Martinez-Gonzalez D, et al. Comparative perspectives that challenge brain warming as the primary function of REM sleep. iScience 2020;23(11):101696.
188. Wehr TA. A brain-warming function for REM sleep. Neurosci Biobehav Rev 1992;16(3):379–97.
189. Sterman MB, Knauss T, Lehmann D, Clemente CD. Circadian sleep and waking patterns in the laboratory cat. Electroencephalogr Clin Neurophysiol 1965;19(5):509–17.
190. Jouvet M, Buda C, Sastre JP. [Is there a bulbar pacemaker responsible for the ultradian rhythm of paradoxical sleep?] (in French). Arch Ital Biol 1995;134(1):39–56.
191. McCoy AN, Tan SY. Otto Loewi (1873–1961): Dreamer and nobel laureate. Singapore Med J 2014; 55(1):3–4.
192. Bast TH. The life and work of Samuel Thomas Von Sömmerring. Ann Med Hist 1924;6(4):369–86.
193. Stoyanov GS, Matev BK. Samuel Thomas von Sömmerring's contributions on the cranial nerves and vomeronasal organ. Cureus 2018;10(6):e2859.
194. Feindel W. The physiologist and the neurosurgeon: The enduring influence of Charles Sherrington on the career of Wilder Penfield. Brain 2007;130(Pt 11):2758–65.
195. Molnar Z, Brown RE. Insights into the life and work of Sir Charles Sherrington. Nat Rev Neurosci 2010;11(6):429–36.
196. Dunbar RIM, Shultz S. Why are there so many explanations for primate brain evolution? Philos Trans R Soc Lond B Biol Sci 2017;372(1727):20160244.
197. Kingsolver JG, Pfennig DW. Individual-level selection as a cause of Cope's rule of phyletic size increase. Evolution 2004;58(7):1608–12.
198. Luppi AI, Rosas FE, Noonan MP, et al. Oxygen and the spark of human brain evolution: Complex interactions of metabolism and cortical expansion across development and evolution. Neuroscientist 2024;30(2):173–98.
199. Harvey PH, Clutton-Brock TH, Mace GM. Brain size and ecology in small mammals and primates. Proc Natl Acad Sci U S A 1980;77(7):4387–9.
200. Jayaraman D, Bae BI, Walsh CA. The genetics of primary microcephaly. Annu Rev Genomics Hum Genet 2018;19:177–200.
201. Evans PD, Gilbert SL, Mekel-Bobrov N, et al. Microcephalin: A gene regulating brain size, continues to evolve adaptively in humans. Science 2005;309(5741):1717–20.

202. Barrickman NL, Bastian ML, Isler K, van Schaik CP. Life history costs and benefits of encephalization: A comparative test using data from long-term studies of primates in the wild. J Hum Evol 2008; 54(5):568–90.
203. Ebinger P, Rohrs M. Volumetric analysis of brain structures, especially of the visual system in wild and domestic turkeys (Meleagris gallopavo). J Hirnforsch 1995;36(2):219–28.
204. Aboitiz F. The evolution of brain size and organization in vertebrates. A program for research. Biol Res 1994;27(1):15–27.
205. Aboitiz F. Does bigger mean better? Evolutionary determinants of brain size and structure. Brain Behav Evol 1996;47(5):225–45.
206. Herculano-Houzel S. The human advantage: A new understanding of how our brain became remarkable. Cambridge, MA: MIT Press, 2016.
207. Andrews P. Last common ancestor of apes and humans: Morphology and environment. Folia Primatol (Basel) 2020;91(2):122–48.
208. King MC, Wilson AC. Evolution at two levels in humans and chimpanzees. Science 1975;188(4184): 107–16.
209. Morange M. What history tells us XXIII. The genetic distance between humans and chimpanzees: What did Mary-Claire King and Allan Wilson really say in 1975? J Biosci 2011;36(1):23–6.
210. Krause J, Lalueza-Fox C, Orlando L, et al. The derived FOXP2 variant of modern humans was shared with Neandertals. Curr Biol 2007;17(21):1908–12.
211. Roberts E, Hampshire DJ, Pattison L, et al. Autosomal recessive primary microcephaly: An analysis of locus heterogeneity and phenotypic variation. J Med Genet 2002;39(10):718–21.
212. Caraffi SG, Pollazzon M, Farooq M, et al. MCPH1: A novel case report and a review of the literature. Genes (Basel) 2022;13(4):634–648.
213. Preuss TM, Wise SP. Evolution of prefrontal cortex. Neuropsychopharmacology 2022;47(1):3–19.
214. Kinney HC, Brody BA, Kloman AS, Gilles FH. Sequence of central nervous system myelination in human infancy. II. Patterns of myelination in autopsied infants. J Neuropathol Exp Neurol 1988;47(3):217–34.
215. Skulachev VP, Holtze S, Vyssokikh MY, et al. Neoteny, prolongation of youth: From naked mole rats to "Naked Apes" (Humans). Physiol Rev 2017;97(2):699–720.
216. Fonseca-Azevedo K, Herculano-Houzel S. Metabolic constraint imposes tradeoff between body size and number of brain neurons in human evolution. Proc Natl Acad Sci U S A 2012;109(45):18571–6.
217. Pagel MD, Harvey PH. Diversity in the brain sizes of newborn mammals. Bio Science 1990;40(2):116–24.
218. Rosch KS, Mostofsky S. Development of the frontal lobe. The Frontal Lobes-Handb Clin Neurol 2019; 163:351–67.
219. Rilling JK, Insel TR. Evolution of the cerebellum in primates: Differences in relative volume among monkeys, apes and humans. Brain Behav Evol 1998;52(6):308–14.
220. Aldridge K. Patterns of differences in brain morphology in humans as compared to extant apes. J Human Evol 2011;60(1):94–105.
221. Herculano-Houzel S. Coordinated scaling of cortical and cerebellar numbers of neurons. Front Neuroanat 2010;4:12.
222. Herculano-Houzel S. The human brain in numbers: A linearly scaled-up primate brain. Front Hum Neurosci 2009;3:31.
223. Herculano-Houzel S. Neuronal scaling rules for primate brains: The primate advantage. Prog Brain Res 2012;195:325–40.
224. Seymour RS, Bosiocic V, Snelling EP. Fossil skulls reveal that blood flow rate to the brain increased faster than brain volume during human evolution. R Soc Open Sci 2016;3(8):160305.
225. Silver I, Erecińska M. Oxygen and ion concentrations in normoxic and hypoxic brain cells. Adv Exp Med Biol 1998;454:7–16.
226. Leonard WR, Robertson ML, Snodgrass JJ, Kuzawa CW. Metabolic correlates of Hominid brain evolution. Comp Biochem Physiol Mol Integr Physiol 2003;136(1):5–15.
227. Attwell D, Buchan AM, Charpak S, et al. Glial and neuronal control of brain blood flow. Nature 2010;468(7321):232–43.
228. Martin RD. Relative brain size and basal metabolic rate in terrestrial vertebrates. Nature 1981; 293(5827):57–60.
229. Somel M, Liu X, Khaitovich P. Human brain evolution: Transcripts, metabolites and their regulators. Nat Rev Neurosci 2013;14(2):112–27.
230. Tauber E, Last KS, Olive PJ, Kyriacou CP. CLOCK gene evolution and functional divergence. J Biol Rhythms 2004;19(5):445–58.
231. Albrecht U. The mammalian circadian clock: A network of gene expression. Front Biosci 2004;9:48–55.

232. Bargiello TA, Young MW. Molecular genetics of a biological clock in Drosophila. Proc Natl Acad Sci U S A 1984;81(7):2142–6.
233. Hastings MH, Maywood ES, Brancaccio M. Generation of circadian rhythms in the suprachiasmatic nucleus. Nat Rev Neurosci 2018;19(8):453–69.
234. Lanfear R, Kokko H, Eyre-Walker A. Population size and the rate of evolution. Trends Ecol Evol 2014;29(1):33–41.
235. Clarke B. Mutation and population size. Heredity (Edinb) 1973;31(3):367–79.
236. Hill RS, Walsh CA. Molecular insights into human brain evolution. Nature 2005;437(7055):64–7.
237. Fu Z, Sajad A, Errington SP, et al. Neurophysiological mechanisms of error monitoring in human and non-human primates. Nat Rev Neurosci 2023;24(3):153–72.
238. Allan Cheyne J, Solman GJ, Carriere JS, Smilek D. Anatomy of an error: A bidirectional state model of task engagement/disengagement and attention-related errors. Cognition 2009;111(1):98–113.
239. Herculano-Houzel S. Life history changes accompany increased numbers of cortical neurons: A new framework for understanding human brain evolution. Prog Brain Res 2019;250:179–216.
240. Deaner RO, Isler K, Burkart J, van Schaik C. Overall brain size, and not encephalization quotient, best predicts cognitive ability across non-human primates. Brain Behav Evol 2007;70(2):115–24.
241. Ragir S. Diet and food preparation: Rethinking early Hominid behavior. Evol Anthropol 2000;9(4):153–5.
242. Wrangham RW. Catching fire: How cooking made us human. New York: Basic Books, 2009.
243. Allen JS, Damasio H, Grabowski TJ, et al. Sexual dimorphism and asymmetries in the gray-white composition of the human cerebrum. Neuroimage 2003;18(4):880–94.
244. Aiello L, Wheeler P. The expensive-tissue hypothesis. Cur Anthropol 1995;36(2):199–221.
245. Herculano-Houzel S. Scaling of brain metabolism with a fixed energy budget per neuron: Implications for neuronal activity, plasticity and evolution. PLoS One 2011;6(3):e17514.
246. Cechetto DF, Weishaupt N. The cerebral cortex in neurodegenerative and neuropsychiatric disorders: Experimental approaches to clinical issues. Amsterdam, Boston: Elsevier/Academic Press, 2017.
247. Rubinov M, Sporns O. Complex network measures of brain connectivity: Uses and interpretations. Neuroimage 2010;52(3):1059–69.
248. Herculano-Houzel S, Messeder DJ, Fonseca-Azevedo K, Pantoja NA. When larger brains do not have more neurons: Increased numbers of cells are compensated by decreased average cell size across mouse individuals. Front Neuroanat 2015;9:64.
249. Boillat Y, Bazin PL, van der Zwaag W. Whole-body somatotopic maps in the cerebellum revealed with 7T fMRI. Neuroimage 2020;211:116624.
250. Kiyatkin EA. Brain temperature homeostasis: Physiological fluctuations and pathological shifts. Front Biosci (Landmark Ed) 2010;15(1):73–92.
251. Kiyatkin EA. Brain temperature and its role in physiology and pathophysiology: Lessons from 20 years of thermorecording. Temperature (Austin) 2019;6(4):271–333.
252. Molnár Z, Clowry GJ, Šestan N, et al. New insights into the development of the human cerebral cortex. J Anat 2019;235(3):432–51.
253. Zeidman P, Maguire EA. Anterior hippocampus: The anatomy of perception, imagination and episodic memory. Nat Rev Neurosci 2016;17(3):173–82.
254. Mesulam MM. Temporopolar regions of the human brain. Brain 2023;146(1):20–41.
255. Marhounová L, Kotrschal A, Kverková K, et al. Artificial selection on brain size leads to matching changes in overall number of neurons. Evolution 2019;73(9):2003–12.
256. Hajdu SI. A note from history: Rudolph Virchow, pathologist, armed revolutionist, politician, and anthropologist. Ann Clin Lab Sci 2005;35(2):203–5.
257. Rather LJ. Virchow and the mechanistic concept of life in 1845: Essay review. Bull Hist Med 1988; 62(1):107–11.
258. Bogerts B. Frantz Nissl, 1860–1919. Am J Psychiatry 1994;151(2):260.
259. Editorial. Frantz Nissl: Neuropathologist. JAMA 1968;205(6):208–9.
260. Karbowski J. Global and regional brain metabolic scaling and its functional consequences. BMC Biol 2007;5:18.
261. Vavilala MS, Lee LA, Lam AM. Cerebral blood flow and vascular physiology. Anesthesiol Clin North Am 2002;20(2):247–64.
262. Wolf TH. Alfred Binet. Chicago: Chicago Uni. Press, 1973.
263. Hofman MA. Evolution of brain size in neonatal and adult placental mammals: A theoretical approach. J Theor Biol 1983;105(2):317–32.
264. Theunissen B. Eugène Dubois and the Ape-Man from Java: The history of the first 'Missing Link' and its discoverer. Dordrecht, Netherlands: Kluwer Academic Publishers, 1990.

265. Jerison HJ. Evolution of the brain and intelligence. New York: Academic Press, Inc, 1973.
266. Gould SJ. Allometry and size in ontogeny and phylogeny. Biol Rev Camb Philos Soc 1966;41(4):587–640.
267. Armstrong E, Falk D. Primate brain evolution: Methods and concepts. New York: Plenum Press, 1982.
268. Holloway RL, Post D. The relativity of relative brain measures and Hominid mosaic evolution. In: Armstrong E, Falk D, eds. Primate brain evolution. New York: Plenum Press; 1982.
269. Holloway RL. Encephalization. In: Trevathan W, ed. Int Encyclo Biologic Anthropol. New Jersy, NJ: John Wiley & Sons, Inc; 2018.
270. Holloway RL. Evolution of the human brain. In: Lock A, Peters CR, eds. Handb-human symbolic evolution. London, UK: Oxford Uni. Press; 1996:74–125.
271. Defelipe J. The evolution of the brain, the human nature of cortical circuits, and intellectual creativity. Front Neuroanat 2011;5:29.
272. Blumenberg B, Fristrup K, Holloway RL, et al. The evolution of the advanced hominid brain. Cur Anthropol 1983;24(5):589–623.
273. Hebb DO. The organization of behavior: A neuropsychological theory. Mahwah, New Jersy: Lawrence Erlbaum Associates, 1949/2002.
274. Bliss TV, Lomo T. Long-lasting potentiation of synaptic transmission in the dentate area of the anaesthetized rabbit following stimulation of the perforant path. J Physiol 1973;232(2):331–56.
275. Cichon J, Gan WB. Branch-specific dendritic Ca(2+) spikes cause persistent synaptic plasticity. Nature 2015;520(7546):180–5.
276. Adolph KE, Cole WG, Komati M, et al. How do you learn to walk? Thousands of steps and dozens of falls per day. Psychol Sci 2012;23(11):1387–94.
277. Atlan G, Terem A, Peretz-Rivlin N, et al. Mapping synaptic cortico-claustral connectivity in the mouse. J Comp Neurol 2017;525(6):1381–402.
278. Urban MC, Richardson JL, Freidenfelds NA. Plasticity and genetic adaptation mediate amphibian and reptile responses to climate change. Evol Appl 2014;7(1):88–103.
279. Adams S. Chinese Sexuality and the bound foot. In: Peakman J, ed. Sexual perversions, 1670–1890. London: Palgrave Macmillan UK; 2009:246–75.
280. Wise NJ, Frangos E, Komisaruk BR. Activation of sensory cortex by imagined genital stimulation: An fMRI analysis. Socio Affect Neurosci Psychol 2016;6:31481.
281. Tan B, Browne CJ, Nobauer T, et al. Drugs of abuse hijack a mesolimbic pathway that processes homeostatic need. bioRxiv 2024;384(6693): eadk6742.
282. MacNicol B. The biology of addiction. Can J Anaesth 2017;64(2):141–48.
283. Buckner RL, Krienen FM. The evolution of distributed association networks in the human brain. Trends Cogn Sci 2013;17(12):648–65.
284. Yu M, Sporns O, Saykin AJ. The human connectome in Alzheimer disease: Relationship to biomarkers and genetics. Nat Rev Neurol 2021;17(9):545–63.
285. Sporns O, Tononi G, Kotter R. The human connectome: A structural description of the human brain. PLoS Comput Biol 2005;1(4):e42.
286. Sporns O. The human connectome: A complex network. Ann N Y Acad Sci 2011;1224:109–25.
287. Sporns O. The human connectome: Origins and challenges. Neuroimage 2013;80:53–61.
288. Kaas JH. The origin and evolution of neocortex: From early mammals to modern humans. Prog Brain Res 2019;250:61–81.
289. Kaas JH. The evolution of the complex sensory and motor systems of the human brain. Brain Res Bull 2008;75(2–4):384–90.
290. Kaiser M. A tutorial in connectome analysis: Topological and spatial features of brain networks. Neuroimage 2011;57(3):892–907.
291. van den Heuvel MP, Kersbergen KJ, de Reus MA, et al. The neonatal connectome during preterm brain development. Cereb Cortex 2015;25(9):3000–13.
292. Glickstein M, Berlucchi G. Classical disconnection studies of the corpus callosum. Cortex 2008; 44(8):914–27.
293. Sato JR, Biazoli CE Jr., Salum GA, et al. Connectome hubs at resting state in children and adolescents: Reproducibility and psychopathological correlation. Dev Cogn Neurosci 2016;20:2–11.
294. Kolk SM, Rakic P. Development of prefrontal cortex. Neuropsychopharmacology 2022;47(1):41–57.
295. Pusey AE, Pintea L, Wilson ML, et al. The contribution of long-term research at Gombe National Park to chimpanzee conservation. Conserv Biol 2007;21(3):623–34.
296. Levy R. The prefrontal cortex: From monkey to man. Brain 2024;147(3):794–815.
297. Donahue CJ, Glasser MF, Preuss TM, et al. Quantitative assessment of prefrontal cortex in humans relative to nonhuman primates. Proc Natl Acad Sci U S A 2018;115(22):E5183–92.

298. Carlen M. What constitutes the prefrontal cortex? Science 2017;358(6362):478–82.
299. Wallace JL, Pollen AA. The genetic symphony underlying evolution of the brain's prefrontal cortex. Nature 2021;598(7881):417–8.
300. Alexandre F. A global framework for a systemic view of brain modeling. Brain Inform 2021;8(1):3.
301. Fuster JM. The prefrontal cortex. 5th ed. Amesterdam: Academic Press, 2015.
302. Segal O, Elkana O. The ventrolateral prefrontal cortex is part of the modular working memory system: A functional neuroanatomical perspective. Front Neuroanat 2023;17:1076095.
303. Fuster JM, Bressler SL. Past makes future: Role of PFC in prediction. J Cogn Neurosci 2015; 27(4):639–54.
304. Fuster JM, Bressler SL. Cognit activation: A mechanism enabling temporal integration in working memory. Trends Cogn Sci 2012;16(4):207–18.
305. Fuster JM. The prefrontal cortex in the neurology clinic. The Frontal Lobes-Handb Clin Neurol. Amesterdam: Elsevier; 2019:3–15.
306. Fuster JM. Cognitive networks (cognits) process and maintain working memory. Front Neural Circuits 2021;15:790691.
307. Singer W. A naturalistic approach to the hard problem of consciousness. Front Syst Neurosci 2019;13:58.
308. Teleanu RI, Niculescu AG, Roza E, et al. Neurotransmitters: Key factors in neurological and neurodegenerative disorders of the central nervous system. Int J Mol Sci 2022;23(11):5954.
309. Cramer MN, Gagnon D, Laitano O, Crandall CG. Human temperature regulation under heat stress in health, disease, and injury. Physiol Rev 2022;102(4):1907–89.
310. Sisodiya SM. Hot brain: Practical climate change advice for neurologists. Pract Neurol 2024;24(1):28–36.
311. Xu X, Rioux TP, Castellani MP. Three dimensional models of human thermoregulation: A review. J Therm Biol 2023;112:103491.
312. Juárez Olguín H, Calderón Guzmán D, Hernández García E, Barragán Mejía G. The role of dopamine and its dysfunction as a consequence of oxidative stress. Oxid Med Cell Longev 2016;2016:9730467.
313. Koob GF, Volkow ND. Neurocircuitry of addiction. Neuropsychopharmacology 2010;35(1):217–38.
314. Ugarte G, Piña R, Contreras D, et al. Attention Deficit-Hyperactivity Disorder (ADHD): From abnormal behavior to impairment in synaptic plasticity. Biology (Basel) 2023;12(9):1241.
315. Vanderhaeghen P, Polleux F. Developmental mechanisms underlying the evolution of human cortical circuits. Nat Rev Neurosci 2023;24(4):213–32.
316. Clagett M. Ancient Egyptian science: A source book. Vol. 2, Philadelphia: Am Philos Soc, 1995.
317. Feldman RP, Goodrich JT. The Edwin Smith surgical papyrus. Childs Nerv Syst 1999;15(6–7):281–4.
318. Atta HM. Edwin Smith surgical papyrus: The oldest known surgical treatise. Am Surg 1999;65(12):1190–2.
319. Livianos-Aldana L, Rojo-Moreno L, Sierra-Sanmiguel P. F.J. Gall and the phrenological movement. Am J Psychiatry 2007;164(3):414.
320. Van Wyhe J. The authority of human nature: The Schädellehre of Franz Joseph Gall. Br J Hist Sci 2002;35(124 Pt 1):17–42.
321. Finger S. Minds behind the brain: A history of the pioneers and their discoveries. New York, NY: Oxford Uni. Press, 2000.
322. Temkin O. Gall and the phrenological movement. Bull Hist Med 1947;21(3):275–321.
323. Silverman BD. Jean Baptiste Bouillaud. Clin Cardiol 1996;19(10):836–7.
324. Luzzatti C, Whitaker H. Jean-Baptiste Bouillaud, Claude-Francois Lallemand, and the role of the frontal lobe: Location and mislocation of language in the early 19th century. Arch Neurol 2001;58(7):1157–62.
325. Michel GF, Babik I, Nelson EL, et al. Evolution and development of handedness: An Evo-Devo approach. Prog Brain Res 2018;238:347–74.
326. Uomini NT, Ruck L. Manual laterality and cognition through evolution: An archeological perspective. Prog Brain Res 2018;238:295–323.
327. Costandi M. Fifty ideas you really need to know: The human brain. London, UK: Quercus Editions Ltd., 2013.
328. Della Sala S, McIntosh RD. Righteous Adam, Sinister Eve. Laterality 2022;27(6):605–15.
329. McManus IC. Right-left and the scrotum in Greek sculpture. Laterality 2004;9(2):189–99.
330. Couzin R. Art and material culture in Medieval and Renaissance Europe. Leiden, The Netherlands: Brill, 2021.
331. Alhassan AB. Left handedness, the Bible and the Quran: Implications for parents and teachers. J Edu Stud 2017;3(1):19–26.
332. Joynt RJ, Benton AL. The memoir of Marc Dax on aphasia. Neurology 1964;14:851–4.
333. Manning L, Thomas-Anterion C. Marc Dax and the discovery of the lateralisation of language in the left cerebral hemisphere. Rev Neurol (Paris) 2011;167(12):868–72.

334. Buckingham HW. The Marc Dax (1770–1837)/Paul Broca (1824–1880) controversy over priority in science: Left hemisphere specificity for seat of articulate language and for lesions that cause aphemia. Clin Linguist Phon 2006;20(7–8):613–9.
335. Amunts K, Zilles K. Architectonic mapping of the human brain beyond Brodmann. Neuron 2015; 88(6):1086–107.
336. Kawamura M. [Korbinian Brodmann's scientific profile, and academic works] (in Japanese). Brain Nerve 2017;69(4):301–12.
337. Mueller T, Kanis-Seyfried U. On the life and work of Korbinian Brodmann (1868–1918). J Hist Neurosci 2019;28(3):307–18.
338. Scerri TS, Brandler WM, Paracchini S, et al. PCSK6 is associated with handedness in individuals with dyslexia. Hum Mol Genet 2011;20(3):608–14.
339. Knecht S, Deppe M, Dräger B, et al. Language lateralization in healthy right-handers. Brain 2000; 123 (Pt 1):74–81.
340. Gilbert AN, Wysocki CJ. Hand preference and age in the United States. Neuropsychologia 1992; 30(7):601–8.
341. McManus IC. The history and geography of human handedness. Language lateralisation and psychosis. Cambridge, MA: Cambridge Uni. Press, 2009.
342. Podolsky E. Cesare Lombroso and criminal man. Med Ann Dist Columbia 1951;20(6):323–5.
343. Chio A, Spreafico R, Avanzini G, et al. Cesare Lombroso, cortical dysplasia, and epilepsy: Keen findings and odd theories. Neurology 2003;61(10):1412–6.
344. Carra G, Barale F. Cesare Lombroso, M.D., 1835–1909. Am J Psychiatry 2004;161(4):624.
345. Hill AV. Gustave Le Bon. Nature 1941;147(3732):573.
346. Tsai SY. Eponym and identity: Benedict Augustin Morel (1809–1873) and Ferdinand Morel (1888–1957). Arch Gen Psychiatry 1968;19(1):104–9.
347. Conti NA. [Benedict Augustin Morel and the origin of the term dementia praecox] (in Spanish). Vertex 2003;14(53):227–31.
348. Guillin V. Theodule Ribot's ambiguous positivism: Philosophical and epistemological strategies in the founding of French scientific psychology. J Hist Behav Sci 2004;40(2):165–81.
349. Lombroso C. Criminal man. Durham and London: Duke Uni. Press, 1896/2007.
350. McManus IC. Half a century of handedness research: Myths, truths, fictions, facts, backwards but mostly forwards. Brain Neurosci Adv 2019;3:2398212818820513.
351. Gutwinski S, Löscher A, Mahler L, et al. Understanding left-handedness. Dtsch Arztebl Int 2011; 108(50):849–53.
352. Shurkin JN. Terman's kids: The groundbreaking study of how the gifted grow up. Boston, MA: Little Brown and Company, 1992.
353. Sears RR. L. M. Terman, pioneer in mental measurement. Science 1957;125(3255):978–9.
354. Corballis MC, Beale IL. Bilateral symmetry and behavior. Psychol Rev 1970;77(5):451–64.
355. Corballis MC. The evolution and genetics of cerebral asymmetry. Philos Trans R Soc Lond B Biol Sci 2009;364(1519):867–79.
356. McManus IC, Drury H. The handedness of Leonardo da Vinci: A tale of the complexities of lateralisation. Brain Cogn 2004;55(2):262–8.
357. Galaburda AM, Geschwind N. Anatomical asymmetries in the adult and developing brain and their implications for function. Adv Pediatr 1981;28:271–92.
358. Papadatou-Pastou M. Handedness and cognitive ability: Using meta-analysis to make sense of the data. Prog Brain Res 2018;238:179–206.
359. Coren S, Porac C. Fifty centuries of right-handedness: The historical record. Science 1977;198(4317):631–2.
360. Annett M. In defence of the right shift theory. Percept Mot Skills 1996;82(1):115–37.
361. Annett M. Handedness and cerebral dominance: The right shift theory. J Neuropsychiatry Clin Neurosci 1998;10(4):459–69.
362. Annett M. Hand preference observed in large healthy samples: Classification, norms and interpretations of increased non-right-handedness by the right shift theory. Br J Psychol 2004;95(Pt 3):339–53.
363. Ackermann H, Riecker A. The contribution(s) of the insula to speech production: A review of the clinical and functional imaging literature. Brain Struct Funct 2010;214(5–6):419–33.
364. Larsson M. Did heart asymmetry play a role in the evolution of human handedness? J Cult Cogn Sci 2017;1:65–76.
365. Faurie C, Raymond M. Handedness, homicide and negative frequency-dependent selection. Proc Biol Sci 2005;272(1558):25–8.
366. Michelini F, Kochy K. Jakob von Uexküll and philosophy. New York: Routledge, 2020.

367. Steele J, Uomini N. Humans, tools and handedness. In: Roux V, Bril B, eds. Stone Knapping: The necessary conditions for a uniquely Hominin behaviour. Cambridge, MA: McDonald Institute for Archaeological Research; 2005.
368. Yeo SS, Jang SH. Medullary decussation of the lateral corticospinal tract. Eur Neurol 2011; 66(5):296–7.
369. Welniarz Q, Dusart I, Roze E. The corticospinal tract: Evolution, development, and human disorders. Dev Neurobiol 2017;77(7):810–29.
370. Luhan JA. Long survival after unilateral stab wound of medulla with unusual pyramidal tract distribution. Arch Neurol 1959;1:427–34.
371. Shinbrot T, Young W. Why decussate? Topological constraints on 3D wiring. Anat Rec (Hoboken) 2008;291(10):1278–92.
372. Banihani SM. Crossing of neuronal pathways: Is it a response to the occurrence of separated parts for the body (limbs, eyes, etc.) during evolution? Med Hypotheses 2010;74(4):741–5.
373. Vulliemoz S, Raineteau O, Jabaudon D. Reaching beyond the midline: Why are human brains cross wired? Lancet Neurol 2005;4(2):87–99.
374. Armand J. The origin, course and terminations of corticospinal fibers in various mammals. Prog Brain Res 1982;57:329–60.
375. Kinsbourne M. Somatic twist: A model for the evolution of decussation. Neuropsychology 2013;27(5): 511–5.
376. Braitenberg V. Taxis, kinesis and decussation. Prog Brain Res 1965;17:210–22.
377. Norcia AM, Gerhard HE. Development of 3D perception in human infants. Annu Rev Vis Sci 2015; 1:569–94.
378. de Lussaneta MHE, Osseb JWM. Axial twist explains the contralateral forebrain and the optic chiasm in vertebrates. Anim Biol 2012;62(2):193–216.
379. de Lussaneta MHE, Osse JWM. Decussation as an axial twist: A comment on Kinsbourne. Neuropsychology 2014;29(5):713–4.
380. Mora C, Velasquez C, Martino J. The neural pathway midline crossing theory: A historical analysis of Santiago Ramon y Cajal's contribution on cerebral localization and on contralateral forebrain organization. Neurosurg Focus 2019;47(3):E10.
381. Fulton JF, Sheehan D. The uncrossed lateral pyramidal tract in higher primates. J Anat 1935;69(Pt 2): 181–7.
382. Verhaart WJ, Kramer W. The uncrossed pyramidal tract. Acta Psychiatr Neurol Scand 1952; 27(1–2):181–200.
383. Frikha R. Klippel-Feil syndrome: A review of the literature. Clin Dysmorphol 2020;29(1):35–7.
384. Granat M. Kallman's syndrome. Fertil Steril 1978;30(2):245.
385. Benarroch E. What are the functions of the superior colliculus and its involvement in neurologic disorders? Neurology 2023;100(16):784–90.
386. Sarnat HB, Flores-Sarnat L. Development of the human olfactory system. In: Dotty RL, ed. Smell and taste - Handb Clin Neurol. Amesterdam, Elsevier; 2019:29–45.
387. McGann JP. Poor human olfaction is a 19th century myth. Science 2017;356(597): eaam7263.
388. Reznik GK. Comparative anatomy, physiology, and function of the upper respiratory tract. Environ Health Perspect 1990;85:171–6.
389. Small DM, Gerber JC, Mak YE, Hummel T. Differential neural responses evoked by orthonasal versus retronasal odorant perception in humans. Neuron 2005;47(4):593–605.
390. Barwich AS. Smellosophy: What the nose tells the mind. Cambridge, MA: Harvard Uni. Press, 2020.
391. De Miguel C, Saniotis A, Cieslik A, Henneberg M. Comparative study of brain size ontogeny: Marsupials and placental mammals. Biology (Basel) 2022;11(6):900.
392. Isa T, Marquez-Legorreta E, Grillner S, Scott EK. The tectum/superior colliculus as the vertebrate solution for spatial sensory integration and action. Curr Biol 2021;31(11):R741–62.
393. Petersen CCH. Sensorimotor processing in the rodent barrel cortex. Nat Rev Neurosci 2019;20(9):533–46.
394. Kaas JH, Qi HX, Stepniewska I. The evolution of parietal cortex in primates. Neuroanatomy and neurophysiology of the parietal lobe. Handb Clin Neurol. Amesterdam: Elsevier;:31–52.
395. Louis ED, Horn S, Roth LA. The neurologic content of S. Weir Mitchell's fiction. Neurology 2006; 66(3):403–7.
396. Bourke J. Silas Weir Mitchell's: The case of George Dedlow. Lancet 2009;373(9672):1332–3.
397. Kline DG. Silas Weir Mitchell and "The strange case of George Dedlow. Neurosurg Focus 2016;41(1):E5.
398. Schneck JM. S. Weir Mitchell, M.D. (1829–1914): Neurologic and psychiatric observations in "The case of George Dedlow". N Y State J Med 1979;79(11):1777–82.

399. Canale DJ. Civil war medicine from the perspective of S. Weir Mitchell'S "The case of George Dedlow". J Hist Neurosci 2002;11(1):11–8.
400. Goler RI. Loss and the persistence of memory: "The case of George Dedlow" and disabled Civil War veterans. Lit Med 2004;23(1):160–83; discussion 201–4.
401. Cascio C, McGlone F, Folger S, et al. Tactile perception in adults with autism: A multidimensional psychophysical study. J Autism Dev Disord 2008;38(1):127–37.
402. Sheffer E. Asperger's children the origins of autism in Nazi Vienna. New York, London: W. W. Norton & Company, 2018.
403. Morrison I, Loken LS, Olausson H. The skin as a social organ. Exp Brain Res 2010;204(3):305–14.
404. Porges SW. Polyvagal theory: A science of safety. Front Integr Neurosci 2022;16:871227.
405. Nakajima M, Schmitt LI, Halassa MM. Prefrontal cortex regulates sensory filtering through a basal ganglia to thalamus pathway. Neuron 2019;103(3):445–58.e10.
406. Craig AD. A new view of pain as a homeostatic emotion. Trends Neurosci 2003;26(6):303–7.
407. Barbalet JM. William James' theory of emotions: Filling in the picture. J Theo Soc Behav 1999; 29(3):251–66.
408. Strigo IA, Craig AD. Interoception, homeostatic emotions and sympathovagal balance. Philos Trans R Soc Lond B Biol Sci 2016;371(1708):20160010.
409. Damasio AR. Emotion in the perspective of an integrated nervous system. Brain Res Brain Res Rev 1998;26(2–3):83–6.
410. Damasio A, Carvalho GB. The nature of feelings: Evolutionary and neurobiological origins. Nat Rev Neurosci 2013;14(2):143–52.
411. Schmitt CM, Schoen S. Interoception: A multi-sensory foundation of participation in daily life. Front Neurosci 2022;16:875200.
412. Ceunen E, Vlaeyen JW, Van Diest I. On the origin of interoception. Front Psychol 2016;7:743.
413. Carvalho GB, Damasio A. Interoception and the origin of feelings: A new synthesis. Bioessays 2021; 43(6):e2000261.
414. Bourque CW, Oliet SH, Richard D. Osmoreceptors, osmoreception, and osmoregulation. Front Neuroendocrinol 1994;15(3):231–74.
415. Butti C, Hof PR. The insular cortex: A comparative perspective. Brain Struct Funct 2010; 214(5–6):477–93.
416. Shine JM, Lewis LD, Garrett DD, Hwang K. The impact of the human thalamus on brain-wide information processing. Nat Rev Neurosci 2023;24(7):416–30.
417. Namkung H, Kim SH, Sawa A. The insula: An underestimated brain area in clinical neuroscience, psychiatry, and neurology. Trends Neurosci 2017;40(4):200–7.
418. Craig AD. Why a soft touch can hurt. J Physiol 2010;588(Pt 1):13.
419. Costandi M. Neuroplasticity. Cambridge, MA: MIT Press, 2016.
420. Cytowic RE. Synesthesia. Cambridge, Massachusetts: MIT Press, 2018.
421. Luria A. The mind of a mnemonist: A little book about a vast memory. Cambridge: Harvard Uni. Press, 1987.
422. Mecacci L. Solomon V. Shereshevsky: The great Russian mnemonist. Cortex 2013;49(8):2260–3.
423. Craig AD. Emotional moments across time: A possible neural basis for time perception in the anterior insula. Philos Trans R Soc Lond B Biol Sci 2009;364(1525):1933–42.
424. Craig AD. How do you feel now? The anterior insula and human awareness. Nat Rev Neurosci 2009; 10(1):59–70.
425. Allman JM, Tetreault NA, Hakeem AY, et al. The von Economo neurons in frontoinsular and anterior cingulate cortex in great apes and humans. Brain Struct Funct 2010;214(5–6):495–517.
426. Damasio A. Self comes to mind: Constructing the conscious brain. Toronto: Pantheon Books, 2010.
427. Costandi M. Body am I: The new science of self-consciousness. Cambridge, Massachusetts: MIT Press, 2022.
428. Huang J, Cooper TG, Satana B, et al. Visual distortion provoked by a stimulus in migraine associated with hyperneuronal activity. Headache 2003;43(6):664–71.
429. Ullsperger M. Performance monitoring in neurological and psychiatric patients. Int J Psychophysiol 2006;59(1):59–69.
430. Rodriguez M, Kross E. Sensory emotion regulation. Trends Cogn Sci 2023;27(4):379–90.
431. Saper CB. The central autonomic nervous system: Conscious visceral perception and autonomic pattern generation. Annu Rev Neurosci 2002;25:433–69.
432. Spence C. Senses of place: Architectural design for the multisensory mind. Cognit Res Princ Implic 2020;5(1):46.

433. Faber DS, Pereda AE. Two forms of electrical transmission between neurons. Front Mol Neurosci 2018;11:427.
434. Ananth MR, Rajebhosale P, Kim R, et al. Basal forebrain cholinergic signalling: Development, connectivity and roles in cognition. Nat Rev Neurosci 2023;24(4):233–51.
435. Fry M, Ferguson AV. The sensory circumventricular organs: Brain targets for circulating signals controlling ingestive behavior. Physiol Behav 2007;91(4):413–23.
436. Duvernoy HM, Risold PY. The circumventricular organs: An atlas of comparative anatomy and vascularization. Brain Res Rev 2007;56(1):119–47.
437. Kiecker C. The origins of the circumventricular organs. J Anat 2018;232(4):540–53.
438. Ganong WF. Circumventricular organs: Definition and role in the regulation of endocrine and autonomic function. Clin Exp Pharmacol Physiol 2000;27(5–6):422–7.
439. Azizi SA. Monoamines: Dopamine, norepinephrine, and serotonin, beyond modulation, "Switches" that alter the state of target networks. Neuroscientist 2022;28(2):121–43.
440. Paulus MP, Stein MB. Interoception in anxiety and depression. Brain Struct Funct 2010; 214(5–6):451–63.
441. Zalc B. The acquisition of myelin: An evolutionary perspective. Brain Res 2016;1641(Pt A):4–10.
442. Damasio A. The strange order of things: Life, feeling, and the making of cultures. New York: Pantheon Books, 2018.
443. Lindell A. Lateralization of the expression of facial emotion in humans. Prog Brain Res 2018;238:249–70.
444. Gilman SL. Seeing bodies in pain: From Hippocrates to Freud. Intl J Psychoanalysis 2011;92(3):661–74.
445. Esfeld M. Aristotle's direct realism in "De Anima". Rev Metaphysics 2000;54(2):321–36.
446. Smith R. The sixth sense: Towards a history of muscular sensation. Gesnerus 2011;68(2):218–71.
447. Sacks O. The river of consciousness. Toronto, Canada: Alfred A. Knopf, A Division of Penguin Random House, 2017.
448. Furness JB, Stebbing MJ. The first brain: Species comparisons and evolutionary implications for the enteric and central nervous systems. Neurogastroenterol Motility 2018;30(2):e13234.
449. Damasio A. Feeling & knowing: Making minds conscious. Cognitive Neurosci 2021;12(2):65–6.
450. Crews DE. Senescence, aging, and disease. J Physiol Anthropol 2007;26(3):365–72.
451. Krout KE, Loewy AD. Parabrachial nucleus projections to midline and intralaminar thalamic nuclei of the rat. J Comp Neurol 2000;428(3):475–94.
452. Travagli RA. The nucleus tractus solitarius: An integrative centre with 'task-matching' capabilities. J Physiol 2007;582(Pt 2):471.
453. Bailey TW, Jin Y, Doyle MW, Andresen MC. Vanilloid-sensitive afferents activate neurons with prominent A-type potassium currents in nucleus tractus solitarius. J Neurosci 2002;22(18):8230–7.
454. Craig AD. Interoception: The sense of the physiological condition of the body. Curr Opin Neurobiol 2003;13(4):500–5.
455. Craig AD. How do you feel? Interoception: The sense of the physiological condition of the body. Nat Rev Neurosci 2002;3(8):655–66.
456. Dukes D, Abrams K, Adolphs R, et al. The rise of affectivism. Nat Hum Behav 2021;5(7):816–20.
457. Medford N, Critchley HD. Conjoint activity of anterior insular and anterior cingulate cortex: Awareness and response. Brain Struct Funct 2010;214(5–6):535–49.
458. Scheibe S, Blanchard-Fields F. Effects of regulating emotions on cognitive performance: What is costly for young adults is not so costly for older adults. Psychol Aging 2009;24(1):217–23.
459. Naqvi NH, Bechara A. The insula and drug addiction: An interoceptive view of pleasure, urges, and decision-making. Brain Struct Funct 2010;214(5–6):435–50.
460. Brune M. The evolutionary psychology of obsessive-compulsive disorder: The role of cognitive metarepresentation. Perspect Biol Med 2006;49(3):317–29.
461. Lad M, Sedley W, Griffiths TD. Sensory loss and risk of dementia. Neuroscientist 2024;30(2):247–59.
462. Craig AD. The sentient self. Brain Struct Funct 2010;214(5–6):563–77.
463. Small DM. Taste representation in the human insula. Brain Struct Funct 2010;214(5–6):551–61.
464. Ibanez A, Gleichgerrcht E, Manes F. Clinical effects of insular damage in humans. Brain Struct Funct 2010;214(5–6):397–410.
465. Lamm C, Singer T. The role of anterior insular cortex in social emotions. Brain Struct Funct 2010; 214(5–6):579–91.
466. Damasio AR. The somatic marker hypothesis and the possible functions of the prefrontal cortex. Philos Trans R Soc Lond B Biol Sci 1996;351(1346):1413–20.
467. Damasio A. Descartes' error: Emotion, reason, and the human brain. New York: Avon Books, 1994.
468. Zeman A. Sherrington's philosophical writings: A zest for life. Brain 2007;130(Pt 8):1984–7.

469. Ullsperger M, Harsay HA, Wessel JR, Ridderinkhof KR. Conscious perception of errors and its relation to the anterior insula. Brain Struct Funct 2010;214(5–6):629–43.
470. Kosillo P, Smith AT. The role of the human anterior insular cortex in time processing. Brain Struct Funct 2010;214(5–6):623–8.
471. Tsao A, Yousefzadeh SA, Meck WH, et al. The neural bases for timing of durations. Nat Rev Neurosci 2022;23(11):646–65.
472. Wittmann M. The inner sense of time: How the brain creates a representation of duration. Nat Rev Neurosci 2013;14(3):217–23.
473. Wittmann M, van Wassenhove V. The experience of time: Neural mechanisms and the interplay of emotion, cognition and embodiment. Philos Trans R Soc Lond B Biol Sci 2009;364(1525):1809–13.
474. Embley TM, Martin W. Eukaryotic evolution, changes and challenges. Nature 2006;440(7084):623–30.
475. Kandel ER. The biology of memory: A forty-year perspective. J Neurosci 2009;29(41):12748–56.
476. Kandel ER, Dudai Y, Mayford MR. The molecular and systems biology of memory. Cell 2014; 157(1):163–86.
477. Smith CU. Darwin's unsolved problem: The place of consciousness in an evolutionary world. J Hist Neurosci 2010;19(2):105–20.
478. Stanford M. The cultural evolution of human nature. Acta Biotheo 2020;68(2):275–85.
479. Damasio AR. How the brain creates the mind. Sci Am 1999;281(6):112–7.
480. Overgaard M, Preston C, Aspell J. The self, its body and its brain. Sci Rep 2023;13(12761):12761.
481. Roh E, Song DK, Kim MS. Emerging role of the brain in the homeostatic regulation of energy and glucose metabolism. Exp Mol Med 2016;48(3):e216.
482. Pylyshyn Z. Return of the mental image: Are there really pictures in the brain? Trends Cogn Sci 2003;7(3):113–8.
483. Chiang MC, Bowen A, Schier LA, et al. Parabrachial complex: A hub for pain and aversion. J Neurosci 2019;39(42):8225–30.
484. Mota-Rojas D, Titto CG, Orihuela A, et al. Physiological and behavioral mechanisms of thermoregulation in mammals. Animals (Basel) 2021;11(6):1733.
485. Dolcos F, Katsumi Y, Moore M, et al. Neural correlates of emotion-attention interactions: From perception, learning, and memory to social cognition, individual differences, and training interventions. Neurosci Biobehav Rev 2020;108:559–601.
486. Šimić G, Tkalčić M, Vukić V, et al. Understanding emotions: Origins and roles of the amygdala. Biomolecules 2021;11(6):823.
487. Kemper TD. How m emotions are there? Wedding the social and the autonomic components. Am J Sociology 1987;3(2):263–89.
488. Petrosini L, Picerni E, Termine A, et al. The cerebellum as an embodying machine. Neuroscientist 2024;30(2):229–46.
489. Quinton. Matter and space: The Cartesian principle. Mind 1964;73(291):332–52.
490. McGinn C. The mysterious flame: Conscious minds in a material world. New York: Basic Books, 2000.
491. Overton WF, Mueller U, Newman JL. Developmental perspectives on embodiment and consciousness. New York: Lawrence Erlbaum Associates, 2008.
492. Chalmers D. Facing up to the problem of consciusness. J Conscious Stud 1995;2(3):200–19.
493. Graziano MSA. Consciousness and the social brain. New York: Oxford Uni. Press, 2013.
494. Graziano MSA. Understanding consciousness. Brain 2021;144(5):1281–3.
495. Kouider S. Neurobiological theories of consciousness. In Branks W. P.: Encyclopedia of consciousness. Amesterdam: Academic Press; 2009:87–100.
496. Modolo J, Hassan M, Wendling F, Benquet P. Decoding the circuitry of consciousness: From local microcircuits to brain-scale networks. Netw Neurosci 2020;4(2):315–37.
497. Damasio A, Damasio H. Feelings are the source of consciousness. Neural Comput 2023;35(3):277–86.
498. Damasio A, Damasio H. Homeostatic feelings and the biology of consciousness. Brain 2022;145(7):2231–5.
499. Kabbara A, El Falou W, Khalil M, et al. The dynamic functional core network of the human brain at rest. Sci Rep 2017;7(1):2936.
500. Sikkens T, Bosman CA, Olcese U. The role of top-down modulation in shaping sensory processing across brain states: Implications for consciousness. Front Syst Neurosci 2019;13:31.
501. Mesulam MM. From sensation to cognition. Brain 1998;121(Pt 6):1013–52.
502. Dennett DC. Quining qualia. In: Block N, Flanagan O, Giizeldere G, eds. The nature of consciousness: Philosophical debates. Cambridge, MA: A Bradford Book, MIT Press; 1997:619–42.
503. Brentano F. Psychology from an empirical standpoint. London/ New York, NY: Routledge, 1874/1995.

504. Fusar-Poli P, Broome M, Stanghellini G. From Brentano to mirror neurons: Bridging phenomenology and clinical neuroscience. Psychiatry Res 2010;183(3):245–6.
505. Meyer A, Hackert B, Weger U. Franz Brentano and the beginning of experimental psychology: Implications for the study of psychological phenomena today. Psychol Res 2018;82(2):245–54.
506. Damasio A, Meyer K. Behind the looking-glass. Nature 2008;454(7201):167–8.
507. Johnson AT. Consciousness in animals. IEEE Pulse 2020;11(2):29–30.
508. Birch J, Schnell AK, Clayton NS. Dimensions of animal consciousness. Trends Cogn Sci 2020; 24(10):789–801.
509. Arbib MA. Co-evolution of human consciousness and language (revisited). J Integr Neurosci 2014; 13(2):187–200.
510. Baptista V, Zheng ZL, Coleman FH, et al. Characterization of neurons of the nucleus tractus solitarius pars centralis. Brain Res 2005;1052(2):139–46.
511. Dunbar R. Neocortex size as a constraint on group size in primates. J Hum Evol 1992;22(6):469–93.
512. Dunbar R. Coevolution of neocortical size, group size and language in humans. Behav Brain Sci 1993; 16(4):681–735.
513. de Ruiter J, Weston G, Lyon SM. Dunbar's number: Group size and brain physiology in humans reexamined. Am Anthropol 2011;113(4):557–68.
514. Wellman B. Is Dunbar's number up? Br J Psychol 2012;103(2):174–6; discussion 180–2.
515. Lindenfors P, Wartel A, Lind J. 'Dunbar's number' deconstructed. Biol Lett 2021;17(5):20210158.
516. Christiansen MH, Chater N. Language as shaped by the brain. Behav Brain Sci 2008;31(5):489–508; discussion 509–58.
517. Whiten A, Byrne RW. Machiavellian intelligence II: Extensions and evaluations. UK: Cambridge Uni. Press, 1997.
518. Byrne RW. Machiavellian intelligence retrospective. J Comp Psychol 2018;132(4):432–6.
519. Hopper LM, van de Waal E, Caldwell CA. Celebrating the continued importance of "Machiavellian Intelligence" 30 years on. J Comp Psychol 2018;132(4):427–31.
520. Allman J, Rosin A, Kumar R, Hasenstaub A. Parenting and survival in anthropoid primates: Caretakers live longer. Proc Natl Acad Sci U S A 1998;95(12):6866–9.
521. Juhasz G, Foldi I, Penke B. Systems biology of Alzheimer's disease: How diverse molecular changes result in memory impairment in AD. Neurochem Int 2011;58(7):739–50.
522. Fullerton SM, Clark AG, Weiss KM, et al. Apolipoprotein E variation at the sequence haplotype level: Implications for the origin and maintenance of a major human polymorphism. Am J Hum Genet 2000;67(4):881–900.
523. Seripa D, D'Onofrio G, Panza F, et al. The genetics of the human APOE polymorphism. Rejuvenation Res 2011;14(5):491–500.
524. Campbell R, MacSweeney M, Waters D. Sign language and the brain: A review. J Deaf Stud Deaf Edu 2008;13(1):3–20.
525. Astrila V, Bardon A, Power SE, Vatakis A. The illusions of time. Cham, Switzerland: Palgrave Macmillan, 2019.
526. White PA. The perceived present: What is it, and what is it there for? Psychon Bull Rev 2020;27(4):583–601.
527. McConwell AK. Biological individuality: Elements in the philosophy of biology. Cambridge, UK: Cambridge Uni. Press, 2023.
528. Nixon GM. Myth and mind: The origin of human consciousness in the discovery of the sacred. J Conscious Explor Res 2010;1(3):289–336.
529. Friedman JM. Myth, history, and political identity. Cult Anthropol 1992;7(2):194–210.
530. Klemm WR. Neural representations of the sense of self. Adv Cogn Psychol 2011;7:16–30.
531. Johnston J, Elliott C. Healthy limb amputation: Ethical and legal aspects. Clin Med (Lond) 2002; 2(5):431–5.
532. Holms GM. Henry Head. Royal Society Biol Memo 1941;3(10):665–89.
533. Adler A. The work of Paul Schilder. Bull N Y Acad Med 1965;41(8):841–53.
534. Jahn M, Steinberg H. Pain asymbolia-discovered around 1930 by Paul F. Schilder, almost forgotten today? (in Deutch). Schmerz 2020;34(2):172–80.
535. Penfield W. The cerebral cortex of man. New York: Macmillan Company, 1950.
536. Penfield W, Boldrey E. Somatic motor and sensory representation in the cerebral cortex of man as studied by electrical stimulation. Brain 1937;60(4):389–443.
537. Penfield W. Ferrier Lecture: Some observations on the cerebral cortex of man. Proc Biol Sci 1947; 134(876):329–47.
538. Cohen S. Origins of growth factors: NGF and EGF. J Biol Chem 2008;283(49):33793–7.

539. Money J, Jobaris R. Apotemnophilia: Two cases of self-demand amputation as a paraphilia. J Sex Res 1977;13(2):115–25.
540. First MB. Desire for amputation of a limb: Paraphilia, psychosis, or a new type of identity disorder. Psychol Med 2005;35(6):919–28.
541. Blom RM, van Wingen GA, van der Wal SJ, et al. The desire for amputation or paralyzation: Evidence for structural brain anomalies in Body Integrity Identity Disorder (BIID). PLoS One 2016;11(11):e0165789.
542. Gerstmann J. Psychological and phenomenological aspects of disorders of the body image. J Nerv Ment Dis 1958;126(6):499–512.
543. Triarhou LC. Josef Gerstmann (1887–1969). J Neurol 2008;255(4):614–5.
544. Damasio A. Feelings of emotion and the self. Ann N Y Acad Sci 2003;1001:253–61.
545. Buckner RL, Carroll DC. Self-projection and the brain. Trends Cogn Sci 2007;11(2):49–57.
546. Damasio A. Mental self: The person within. Nature 2003;423(6937):227.
547. Blanke O, Metzinger T, Lenggenhager B. How does the brain localize the self? Science 2008; 317(5841):1239–47.
548. Strehler BL. Where is the self? A neuroanatomical theory of consciousness. Synapse 1991;7(1):44–91.
549. Gazzaniga MS. The mind's past. Berkeley, California: Uni. California Press, 1998.
550. Ehrsson HH. The experimental induction of out of body experiences. Science 2007;317(5841):1048.
551. Damasio AR. Investigating the biology of consciousness. Philos Trans R Soc Lond B Biol Sci 1998; 353(1377):1879–82.
552. Araujo HF, Kaplan J, Damasio H, Damasio A. Neural correlates of different self domains. Brain Behav 2015;5(12):e00409.
553. Fivush R. The development of autobiographical memory. Annu Rev Psychol 2011;62:559–82.
554. Merker B. Consciousness without a cerebral cortex: A challenge for neuroscience and medicine. Behav Brain Sci 2007;30(1):63–81; discussion 81–134.
555. Hagmann P, Cammoun L, Gigandet X, et al. Mapping the structural core of human cerebral cortex. PLoS Biology 2008;6(7):e159.
556. Cauda F, Geminiani G, D'Agata F, et al. Functional connectivity of the posteromedial cortex. PLoS One 2010;5(9):e1-3107.
557. Buckner RL, Andrews-Hanna JR, Schacter DL. The brain's default network: Anatomy, function, and relevance to disease. Ann N Y Acad Sci 2008;1124:1–38.
558. Botvinick M, Cohen J. Rubber hands 'feel' touch that eyes see. Nature 1998;391(6669):756.
559. Crea S, D'Alonzo M, Vitiello N, Cipriani C. The rubber foot illusion. J Neuroeng Rehabil 2015;12:77.
560. Chancel M, Ehrsson HH, Ma WJ. Uncertainty-based inference of a common cause for body ownership. eLife 2022;11:e77221.
561. Crucianelli L, Reader AT, Ehrsson HH. Subcortical contributions to the sense of body ownership. Brain 2024;147(2):390–405.
562. Romano D, Maravita A. The dynamic nature of the sense of ownership after brain injury: Clues from asomatognosia and somatoparaphrenia. Neuropsychologia 2019;132:107119.
563. Rao YL, Ganaraja B, Murlimanju BV, et al. Hippocampus and its involvement in Alzheimer's disease: A review. 3 Biotech 2022;12(2):55.
564. Igarashi KM. Entorhinal cortex dysfunction in Alzheimer's disease. Trends Neurosci 2023;46(2):124–36.
565. McDade EM. Alzheimer disease. Continuum (Minneap Minn) 2022;28(3):648–75.
566. Philippi CL, Feinstein JS, Khalsa SS, et al. Preserved self-awareness following extensive bilateral brain damage to the insula, anterior cingulate, and medial prefrontal cortices. PLoS One 2012;7(8):e38413.
567. Chalmers D. The conscious mind: In search of a fundamental theory. New York: Oxford Uni. Press, 1996.
568. Li M, Zhang W, Zhou X. Identification of genes involved in the evolution of human intelligence through combination of inter-species and intra-species genetic variations. PeerJ 2020;8:e8912.
569. Reynolds EH. John Hughlings Jackson and Thomas Laycock: Brain and mind. Brain 2020;143(2):711–4.
570. Smith CU. Evolution and the problem of mind: John Hughlings Jackson. J Hist Biol 1982;15(2):241–62.
571. Smith CU. Philosophy's loss, neurology's gain: The endeavor of John Hughlings-Jackson. Perspect Biol Med 2012;55(1):81–91.
572. Brothers L. Brain mechanisms of social cognition. J Psychopharmacol 1996;10(1):2–8.
573. Cornwell SE. Language classification in the Ethnologue and its consequences. 2019.
574. Ethnologue. Ethnologue. SIL International, 2023. https://www.ethnologue.com/.
575. Nölle J, Hrtmann S, Tinits P. Language evolution research in the year 2020: A survey of new directions. Lang Dynamics Change 2020;10:3–26.

576. Markov I, Kharitonova K, Grigorenko EL. Language: Its origin and ongoing evolution. J Intell 2023; 11(4):61.
577. Lofting H. The story of Doctor Dolittle. Dover ed. Mineola, N.Y.: Dover Publications, 2004.
578. Seyfarth RM, Cheney DL, Marler P. Monkey responses to three different alarm calls: Evidence of predator classification and semantic communication. Science 1980;210(4471):801–3.
579. Lenneberg E. Biological foundations of language. New York: John Wiley and Sons, Inc, 1967.
580. Levinson SC. Turn-taking in human communication: Origins and implications for language processing. Trends Cogn Sci 2016;20(1):6–14.
581. Hockett CF. Animal languages and human language. Hum Biol 1959;31(1):32–9.
582. Hockett CF. The origin of speech. Sci Am 1960;203:89–96.
583. Hickok G, Bellugi U, Klima ES. Sign language in the brain. Sci Am 2001;284(6):58–65.
584. Chomsky N. Syntactic structures. Berlin: de Gruyter Mouton, 2002.
585. Knight C. Decoding Chomsky: Science and revolutionary politics. Totton, Hampshire, UK: Yale Uni. Press, 2016.
586. Jackendoff R, Audring J. Morphology and memory: Toward an integrated theory. Top Cogn Sci 2020; 12(1):170–96.
587. Jackendoff R. In defense of theory. Cogn Sci 2017;41(Suppl 2):185–212.
588. Simonyan K, Ackermann H, Chang EF, Greenlee JD. New developments in understanding the complexity of human speech production. J Neurosci 2016;36(45):11440–8.
589. Brentari D, Coppola M. What sign language creation teaches us about language. Wiley Interdiscip Rev Cogn Sci 2013;4(2):201–11.
590. Holm J. An introduction to Pidgins and Creoles. Cambridge, UK: Cambridge Uni. Press, 2004.
591. Watson B. Edgar Rice Burroughs: The man who created Tarzan. New York: New Word City, 2014.
592. Hayes KJ, Hayes C. Imitation in a home-raised chimpanzee. J Comp Physiol Psychol 1952;45(5):450–9.
593. Weichhold GM, Bark JE, Korte W, et al. DNA analysis in the case of Kaspar Hauser. Int J Legal Med 1998;111(6):287–91.
594. Hotchkiss J. The jungle of Eden: Kipling, Wolf boys and the colonial imagination. Victorian Liter Cult 2001;35(3):435–49.
595. Simon N. Kaspar Hauser's recovery and autopsy: A perspective on neurological and sociological requirements for language development. J Autism Child Schizophr 1978;8(2):209–17.
596. Snow CE, Hoefnagel-Höhle M. The critical period for language acquisition: Evidence from second language learning. Child Dev 1978;49(4):1114–28.
597. Rosselli M, Ardila A, Matute E, Vélez-Uribe I. Language development across the life span: A neuropsychological-neuroimaging perspective. Neuroscience J 2014;2014:585237.
598. Torero Ibad R, Rheey J, Mrejen S, et al. Otx2 promotes the survival of damaged adult retinal ganglion cells and protects against excitotoxic loss of visual acuity in vivo. J Neurosci 2011;31(14):5495–503.
599. Magrath RD, Platzen D, Kondo J. From nestling calls to fledgling silence: Adaptive timing of change in response to aerial alarm calls. Proc Biol Sci 2006;273(1599):2335–41.
600. Broce G. Herder and ethnography. J Hist Behav Sci 1986;22(2):150–70.
601. Gode A, Herder JG, Moran JH, Rousseau JJ. On the origin of language. New York: Uni. Chicago Press, 1986.
602. Arbib MA. Toward the language-ready brain: Biological evolution and primate comparisons. Psychon Bull Rev 2017;24(1):142–50.
603. Arbib M, Bota M. Language evolution: Neural homologies and neuroinformatics. Neural Netw 2003; 16(9):1237–60.
604. Arbib MA. How the brain got language. New York: Oxford Uni. Press, 2012.
605. Buckingham HW, Christman SS. Charles Darwin and the evolution of human grammatical systems. J Hist Neurosci 2010;19(2):121–39.
606. O'Shannessy C. Light Warlpiri: A new language. Austral J Linguistics 2005;25(1):31–57.
607. Rakhmonbek-kizi YN, Glulomjon-ug'li SA. What comes first: A fact or a topic? Int J Integ Edu 2022;5(9):162–65.
608. Pinker S. Rules of language. Science 1991;253(5019):530–5.
609. Pinker S. On language. J Cogn Neurosci 1994;6(1):92–8.
610. Pinker S. The language instinct: How the mind creates language. Penguin Books, 1995.
611. Everett DL. Cultural constraints on grammar and cognition in Pirahã: Another look at the design features of human language. Curr Anthropol 2005;46(4):621–46.
612. Holloway RL. Language and tool making are similar cognitive processes. Behav Brain Sci 2012;35(4):226.
613. Vaesen K. The cognitive bases of human tool use. Behav Brain Sci 2012;35:203–62.

614. Calvin WH. A stone's throw and its launch window: Timing precision and its implications for language and hominid brains. J Theor Biol 1983;104(1):121–35.
615. Pearce E, Launay J, van Duijn M, et al. Singing together or apart: The effect of competitive and cooperative singing on social bonding within and between sub-groups of a university Fraternity. Psychol Music 2016;44(6):1255–73.
616. Deacon TW. The symbolic species: The co-evolution of language and the brain. New York: W. W. Norton & Company, Inc, 1997.
617. Deacon TW. Evolutionary perspectives on language and brain plasticity. J Commun Disord 2000; 33(4):273–90; quiz 290–1.
618. Aarsleff H. Condillac: Essay on the origin of human knowledge. Cambridge, UK: Cambridge Uni. Press, 2001.
619. Corballis MC. Recursion, language, and starlings. Cogn Sci 2007;31(4):697–704.
620. Corballis MC. The gestural origins of language. Wiley Interdiscip Rev Cogn Sci 2010;1(1):2–7.
621. Corballis MC. From mouth to hand: Gesture, speech, and the evolution of right-handedness. Behav Brain Sci 2003;26(2):199–208; discussion 208–60.
622. MacSweeney M, Capek CM, Campbell R, Woll B. The signing brain: The neurobiology of sign language. Trends Cogn Sci 2008;12(11):432–40.
623. Fitch WT. The evolution of syntax: An exaptationist perspective. Front Evol Neurosci 2011;3:9.
624. Pinker S, Bloom P. Natural language and natural selection. Behav Brain Sci 1990;13(4):707–27.
625. Livio M. The golden ratio: The story of Phi, the world's most astonishing number. New York: Broadway Books, 2003.
626. Martins MD. Distinctive signatures of recursion. Philos Trans R Soc Lond B Biol Sci 2012;367(1598): 2055–64.
627. Coolidge FL, Overmann KA, Wynn T. Recursion: What is it, who has it, and how did it evolve? Wiley Interdiscip Rev Cogn Sci 2011;2(5):547–54.
628. Luuk E, Luuk H. The redundancy of recursion and infinity for natural language. Cogn Process 2011; 12(1):1–11.
629. Corballis MC. The recursive mind: The origins of human language, thought, and civilization. Princeton, New Jersey: Princeton Uni. Press, 2011.
630. Vicari G, Adenzato M. Is recursion language-specific? Evidence of recursive mechanisms in the structure of intentional action. Conscious Cogn 2014;26:169–88.
631. Johnson-Laird PN, Bucciarelli M, Mackiewicz R, Khemlani SS. Recursion in programs, thought, and language. Psychon Bull Rev 2022;29(2):430–54.
632. Gong T, Shuai L, Wu Y. Rethinking foundations of language from a multidisciplinary perspective. Phys Life Rev 2018;26–27:120–38.
633. Chu H. Foundations of language and the recursion-only hypothesis: Comment on "Rethinking foundations of language from a multidisciplinary perspective" by Tao Gong, Lan Shuai and Yicheng Wu. Phys Life Rev 2018;26–27:160–3.
634. Fedurek P, Slocombe KE. Primate vocal communication: A useful tool for understanding human speech and language evolution? Hum Biol 2011;83(2):153–73.
635. Beecher MD. Why are no animal communication systems simple languages? Front Psychol 2021; 12:602635.
636. Liu ST, Montes-Lourido P, Wang X, Sadagopan S. Optimal features for auditory categorization. Nature Commun 2019;10(1):1302.
637. Samhita L, Gross HJ. The "Clever Hans Phenomenon" revisited. Commun Integr Biol 2013;6(6):e27122.
638. Lofting H. Doctor Dolittle Circus. Fairfield, IA: Dell, 1967.
639. Cho PW, Szkudlarek E, Tabor W. Discovery of a recursive principle: An artificial grammar investigation of human learning of a counting recursion language. Front Psychol 2016;7:867.
640. von Frisch K, Chadwick LE, Seeley TD. The dance language and orientation of bees. Cambridge, MA: Harvard Uni. Press, 1993.
641. Bringmann WG, Balance WD, Evans RB. Wilhelm Wundt 1832–1920: A brief biographical sketch. J Hist Behav Sci 1975;11(3):287–97.
642. Hewes GW. Primate communication and the gestural origin of language. Curr Anthropol 1974; 14(1–2):5–24.
643. Armstrong DF, Stokoe WC, Wilcox SE. Gesture and the nature of language. Cambridge, MA: Cambridge Uni. Press, 1995.
644. Lieberman P. The biology and evolution of language. Cambridge, MA: Harvard Uni. Press, 1984.
645. MacNeilage PF. The origin of speech. Oxford, UK: Oxford Uni. Press, 2008.

646. Jürgens U. The role of the periaqueductal grey in vocal behaviour. Behav Brain Res 1994;62:107–17.
647. Kuypers HGJM. Corticobulbar connections to the pons and lower brainstem in man: An anatomical study. Brain 1958;81:364–88.
648. Gallese V, Fadiga L, Fogassi L, Rizzolatti G. Action recognition in the premotor cortex. Brain 1996; 119(Pt 2):593–609.
649. di Pellegrino G, Fadiga L, Fogassi L, et al. Understanding motor events: A neurophysiological study. Exp Brain Res 1992;91(1):176–80.
650. Ramachandran VS, Rogers-Ramachandran D. Phantom limbs and neural plasticity. Arch Neurol 2000;57(3):317–20.
651. Rizzolatti G, Arbib MA. From grasping to speech: Imitation might provide a missing link: Reply. Trends Neurosci 1999;22(4):152.
652. Rizzolatti G, Fadiga L, Fogassi L, Gallese V. Resonance behaviors and mirror neurons. Arch Ital Biol 1999;137(2–3):85–100.
653. Bonini L, Rotunno C, Arcuri E, Gallese V. Mirror neurons 30 years later: Implications and applications. Trends Cogn Sci 2022;26(9):767–81.
654. Heyes C, Catmur C. What happened to mirror neurons? Perspect Psychol Sci 2022;17(1):153–68.
655. Acharya S, Shukla S. Mirror neurons: Enigma of the metaphysical modular brain. J Nat Sci Biol Med 2012;3(2):118–24.
656. Aboitiz F. Gestures, vocalizations, and memory in language origins. Front Evol Neurosci 2012;4:2.
657. Sandler W. The body as evidence for the nature of language. Front Psychol 2018;9:1782.
658. Petkov CI, Jarvis ED. Birds, primates, and spoken language origins: Behavioral phenotypes and neurobiological substrates. Front Evol Neurosci 2012;4:12.
659. Jarvis ED. Learned birdsong and the neurobiology of human language. Ann N Y Acad Sci 2004; 1016:749–77.
660. Liberman AM, Mattingly IG. The motor theory of speech perception revised. Cognition 1985;21(1):1–36.
661. Liberman AM, Cooper FS, Shankweiler DP, Studdert-Kennedy M. Perception of the speech code. Psychol Rev 1967;74(6):431–61.
662. MacDonald J, McGurk H. Visual influences on speech perception processes. Percept Psychophys 1978; 24(3):253–7.
663. McGurk H, MacDonald J. Hearing lips and seeing voices. Nature 1976;264(5588):746–8.
664. Ferrari PF, Bonini L, Fogassi L. From monkey mirror neurons to primate behaviours: Possible 'direct' and 'indirect' pathways. Philos Trans R Soc Lond B Biol Sci 2009;364(1528):2311–23.
665. Menyuk P. Sign language and the brain: What the hands reveal about the brain. Science 1987;238(4828):833.
666. Roberts AI, Roberts SG, Vick SJ. The repertoire and intentionality of gestural communication in wild chimpanzees. Anim Cogn 2014;17(2):317–36.
667. Genty E, Breuer T, Hobaiter C, Byrne RW. Gestural communication of the Gorilla (Gorilla Gorilla): Repertoire, intentionality and possible origins. Anim Cogn 2009;12(3):527–46.
668. Johnson RT, O'Neill MC, Umberger BR. The effects of posture on the three-dimensional gait mechanics of human walking in comparison with walking in bipedal chimpanzees. J Exp Biol 2022; 225(5):jeb243272.
669. Alexander RM. Bipedal animals, and their differences from humans. J Anat 2004;204(5):321–30.
670. Christopoulou C, Bonvillian JD. Sign language, pantomime, and gestural processing in aphasic persons: A review. J Commun Disord 1985;18(1):1–20.
671. Emmorey K, McCullough S, Mehta S, et al. Sign language and pantomime production differentially engage frontal and parietal cortices. Lang Cogn Process 2011;26(7):878–901.
672. Gimeno-Martínez M, Baus C. Iconicity in sign language production: Task matters. Neuropsychologia 2022;167:108166.
673. Loos C, German A, Meier RP. Simultaneous structures in sign languages: Acquisition and emergence. Front Psychol 2022;13:992589.
674. Joseph JE. Saussure. Oxford, Uk: Oxford Uni. Press, 2012.
675. Hickok G, Poeppel D. Dorsal and ventral streams: A framework for understanding aspects of the functional anatomy of language. Cognition 2004;92(1–2):67–99.
676. Hickok G, Poeppel D. Towards a functional neuroanatomy of speech perception. Trends Cogn Sci 2000; 4(4):131–38.
677. Mishkin M, Ungerleider LG, Macko KA. Object vision and spatial vision: Two cortical pathways. Trend Neurosci 1983;6:414–17.
678. Mishkin M, Ungerleider LG. Contribution of striate inputs to the visuospatial functions of parieto-preoccipital cortex in monkeys. Behav Brain Res 1982;6(1):57–77.

679. Brauer J, Anwander A, Friederici AD. Neuroanatomical prerequisites for language functions in the maturing brain. Cereb Cortex 2011;21(2):459–66.
680. Westerlund M, Kastner I, Al Kaabi M, Pylkkanen L. The LATL as locus of composition: MEG evidence from English and Arabic. Brain Lang 2015;141:124–34.
681. Saur D, Kreher BW, Schnell S, et al. Ventral and dorsal pathways for language. Proc Natl Acad Sci U S A 2008;105(46):18035–40.
682. Bach-y-Rita P, Collins CC, White B, et al. A tactile vision substitution system. Am J Optom Arch Am Acad Optom 1969;46(2):109–11.
683. Bach-y-Rita P, Collins CC, Saunders FA, et al. Vision substitution by tactile image projection. Nature 1969;221(5184):963–4.
684. Boraud T, Forkel SJ. Paul Broca: From fame to shame? Brain 2022;145(3):801–4.
685. Geschwind N. Wernicke's contribution to the study of aphasia. 1966:449–63 (4th annual meeting of the academy of aphasia in Chicago).
686. Pillmann F. Carl Wernicke (1848–1905). J Neurol 2003;250(11):1390–1.
687. Saur D. Beyond ventral and dorsal streams: Thalamo-cortical connections for subcortical language integration. Brain 2024;147(6):1927–8.
688. Lichtheim L. On aphasia. Brain 1885;7(4):434–85.
689. Graves RE. The legacy of the Wernicke-Lichtheim model. J Hist Neurosci 1997;6(1):3–20.
690. Eling P. Lichtheim's golden shot. Cortex 2011;47(4):501–8.
691. Catani M, Mesulam M. The arcuate fasciculus and the disconnection theme in language and aphasia: History and current state. Cortex 2008;44(8):953–61.
692. Fitch WT. The evolution of language: A comparative review. Biol Philosophy 2005;20(2):193–230.
693. Bickerton D. Unified cognitive theory: You can't get there from here. Behav Brain Sci 1992;15(3): 437–8.
694. Bickerton D. Afferent isn't efferent, and language isn't logic, either. Behav Brain Sci 2003;26(3):286–7.
695. Grant A. The Oxford handbook of language contact. New York, NY: Oxford Uni. Press, 2019:1 online resource.
696. Hickok G. The functional neuroanatomy of language. Phys Life Rev 2009;6(3):121–43.
697. Elliott EA, Jacobs AM. Facial expressions, emotions, and sign languages. Front Psychol 2013;4:115.
698. Summers RJ, Bailey PJ, Roberts B. Effects of the rate of formant-frequency variation on the grouping of formants in speech perception. J Assoc Res Otolaryngol 2012;13(2):269–80.
699. Feinberg DR, Jones BC, Little AC, et al. Manipulations of fundamental and formant frequencies influence the attractiveness of human male voices. Anim Behaviour 2005;69(3):561–8.
700. Bakhshaee H, Young J, Yang JC, et al. Determination of strain field on the superior surface of excised larynx vocal folds using DIC. J Voice 2013;27(6):659–67.
701. Esteve-Gibert N, Prieto P. Infants temporally coordinate gesture-speech combinations before they produce their first words. Speech Communication 2013;57:301–16.
702. Asch SE. Wolfgang Kohler: 1887–1967. Am J Psychol 1968;81(1):110–9.
703. Ruiz G, Sanchez N. Wolfgang Kohler's "The mentality of apes" and the animal psychology of his time. Span J Psychol 2014;17:E69.
704. Kohler W. Gestalt psychology: An introduction to new concepts in modern psychology. New York: Liverright Publishing Corporation, 1947/1992.
705. Kibbee DA. History of linguistics 2005: Selected papers from the tenth international conference on the history of the language sciences. Amesterdam: John Benjamins Publishing Company, 2007.
706. Knoll E. The science of language and the evolution of mind: Max Muller's quarrel with Darwinism. J Hist Behav Sci 1986;22(1):3–22.
707. Greenberg SA. Henry Head (1861–1940). J Neurol 2004;251(9):1158–9.
708. Pearce JM. Henry Head (1861–1940). J Neurol Neurosurg Psychiatry 2000;69(5):578.
709. Swash M. Henry Head's lifelong studies of cutaneous sensation. J Med Biogr 2022;30(1):57–63.
710. Bartlett FC. Remembering: A study in experimental and social psychology. Cambridge: Cambridge Uni. Press, 1932/1995.
711. Corballis MC. The truth about language: What it is and where it came from. Chicago: Chicago Uni. Press, 2017.
712. Gentilucci M, Corballis MC. From manual gesture to speech: A gradual transition. Neurosci Biobehav Rev 2006;30(7):949–60.
713. Savage-Rumbaugh S, Shanker SG, Taylor TJ. Apes, language, and the human mind. London, UK: Oxford Uni. Press, 2001.
714. Rizzolatti G, Arbib MA. Language within our grasp. Trends Neurosci 1998;21(5):188–94.

715. Negus V. The evolutionary history of man from the evidence of the nose and larynx. AMA Arch Otolaryngol 1957;66(4):414–29.
716. Bateman G. In memoriam: Sir Victor Ewings Negus. Ann R Coll Surg Engl 1974;55(4):201–2.
717. Behrman A, Finan D. Speech and voice science. 4th ed. San Diego, CA: Plural Publishing, 2021.
718. Hiiemae KM, Palmer JB. Tongue Movements in feeding and speech. Crit Rev Oral Biol Med 2003; 14(6):413–29.
719. Zilberman U, Smith P. A comparison of tooth structure in Neanderthals and early Homo sapiens sapiens: A radiographic study. J Anat 1992;180(Pt 3):387–93.
720. Arensburg B, Schepartz LA, Tillier AM, et al. A reappraisal of the anatomical basis for speech in Middle Palaeolithic Hominids. Am J Phys Anthropol 1990;83(2):137–46.
721. Fisher SE. Evolution of language: Lessons from the genome. Psychon Bull Rev 2017;24(1):34–40.
722. Hurst JA, Baraitser M, Auger E, et al. An extended family with a dominantly inherited speech disorder. Dev Med Child Neurol 1990;32(4):352–5.
723. Gopnik M, Crago MB. Familial aggregation of a developmental language disorder. Cognition 1991; 39(1):1–50.
724. Gopnik M. Language deficits and genetic factors. Trends Cogn Sci 1997;1(1):5–9.
725. Lai CS, Fisher SE, Hurst JA, et al. A forkhead-domain gene is mutated in a severe speech and language disorder. Nature 2001;413(6855):519–23.
726. Enard W, Khaitovich P, Klose J, et al. Intra- and interspecific variation in primate gene expression patterns. Science 2002;296(5566):340–3.
727. Graham SA, Fisher SE. Understanding language from a genomic perspective. Annu Rev Genet 2015; 49:131–60.
728. Ghazanfar AA, Takahashi DY, Mathur N, Fitch WT. Cineradiography of monkey lip-smacking reveals putative precursors of speech dynamics. Curr Biol 2012;22(13):1176–82.
729. MacNeilage PF. The frame/content theory of evolution of speech production. Behav Brain Sci 1998; 21(4):499–511; discussion 511–46.
730. Fitch HL, Halwes T, Erickson DM, Liberman AM. Perceptual equivalence of two acoustic cues for stop-consonant manner. Percept Psychophys 1980;27(4):343–50.
731. DeGusta D, Gilbert WH, Turner SP. Hypoglossal canal size and hominid speech. Proc Natl Acad Sci U S A 1999;96(4):1800–4.
732. Cross N. Navajo code talkers: A guide to first-person narratives in the veterans history project. Library of Congress. 2024. (https://guides.loc.gov/navajo-code-talkers).
733. Meadows WC. The Comanche code talkers of World War II. Austin, TX: Uni. of Texas Press, 2002.
734. Hunter D, DeFelipe J, Mehta AR, Conway BR. Art, intuition, and identity in Ramón y Cajal. Neuroscientist 2024; 30(4):1–16.
735. Kilstrup M. Naturalizing semiotics: The triadic sign of Charles Sanders Peirce as a systems property. Prog Biophys Mol Biol 2015;119(3):563–75.
736. Tomasello M. The cultural origins of the human cognition. Cambridge, MA: MIT Press, 2000.
737. Tomasello M. Origins of human communication. Cambridge, MA: MIT Press, 2008.
738. Everett DL. Language: The cultural tool. New York: Pantheon Books, 2012.
739. Huebner GM, Gegenfurtner KR. Conceptual and visual features contribute to visual memory for natural images. PLoS One 2012;7(6):e37575.
740. Corballis MC. From hand to mouth: The origins of language. Princeton, New Jersey: Princeton Uni. Press, 2002.
741. Gazzaniga MS. Cerebral specialization and interhemispheric communication: Does the corpus callosum enable the human condition? Brain 2000;123(Pt 7):1293–326.
742. Kundera M, Asher L. Ignorance. 1st ed. New York: Harper Collins Publishers, 2002.
743 Lumen Learning: Child development. LibreTexts. 2023. www.libretexts.org/
744. Unit EI. Country profile: Venezuela. London, UK: The Unit, 1993.
745. Deane PD. Grammar in mind and brain: Explorations in cognitive syntax. Berlin; New York: Mouton de Gruyter, 1992.
746. O'Keefe J, Nadel L. The hippocampus as a cognitive map. Oxford: Clarendon Press, 1978.
747. O'Keefe J, Burgess N, Donnett JG, et al. Place cells, navigational accuracy, and the human hippocampus. Philos Trans R Soc Lond B Biol Sci 1998;353(1373):1333–40.
748. O'Keefe J, Dostrovsky J. The hippocampus as a spatial map: Preliminary evidence from unit activity in the freely-moving rat. Brain Res 1971;34(1):171–5.
749. Squire LR. The legacy of patient H.M. for neuroscience. Neuron 2009;61(1):6–9.
750. Carruthers P. Evolution of working memory. Proc Natl Acad Sci U S A 2013;110(Suppl 2):10371–8.

751. Corkin S. Permanent present tense: The unforgettable life of the amnesic patient. H.M. New York: Basic Books, 2013.
752. Rosenbaum RS, Winocur G, Moscovitch M. New views on old memories: Re-evaluating the role of the hippocampal complex. Behav Brain Res 2001;127(1–2):183–97.
753. Opitz B. Memory function and the hippocampus. Front Neurol Neurosci 2014;34:51–9.
754. Kandel ER. The molecular biology of memory storage: A dialogue between genes and synapses. Science 2001;294(5544):1030–8.
755. Segal M. Changing views of Cajal's neuron: The case of the dendritic spine. Prog Brain Res 2002; 136:101–7.
756. Lashley KS, Chow KL, Semmes J. An examination of the electrical field theory of cerebral integration. Psychol Rev 1951;58(2):123–36.
757. Edelman GM. The remembered present: A biological theory of consciousness. New York: Basic Books, 1989.
758. Stouffer KM, Grande X, Düzel E, et al. Amidst an amygdala renaissance in Alzheimer's disease. Brain 2024;147(3):816–29.
759. Lum JA, Conti-Ramsden G. Long-term memory: A review and meta-analysis of studies of declarative and procedural memory in specific language impairment. Top Lang Disord 2013;33(4):282–97.
760. Haidle MN. Working memory capacity and the evolution of modern cognitive potential. Cur Anthropol 2010;51(Supp 1):S149–S166.
761. Moscovitch M, Rosenbaum RS, Gilboa A, et al. Functional neuroanatomy of remote episodic, semantic and spatial memory: A unified account based on multiple trace theory. J Anat 2005; 207(1):35–66.
762. Voss JL, Bridge DJ, Cohen NJ, Walker JA. A closer look at the hippocampus and memory. Trends Cogn Sci 2017;21(8):577–88.
763. Jensen JV. Return to the Wilberforce-Huxley debate. Br J Hist Sci 1988;21(2):161–79.
764. Dumont JR, Taube JS. The neural correlates of navigation beyond the hippocampus. Prog Brain Res 2015;219:83–102.
765. Sanders H, Rennó-Costa C, Idiart M, Lisman J. Grid cells and place cells: An integrated view of their navigational and memory function. Trends Neurosci 2015;38(12):763–75.
766. Moser MB, Rowland DC, Moser EI. Place cells, grid cells, and memory. Cold Spring Harb Perspect Biol 2015;7(2):a021808.
767. Wheeler MA, Stuss DT, Tulving E. Toward a theory of episodic memory: The frontal lobes and autonoetic consciousness. Psychol Bull 1997;121(3):331–54.
768. Corballis MC. The uniqueness of human recursive thinking: The ability to think about thinking may be the critical attribute that distinguishes us from all other species. Am Scientist 2007;95(3):240–48.
769. Bor D, Billington J, Baron-Cohen S. Savant memory for digits in a case of synaesthesia and Asperger syndrome is related to hyperactivity in the lateral prefrontal cortex. Neurocase 2007;13(5):311–9.
770. Brancaccio MT. Between Charcot and Bernheim: The debate on hypnotism in fin-de-siecle Italy. Notes Rec R Soc Lond 2017;71(2):157–77.
771. Patihis L, Burton HJY. False memories in therapy and hypnosis before 1980. Psycho Conscious Theo Res Pract, 2015;2(2):153–69.
772. Loftus EF, Ketcham K. Witness for the defense: The accused, the eyewitness, and the expert who puts memory on trial. 1st ed. New York: St. Martin's Press, 1991.
773. Suarez M, Beato MS. The role of language proficiency in false memory: A mini review. Front Psychol 2021;12:659434.
774. Chung D. Conscious, explicit memory, implicit memory, and instincts as the four-component mind for the mental origins of psychotherapy and personality. J Behav Brain Sci 2020;10(11):471–516.
775. Pavlopoulos E, Jones S, Kosmidis S, et al. Molecular mechanism for age-related memory loss: The histone-binding protein RbAp48. Sci Transl Med 2013;5(200):200ra115.
776. Dickerson BC, Brickhouse M, McGinnis S, Wolk DA. Alzheimer's disease: The influence of age on clinical heterogeneity through the human brain connectome. Alzheimers Dement (Amst) 2017;6:122–35.
777. Steiger JH, Yuille JC. Long-term memory and mental rotation. Can J Psychol 1983;37(3):367–89.
778. Venezky RL. English Orthography: Its graphical structure and its relation to sound. Reading Res Q 1967;2(3):75–105.
779. Flaherty M. Sign language and Chinese characters on visual-spatial memory: A literature review. Percept Mot Skills 2003;97(3 Pt 1):797–802.
780. Best CT. The diversity of tone languages and the roles of pitch variation in non-tone languages: Considerations for tone perception research. Front Psychol 2019;10:364.

781. Güldemann T, Stoneking M. A historical appraisal of clicks: A linguistic and genetic population perspective. Annu Rev Anthropol 2008;37:93–109.
782. Pakendorf B, Stoneking M. The genomic prehistory of peoples speaking Khoisan languages. Hum Mol Genet 2021;30(R1):R49–55.
783. Beck HP, Levinson S, Irons G. Finding little Albert: A journey to John B. Watson's infant laboratory. Am Psychol 2009;64(7):605–14.
784. Watson JB. Psychology: From the standpoint of a behaviorist. London; Dover, N.H: F. Pinter, 1924/1983.
785. Perlovsky L, Ilin R. Brain: Conscious and unconscious mechanisms of cognition, emotions, and language. Brain Sci 2012;2(4):790–834.
786. Savage-Rumbaugh S, Rumbaugh DM, Boysen S. Do apes use language? One research group considers the evidence for representational ability in apes. Am Scientist 1980;68(1):49–61.
787. Muindi FJ, Ramachandran L, Tsai JW. Human narratives in science: The power of storytelling. Trends Mol Med 2020;26(3):249–51.
788. Nelson K. Developmental narratives of the experiencing child. Child Dev Perspect 2010;4(1):42–7.
789. Nelson K. Narrative practices and folk psychology: A perspective from developmental psychology. J Conscious Stud 2009;16(6–7):69–93.
790. Niles J. Homo Narrans: The poetics and anthropology of oral literature. Philadelphia, Pennsylvania: Uni. Pennsylvania Press, 1999.
791. Bond BJ. Following your "Friend": Social media and the strength of adolescents' parasocial relationships with media personae. Cyberpsychol Behav Soc Netw 2016;19(11):656–60.
792. Frank AW. The wounded storyteller: Body, illness, and ethics. Chicago, USA: Chicago Uni. Press, 1995.
793. Coen E. The storytelling arms race: Origin of human intelligence and the scientific mind. Heredity (Edinb) 2019;123(1):67–78.
794. Piantadosi ST. Zipf's word frequency law in natural language: A critical review and future directions. Psychon Bull Rev 2014;21(5):1112–30.
795. Zipf GK. Selected studies of the principle of relative frequency in language. Cambridge, MA: Harvard Uni. Press, 1932.
796. Baayen RH. Word frequency distributions. Dordrecht, Netherland: Kluwer Academic Publishers, 2001.
797. Adelaar WFH, Muysken PC. The languages of the Andes. Cambridge, UK: Cambridge Uni. Press, 2004.
798. Everett DL. Piraha culture and gramme: A response to some criticisms. Language 2009;85(2):405–42.
799. Dixon RMW, Aikhenvald AY. The Amazonian languages. Cambridge, UK: Cambridge Uni. Press, 1999.
800. Dunbar R. How religion evolved: And why it endures? New York: Oxford Uni. Press, 2022.
801. Lewis-Williams JD. Conceiving god: The cognitive origin and evolution of religion. Thames Hudson, 2010.
802. Goldman AH. The appeal of the mystery. J Aesthetics Art Criticism 2011;69(3):261–72.
803. Pulimeno M, Piscitelli P, Colazzo S. Children's literature to promote students' global development and wellbeing. Health Promot Perspect 2020;10(1):13–23.
804. Gross M. The evolution of writing. Curr Biol 2012;22(23):981–5.
805. Boas F. The mind of primitive man. USA: The MacMillan Company, 1911/1944.
806. Broca P. Comparative anatomy of the cerebral convolutions: The great limbic lobe and the limbic fissure in the mammalian series. J Comp Neurol 1878/2015;523(17):2501–54.
807. Muller-Wille S. Claude Levi-Strauss on race, history, and genetics. Biosocieties 2010;5(3):330–47.
808. Unseth P. The international impact of Sequoyah's Cherokee syllabary. Written Lang Literacy 2016; 19(1):75–93.
809. Everett DL. An evaluation of universal grammar and the phonological mind. Front Psychol 2016;7:15.
810. Schutz AJ. The Fijian language. Honolulu, Hawaii: Uni. of Hawaii Press, 1985.
811. Suddendorf T, Addis DR, Corballis MC. Mental time travel and the shaping of the human mind. Philos Trans R Soc Lond B Biol Sci 2009;364(1521):1317–24.
812. Suddendorf T, Corballis MC. The evolution of foresight: What is mental time travel, and is it unique to humans? Behav Brain Sci 2007;30(3):299–313; discussion 313–51.
813. Suddendorf T, Corballis MC. Behavioural evidence for mental time travel in non-human animals. Behav Brain Res 2010;215(2):292–8.
814. Baird B, Smallwood J, Schooler JW. Back to the future: Autobiographical planning and the functionality of mind-wandering. Conscious Cogn 2011;20(4):1604–11.
815. Mooneyham BW, Schooler JW. The costs and benefits of mind-wandering: A review. Can J Exp Psychol 2013;67(1):11–8.

816. Killingsworth MA, Gilbert DT. A wandering mind is an unhappy mind. Science 2010;330(6006):932.
817. Wilson BA, Kopelman M, Kapur N. Prominent and persistent loss of past awareness in amnesia: Delusion, impaired consciousness or coping strategy? Neuropsychol Rehabil 2008;18(5–6):527–40.
818. Kohler W. The mentality of apes. New York: Harcourt, Brace and Company, Inc, 1925.
819. Herzfeld C. The great apes: A short history. New Haven, London: Yale Uni. Press, 2017.
820. Teuber ML. The founding of the primate station, Tenerife, Canary Islands. Am J Psychol 1994;107(4): 551–81.
821. Hutchins HE, Lanner RM. The central role of Clark's nutcracker in the dispersal and establishment of whitebark pine. Oecologia 1982;55(2):192–201.
822. Qadri MAJ, Leonard K, Cook RG, Kelly DM. Examination of long-term visual memorization capacity in the Clark's nutcracker (Nucifraga columbiana). Psychon Bull Rev 2018;25(6):2274–80.
823. Paxton R, Hampton RR. Tests of planning and the Bischof-Kohler hypothesis in Rhesus monkeys (Macaca mulatta). Behav Processes 2009;80(3):238–46.
824. Naqshbandi M, Roberts WA. Anticipation of future events in squirrel monkeys (Saimiri sciureus) and rats (Rattus norvegicus): Tests of the Bischof-Kohler hypothesis. J Comp Psychol 2006;120(4):345–57.
825. Inbadas H. Indian philosophical foundations of spirituality at the end of life. Mortality (Abingdon) 2018;23(4):320–33.
826. Baron L, Shemie SD, Teitelbaum J, Doig CJ. Brief review: History, concept and controversies in the neurological determination of death. Can J Anaesth 2006;53(6):602–8.
827. Bernat JL. Contemporary controversies in the definition of death. Prog Brain Res 2009;177:21–31.
828. Kondziella D. The neurology of death and the dying brain: A pictorial essay. Front Neurol 2020;11:736.
829. Markus H, Nurius P. Possible selves. Am Psychol 1986;41(9):954–69.
830. Hasselmo ME. If I had a million neurons: Potential tests of cortico-hippocampal theories. Prog Brain Res 2015;219:1–19.
831. Rolls ET. Diluted connectivity in pattern association networks facilitates the recall of information from the hippocampus to the neocortex. Prog Brain Res 2015;219:21–43.
832. Apperly I. Mindreaders: The cognitive basis of theory of mind. New York: Psychology Press, 2011.
833. Anderson SR. Doctor Dolittle's delusion: Animals and the uniqueness of human language. New Haven: Yale Uni. Press, 2004.
834. Randi J. The truth about Uri Geller. Falls Church, VA: James Randi Educational Foundation, 2011.
835. Morss JR. The construction of perspectives: Piaget's alternative to spatial egocentrism. Intl J Behav Dev 1987;10(3):263–79.
836. Baron-Cohen S, Leslie AM, Frith U. Does the autistic child have a "Theory of mind"? Cognition 1985;21(1):37–46.
837. Donald M. Origins of the modern mind: Three stages in the evolution of culture and cognition. Cambridge, MA: Harvard Uni. Press, 1991.
838. Martin C. What the original "Pinocchio" really says about lying. Condé Nast, Advance Publications. 2015. (https://www.newyorker.com/books/page-turner/original-pinocchio-really-says-lying).
839. Scott-Phillips TC. Speaking our minds: Why human communication is different, and how language evolved to make it special. Houndmills, Basingstoke, Hampshire; New York, NY: Palgrave Macmillan, 2015.
840. Scott-Phillips TC, Kirby S. Language evolution in the laboratory. Trends Cogn Sci 2010;14(9):411–7.
841. Scott-Phillips TC. Speaking our minds: Why human communication is different, and how language evolved to make it special. Houndmills, Basingstoke: Hampshire Palgrave Macmillan, 2015.
842. Dennett DC. The cultural evolution of words and other thinking tools. Cold Spring Harb Symp Quant Biol 2009;74:435–41.
843. Whiten A, Erdal D. The human socio-cognitive niche and its evolutionary origins. Philos Trans R Soc Lond B Biol Sci 2012;367(1599):2119–29.
844. Ogura T, Maki M, Nagata S, Nakamura S. Dogs (Canis familiaris) gaze at our hands: A preliminary eye-tracker experiment on selective attention in dogs. Animals (Basel) 2020;10(5):755.
845. Hare B, Ferrans M. Is cognition the secret to working dog success? Anim Cogn 2021;24(2):231–7.
846. Neale S. Paul Grice and the philosophy of language. Linguist Phil 1992;15:509–99.
847. Thierry G. Neurolinguistic relativity: How language flexes human perception and cognition. Lang Learn 2016;66(3):690–713.
848. Dunbar RIM. Grooming, gossip, and the evolution of language. Cambridge, MA: Harvard Uni. Press, 1996.

Index

Note: Pages in **bold** refer to tables.

N

O

P

W

X

Y

Z